Gretl - Gnu Regression, Econometrics and Time-series Library

A catalogue record for this book is available from the Hong Kong Public Libraries.

Published in Hong Kong by Samurai Media Limited.

Email: info@samuraimedia.org

ISBN 978-988-8406-27-2

Background Cover Image by https://www.flickr.com/people/webtreatsetc/

Contents

1 **Introduction** 1

1.1 Features at a glance 1

1.2 Acknowledgements 1

1.3 Installing the programs 2

I Running the program 3

2 **Getting started** 4

2.1 Let's run a regression 4

2.2 Estimation output 6

2.3 The main window menus 6

2.4 Keyboard shortcuts 10

2.5 The gretl toolbar 10

3 **Modes of working** 11

3.1 Command scripts 11

3.2 Saving script objects 12

3.3 The gretl console 13

3.4 The Session concept 14

4 **Data files** 17

4.1 Data file formats 17

4.2 Databases 17

4.3 Creating a dataset from scratch 18

4.4 Structuring a dataset 20

4.5 Panel data specifics 21

4.6 Missing data values 25

4.7 Maximum size of data sets 26

4.8 Data file collections 26

4.9 Assembling data from multiple sources 28

5 **Sub-sampling a dataset** 29

5.1 Introduction 29

5.2 Setting the sample 29

5.3 Restricting the sample 30

5.4 Panel data 31

5.5 Resampling and bootstrapping 32

6 Graphs and plots **34**

6.1 Gnuplot graphs 34

6.2 Plotting graphs from scripts 37

6.3 Boxplots 40

7 Joining data sources **42**

7.1 Introduction 42

7.2 Basic syntax 42

7.3 Filtering 43

7.4 Matching with keys 44

7.5 Aggregation 47

7.6 String-valued key variables 47

7.7 Importing multiple series 48

7.8 A real-world case 49

7.9 The representation of dates 51

7.10 Time-series data 52

7.11 Special handling of time columns 54

7.12 Panel data 55

7.13 Memo: `join` options 57

8 Realtime data **60**

8.1 Introduction 60

8.2 Atomic format for realtime data 60

8.3 More on time-related options 62

8.4 Getting a certain data vintage 62

8.5 Getting the n-th release for each observation period 63

8.6 Getting the values at a fixed lag after the observation period 64

8.7 Getting the revision history for an observation 65

9 Special functions in genr **68**

9.1 Introduction 68

9.2 Long-run variance 68

9.3 Cumulative densities and p-values 68

9.4 Retrieving internal variables 69

9.5 The discrete Fourier transform 70

10 Gretl data types **73**

10.1 Introduction 73

10.2 Series 73

10.3 Scalars 74

10.4 Matrices 74

10.5 Lists 74

10.6 Strings 74

10.7 Bundles 74
10.8 Arrays 77
10.9 The life cycle of gretl objects 79

11 Discrete variables **82**
11.1 Declaring variables as discrete 82
11.2 Commands for discrete variables 83

12 Loop constructs **87**
12.1 Introduction 87
12.2 Loop control variants 87
12.3 Progressive mode 90
12.4 Loop examples 90

13 User-defined functions **94**
13.1 Defining a function 94
13.2 Calling a function 96
13.3 Deleting a function 97
13.4 Function programming details 98
13.5 Function packages 104
13.6 Memo: updating old-style functions 107

14 Named lists and strings **108**
14.1 Named lists 108
14.2 Named strings 112

15 Matrix manipulation **117**
15.1 Creating matrices 117
15.2 Empty matrices 118
15.3 Selecting sub-matrices 119
15.4 Matrix operators 120
15.5 Matrix-scalar operators 121
15.6 Matrix functions 121
15.7 Matrix accessors 128
15.8 Namespace issues 129
15.9 Creating a data series from a matrix 130
15.10 Matrices and lists 130
15.11 Deleting a matrix 130
15.12 Printing a matrix 131
15.13 Example: OLS using matrices 132

16 Cheat sheet **133**
16.1 Dataset handling 133
16.2 Creating/modifying variables 135

16.3 Neat tricks . . . 139

II Econometric methods 142

17 Robust covariance matrix estimation 143
17.1 Introduction . . . 143
17.2 Cross-sectional data and the HCCME . . . 144
17.3 Time series data and HAC covariance matrices . . . 145
17.4 Special issues with panel data . . . 149
17.5 The cluster-robust estimator . . . 150

18 Panel data 151
18.1 Estimation of panel models . . . 151
18.2 Autoregressive panel models . . . 157

19 Dynamic panel models 159
19.1 Introduction . . . 159
19.2 Usage . . . 162
19.3 Replication of DPD results . . . 164
19.4 Cross-country growth example . . . 167
19.5 Auxiliary test statistics . . . 169
19.6 Memo: `dpanel` options . . . 169

20 Nonlinear least squares 170
20.1 Introduction and examples . . . 170
20.2 Initializing the parameters . . . 170
20.3 NLS dialog window . . . 171
20.4 Analytical and numerical derivatives . . . 171
20.5 Controlling termination . . . 172
20.6 Details on the code . . . 172
20.7 Numerical accuracy . . . 172

21 Maximum likelihood estimation 175
21.1 Generic ML estimation with gretl . . . 175
21.2 Gamma estimation . . . 176
21.3 Stochastic frontier cost function . . . 178
21.4 GARCH models . . . 179
21.5 Analytical derivatives . . . 182
21.6 Debugging ML scripts . . . 184
21.7 Using functions . . . 184
21.8 Advanced use of `mle`: functions, analytical derivatives, algorithm choice . . . 187

22 GMM estimation 191

22.1 Introduction and terminology . . . 191
22.2 GMM as Method of Moments . . . 192
22.3 OLS as GMM . . . 195
22.4 TSLS as GMM . . . 196
22.5 Covariance matrix options . . . 196
22.6 A real example: the Consumption Based Asset Pricing Model . . . 198
22.7 Caveats . . . 199

23 Model selection criteria **203**
23.1 Introduction . . . 203
23.2 Information criteria . . . 203

24 Time series filters **205**
24.1 Fractional differencing . . . 205
24.2 The Hodrick-Prescott filter . . . 205
24.3 The Baxter and King filter . . . 206
24.4 The Butterworth filter . . . 207

25 Univariate time series models **209**
25.1 Introduction . . . 209
25.2 ARIMA models . . . 209
25.3 Unit root tests . . . 214
25.4 Cointegration tests . . . 218
25.5 ARCH and GARCH . . . 219

26 Vector Autoregressions **222**
26.1 Notation . . . 222
26.2 Estimation . . . 223
26.3 Structural VARs . . . 226

27 Cointegration and Vector Error Correction Models **228**
27.1 Introduction . . . 228
27.2 Vector Error Correction Models as representation of a cointegrated system . . . 229
27.3 Interpretation of the deterministic components . . . 230
27.4 The Johansen cointegration tests . . . 232
27.5 Identification of the cointegration vectors . . . 233
27.6 Over-identifying restrictions . . . 235
27.7 Numerical solution methods . . . 241

28 Multivariate models **244**
28.1 The system command . . . 244
28.2 Restriction and estimation . . . 246
28.3 System accessors . . . 247

29 **Forecasting** **249**
29.1 Introduction . . . 249
29.2 Saving and inspecting fitted values . . . 249
29.3 The `fcast` command . . . 249
29.4 Univariate forecast evaluation statistics . . . 251
29.5 Forecasts based on VAR models . . . 252
29.6 Forecasting from simultaneous systems . . . 252

30 **The Kalman Filter** **253**
30.1 Preamble . . . 253
30.2 Notation . . . 253
30.3 Intended usage . . . 255
30.4 Overview of syntax . . . 255
30.5 Defining the filter . . . 255
30.6 The `kfilter` function . . . 258
30.7 The `ksmooth` function . . . 259
30.8 The `ksimul` function . . . 260
30.9 Example 1: ARMA estimation . . . 260
30.10 Example 2: local level model . . . 261

31 **Numerical methods** **265**
31.1 BFGS . . . 265
31.2 Newton–Raphson . . . 266
31.3 Simulated Annealing . . . 268
31.4 Computing a Jacobian . . . 269

32 **Discrete and censored dependent variables** **273**
32.1 Logit and probit models . . . 273
32.2 Ordered response models . . . 276
32.3 Multinomial logit . . . 277
32.4 Bivariate probit . . . 277
32.5 Panel estimators . . . 280
32.6 The Tobit model . . . 281
32.7 Interval regression . . . 281
32.8 Sample selection model . . . 282
32.9 Count data . . . 285
32.10 Duration models . . . 286

33 **Quantile regression** **293**
33.1 Introduction . . . 293
33.2 Basic syntax . . . 293
33.3 Confidence intervals . . . 294
33.4 Multiple quantiles . . . 294

33.5 Large datasets 295

34 Nonparametric methods **297**

34.1 Locally weighted regression (loess) 297

34.2 The Nadaraya–Watson estimator 298

III Technical details **302**

35 Gretl and TEX **303**

35.1 Introduction 303

35.2 TEX-related menu items 303

35.3 Fine-tuning typeset output 305

35.4 Installing and learning TEX 307

36 Gretl and R **308**

36.1 Introduction 308

36.2 Starting an interactive R session 308

36.3 Running an R script 311

36.4 Taking stuff back and forth 311

36.5 Interacting with R from the command line 315

36.6 Performance issues with R 317

36.7 Further use of the R library 317

37 Gretl and Ox **318**

37.1 Introduction 318

37.2 Ox support in gretl 318

37.3 Illustration: replication of DPD model 320

38 Gretl and Octave **322**

38.1 Introduction 322

38.2 Octave support in gretl 322

38.3 Illustration: spectral methods 323

39 Gretl and Stata **326**

40 Gretl and Python **327**

40.1 Introduction 327

40.2 Python support in gretl 327

40.3 Illustration: linear regression with multicollinearity 327

41 Gretl and Julia **329**

41.1 Introduction 329

41.2 Julia support in gretl 329

41.3 Illustration 329

42 **Troubleshooting gretl** 330
42.1 Bug reports . . . 330
42.2 Auxiliary programs . . . 331

43 **The command line interface** 332

IV **Appendices** 333

A **Data file details** 334
A.1 Basic native format . . . 334
A.2 Binary data file format . . . 334
A.3 Native database format . . . 335

B **Data import via ODBC** 336
B.1 ODBC support . . . 336
B.2 ODBC base concepts . . . 336
B.3 Syntax . . . 337
B.4 Examples . . . 339

C **Building gretl** 342
C.1 Installing the prerequisites . . . 342
C.2 Getting the source: release or git . . . 343
C.3 Configure the source . . . 344
C.4 Build and install . . . 344

D **Numerical accuracy** 347

E **Related free software** 348

F **Listing of URLs** 349

Bibliography 350

Chapter 1

Introduction

1.1 Features at a glance

Gretl is an econometrics package, including a shared library, a command-line client program and a graphical user interface.

User-friendly Gretl offers an intuitive user interface; it is very easy to get up and running with econometric analysis. Thanks to its association with the econometrics textbooks by Ramu Ramanathan, Jeffrey Wooldridge, and James Stock and Mark Watson, the package offers many practice data files and command scripts. These are well annotated and accessible. Two other useful resources for gretl users are the available documentation and the gretl-users mailing list.

Flexible You can choose your preferred point on the spectrum from interactive point-and-click to complex scripting, and can easily combine these approaches.

Cross-platform Gretl's "home" platform is Linux but it is also available for MS Windows and Mac OS X, and should work on any unix-like system that has the appropriate basic libraries (see Appendix C).

Open source The full source code for gretl is available to anyone who wants to critique it, patch it, or extend it. See Appendix C.

Sophisticated Gretl offers a full range of least-squares based estimators, either for single equations and for systems, including vector autoregressions and vector error correction models. Several specific maximum likelihood estimators (e.g. probit, ARIMA, GARCH) are also provided natively; more advanced estimation methods can be implemented by the user via generic maximum likelihood or nonlinear GMM.

Extensible Users can enhance gretl by writing their own functions and procedures in gretl's scripting language, which includes a wide range of matrix functions.

Accurate Gretl has been thoroughly tested on several benchmarks, among which the NIST reference datasets. See Appendix D.

Internet ready Gretl can fetch materials such databases, collections of textbook datafiles and add-on packages over the internet.

International Gretl will produce its output in English, French, Italian, Spanish, Polish, Portuguese, German, Basque, Turkish, Russian, Albanian or Greek depending on your computer's native language setting.

1.2 Acknowledgements

The gretl code base originally derived from the program ESL ("Econometrics Software Library"), written by Professor Ramu Ramanathan of the University of California, San Diego. We are much in debt to Professor Ramanathan for making this code available under the GNU General Public Licence and for helping to steer gretl's early development.

We are also grateful to the authors of several econometrics textbooks for permission to package for gretl various datasets associated with their texts. This list currently includes William Greene, author of *Econometric Analysis*; Jeffrey Wooldridge (*Introductory Econometrics: A Modern Approach*); James Stock and Mark Watson (*Introduction to Econometrics*); Damodar Gujarati (*Basic*

Econometrics); Russell Davidson and James MacKinnon (*Econometric Theory and Methods*); and Marno Verbeek (*A Guide to Modern Econometrics*).

GARCH estimation in gretl is based on code deposited in the archive of the *Journal of Applied Econometrics* by Professors Fiorentini, Calzolari and Panattoni, and the code to generate *p*-values for Dickey–Fuller tests is due to James MacKinnon. In each case we are grateful to the authors for permission to use their work.

With regard to the internationalization of gretl, thanks go to Ignacio Díaz-Emparanza (Spanish), Michel Robitaille and Florent Bresson (French), Cristian Rigamonti (Italian), Tadeusz Kufel and Pawel Kufel (Polish), Markus Hahn and Sven Schreiber (German), Hélio Guilherme and Henrique Andrade (Portuguese), Susan Orbe (Basque), Talha Yalta (Turkish) and Alexander Gedranovich (Russian).

Gretl has benefitted greatly from the work of numerous developers of free, open-source software: for specifics please see Appendix C. Our thanks are due to Richard Stallman of the Free Software Foundation, for his support of free software in general and for agreeing to "adopt" gretl as a GNU program in particular.

Many users of gretl have submitted useful suggestions and bug reports. In this connection particular thanks are due to Ignacio Díaz-Emparanza, Tadeusz Kufel, Pawel Kufel, Alan Isaac, Cri Rigamonti, Sven Schreiber, Talha Yalta, Andreas Rosenblad, and Dirk Eddelbuettel, who maintains the gretl package for Debian GNU/Linux.

1.3 Installing the programs

Linux

On the Linux[1] platform you have the choice of compiling the gretl code yourself or making use of a pre-built package. Building gretl from the source is necessary if you want to access the development version or customize gretl to your needs, but this takes quite a few skills; most users will want to go for a pre-built package.

Some Linux distributions feature gretl as part of their standard offering: Debian, Ubuntu and Fedora, for example. If this is the case, all you need to do is install gretl through your package manager of choice. In addition the gretl webpage at `http://gretl.sourceforge.net` offers a "generic" package in `rpm` format for modern Linux systems.

If you prefer to compile your own (or are using a unix system for which pre-built packages are not available), instructions on building gretl can be found in Appendix C.

MS Windows

The MS Windows version comes as a self-extracting executable. Installation is just a matter of downloading `gretl_install.exe` and running this program. You will be prompted for a location to install the package.

Mac OS X

The Mac version comes as a gzipped disk image. Installation is a matter of downloading the image file, opening it in the Finder, and dragging `Gretl.app` to the Applications folder. However, when installing for the first time two prerequisite packages must be put in place first; details are given on the gretl website.

[1] In this manual we use "Linux" as shorthand to refer to the GNU/Linux operating system. What is said herein about Linux mostly applies to other unix-type systems too, though some local modifications may be needed.

Part I

Running the program

Chapter 2

Getting started

2.1 Let's run a regression

This introduction is mostly angled towards the graphical client program; please see Chapter 43 below and the *Gretl Command Reference* for details on the command-line program, gretlcli.

You can supply the name of a data file to open as an argument to gretl, but for the moment let's not do that: just fire up the program.[1] You should see a main window (which will hold information on the data set but which is at first blank) and various menus, some of them disabled at first.

What can you do at this point? You can browse the supplied data files (or databases), open a data file, create a new data file, read the help items, or open a command script. For now let's browse the supplied data files. Under the File menu choose "Open data, Sample file". A second notebook-type window will open, presenting the sets of data files supplied with the package (see Figure 2.1). Select the first tab, "Ramanathan". The numbering of the files in this section corresponds to the chapter organization of Ramanathan (2002), which contains discussion of the analysis of these data. The data will be useful for practice purposes even without the text.

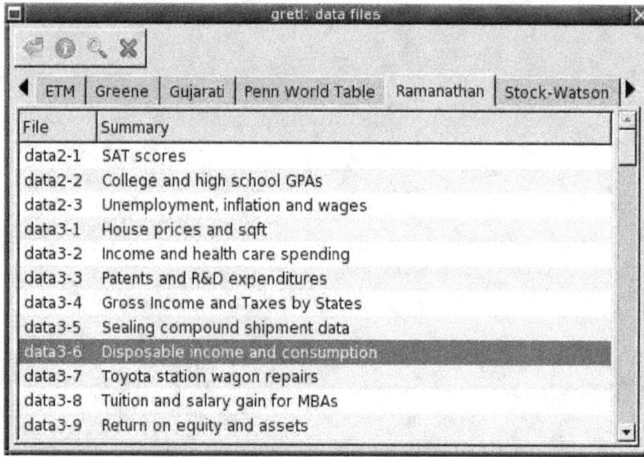

Figure 2.1: Practice data files window

If you select a row in this window and click on "Info" this opens a window showing information on the data set in question (for example, on the sources and definitions of the variables). If you find a file that is of interest, you may open it by clicking on "Open", or just double-clicking on the file name. For the moment let's open `data3-6`.

☞ In gretl windows containing lists, double-clicking on a line launches a default action for the associated list entry: e.g. displaying the values of a data series, opening a file.

This file contains data pertaining to a classic econometric "chestnut", the consumption function. The data window should now display the name of the current data file, the overall data

[1] For convenience we refer to the graphical client program simply as gretl in this manual. Note, however, that the specific name of the program differs according to the computer platform. On Linux it is called `gretl_x11` while on MS Windows it is `gretl.exe`. On Linux systems a wrapper script named `gretl` is also installed — see also the *Gretl Command Reference*.

range and sample range, and the names of the variables along with brief descriptive tags—see Figure 2.2.

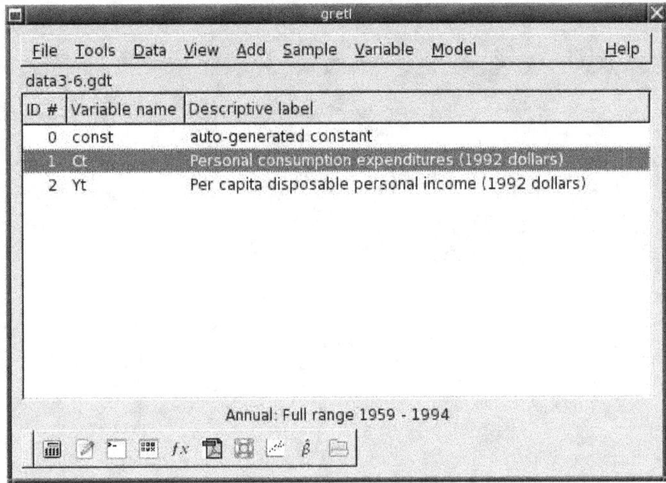

Figure 2.2: Main window, with a practice data file open

OK, what can we do now? Hopefully the various menu options should be fairly self explanatory. For now we'll dip into the Model menu; a brief tour of all the main window menus is given in Section 2.3 below.

Gretl's Model menu offers numerous various econometric estimation routines. The simplest and most standard is Ordinary Least Squares (OLS). Selecting OLS pops up a dialog box calling for a *model specification*—see Figure 2.3.

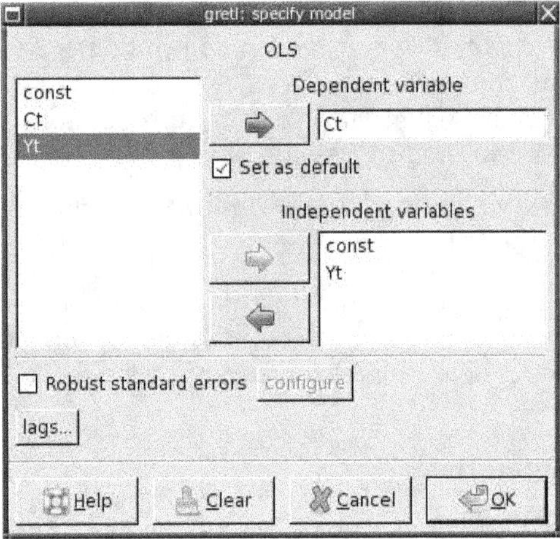

Figure 2.3: Model specification dialog

To select the dependent variable, highlight the variable you want in the list on the left and click the arrow that points to the Dependent variable slot. If you check the "Set as default" box this variable will be pre-selected as dependent when you next open the model dialog box. Shortcut: double-clicking on a variable on the left selects it as dependent and also sets it as the default. To select independent variables, highlight them on the left and click the green arrow (or right-click the highlighted variable); to remove variables from the selected list, use the rad arrow. To select several variable in the list box, drag the mouse over them; to select several non-contiguous variables, hold down the `Ctrl` key and click on the variables you want. To run a regression with consumption as the dependent variable and income as independent, click `Ct` into the Dependent slot and add `Yt` to the Independent variables list.

2.2 Estimation output

Once you've specified a model, a window displaying the regression output will appear. The output is reasonably comprehensive and in a standard format (Figure 2.4).

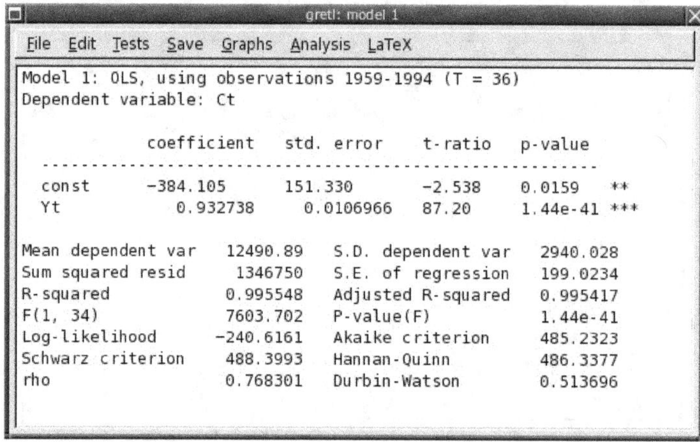


```
Model 1: OLS, using observations 1959-1994 (T = 36)
Dependent variable: Ct

             coefficient    std. error    t-ratio    p-value
  -----------------------------------------------------------
  const      -384.105       151.330       -2.538     0.0159    **
  Yt            0.932738      0.0106966   87.20      1.44e-41  ***

Mean dependent var   12490.89   S.D. dependent var   2940.028
Sum squared resid     1346750   S.E. of regression   199.0234
R-squared            0.995548   Adjusted R-squared   0.995417
F(1, 34)             7603.702   P-value(F)           1.44e-41
Log-likelihood      -240.6161   Akaike criterion     485.2323
Schwarz criterion    488.3993   Hannan-Quinn         486.3377
rho                  0.768301   Durbin-Watson        0.513696
```

Figure 2.4: Model output window

The output window contains menus that allow you to inspect or graph the residuals and fitted values, and to run various diagnostic tests on the model.

For most models there is also an option to print the regression output in LaTeX format. See Chapter 35 for details.

To import gretl output into a word processor, you may copy and paste from an output window, using its Edit menu (or Copy button, in some contexts) to the target program. Many (not all) gretl windows offer the option of copying in RTF (Microsoft's "Rich Text Format") or as LaTeX. If you are pasting into a word processor, RTF may be a good option because the tabular formatting of the output is preserved.[2] Alternatively, you can save the output to a (plain text) file then import the file into the target program. When you finish a gretl session you are given the option of saving all the output from the session to a single file.

Note that on the gnome desktop and under MS Windows, the File menu includes a command to send the output directly to a printer.

☞ When pasting or importing plain text gretl output into a word processor, select a monospaced or typewriter-style font (e.g. Courier) to preserve the output's tabular formatting. Select a small font (10-point Courier should do) to prevent the output lines from being broken in the wrong place.

2.3 The main window menus

Reading left to right along the main window's menu bar, we find the File, Tools, Data, View, Add, Sample, Variable, Model and Help menus.

- File menu
  - Open data: Open a native gretl data file or import from other formats. See Chapter 4.
  - Append data: Add data to the current working data set, from a gretl data file, a comma-separated values file or a spreadsheet file.

[2] Note that when you copy as RTF under MS Windows, Windows will only allow you to paste the material into applications that "understand" RTF. Thus you will be able to paste into MS Word, but not into notepad. Note also that there appears to be a bug in some versions of Windows, whereby the paste will not work properly unless the "target" application (e.g. MS Word) is already running prior to copying the material in question.

- Save data: Save the currently open native gretl data file.
- Save data as: Write out the current data set in native format, with the option of using gzip data compression. See Chapter 4.
- Export data: Write out the current data set in Comma Separated Values (CSV) format, or the formats of GNU R or GNU Octave. See Chapter 4 and also Appendix E.
- Send to: Send the current data set as an e-mail attachment.
- New data set: Allows you to create a blank data set, ready for typing in values or for importing series from a database. See below for more on databases.
- Clear data set: Clear the current data set out of memory. Generally you don't have to do this (since opening a new data file automatically clears the old one) but sometimes it's useful.
- Script files: A "script" is a file containing a sequence of gretl commands. This item contains entries that let you open a script you have created previously ("User file"), open a sample script, or open an editor window in which you can create a new script.
- Session files: A "session" file contains a snapshot of a previous gretl session, including the data set used and any models or graphs that you saved. Under this item you can open a saved session or save the current session.
- Databases: Allows you to browse various large databases, either on your own computer or, if you are connected to the internet, on the gretl database server. See Section 4.2 for details.
- Exit: Quit the program. You'll be prompted to save any unsaved work.

• Tools menu

- Statistical tables: Look up critical values for commonly used distributions (normal or Gaussian, *t*, chi-square, *F* and Durbin-Watson).
- P-value finder: Look up p-values from the Gaussian, *t*, chi-square, *F*, gamma, binomial or Poisson distributions. See also the `pvalue` command in the *Gretl Command Reference*.
- Distribution graphs: Produce graphs of various probability distributions. In the resulting graph window, the pop-up menu includes an item "Add another curve", which enables you to superimpose a further plot (for example, you can draw the *t* distribution with various different degrees of freedom).
- Test statistic calculator: Calculate test statistics and p-values for a range of common hypothesis tests (population mean, variance and proportion; difference of means, variances and proportions).
- Nonparametric tests: Calculate test statistics for various nonparametric tests (Sign test, Wilcoxon rank sum test, Wilcoxon signed rank test, Runs test).
- Seed for random numbers: Set the seed for the random number generator (by default this is set based on the system time when the program is started).
- Command log: Open a window containing a record of the commands executed so far.
- Gretl console: Open a "console" window into which you can type commands as you would using the command-line program, gretlcli (as opposed to using point-and-click).
- Start Gnu R: Start R (if it is installed on your system), and load a copy of the data set currently open in gretl. See Appendix E.
- Sort variables: Rearrange the listing of variables in the main window, either by ID number or alphabetically by name.
- Function packages: Handles "function packages" (see Section 13.5), which allow you to access functions written by other users and share the ones written by you.
- NIST test suite: Check the numerical accuracy of gretl against the reference results for linear regression made available by the (US) National Institute of Standards and Technology.

- Preferences: Set the paths to various files gretl needs to access. Choose the font in which gretl displays text output. Activate or suppress gretl's messaging about the availability of program updates, and so on. See the *Gretl Command Reference* for further details.

- Data menu
  - Select all: Several menu items act upon those variables that are currently selected in the main window. This item lets you select all the variables.
  - Display values: Pops up a window with a simple (not editable) printout of the values of the selected variable or variables.
  - Edit values: Opens a spreadsheet window where you can edit the values of the selected variables.
  - Add observations: Gives a dialog box in which you can choose a number of observations to add at the end of the current dataset; for use with forecasting.
  - Remove extra observations: Active only if extra observations have been added automatically in the process of forecasting; deletes these extra observations.
  - Read info, Edit info: "Read info" just displays the summary information for the current data file; "Edit info" allows you to make changes to it (if you have permission to do so).
  - Print description: Opens a window containing a full account of the current dataset, including the summary information and any specific information on each of the variables.
  - Add case markers: Prompts for the name of a text file containing "case markers" (short strings identifying the individual observations) and adds this information to the data set. See Chapter 4.
  - Remove case markers: Active only if the dataset has case markers identifying the observations; removes these case markers.
  - Dataset structure: invokes a series of dialog boxes which allow you to change the structural interpretation of the current dataset. For example, if data were read in as a cross section you can get the program to interpret them as time series or as a panel. See also section 4.4.
  - Compact data: For time-series data of higher than annual frequency, gives you the option of compacting the data to a lower frequency, using one of four compaction methods (average, sum, start of period or end of period).
  - Expand data: For time-series data, gives you the option of expanding the data to a higher frequency.
  - Transpose data: Turn each observation into a variable and vice versa (or in other words, each row of the data matrix becomes a column in the modified data matrix); can be useful with imported data that have been read in "sideways".

- View menu
  - Icon view: Opens a window showing the content of the current session as a set of icons; see section 3.4.
  - Graph specified vars: Gives a choice between a time series plot, a regular X–Y scatter plot, an X–Y plot using impulses (vertical bars), an X–Y plot "with factor separation" (i.e. with the points colored differently depending to the value of a given dummy variable), boxplots, and a 3-D graph. Serves up a dialog box where you specify the variables to graph. See Chapter 6 for details.
  - Multiple graphs: Allows you to compose a set of up to six small graphs, either pairwise scatter-plots or time-series graphs. These are displayed together in a single window.
  - Summary statistics: Shows a full set of descriptive statistics for the variables selected in the main window.

  - Correlation matrix: Shows the pairwise correlation coefficients for the selected variables.
  - Cross Tabulation: Shows a cross-tabulation of the selected variables. This works only if at least two variables in the data set have been marked as discrete (see Chapter 11).
  - Principal components: Produces a Principal Components Analysis for the selected variables.
  - Mahalanobis distances: Computes the Mahalanobis distance of each observation from the centroid of the selected set of variables.
  - Cross-correlogram: Computes and graphs the cross-correlogram for two selected variables.

- Add menu Offers various standard transformations of variables (logs, lags, squares, etc.) that you may wish to add to the data set. Also gives the option of adding random variables, and (for time-series data) adding seasonal dummy variables (e.g. quarterly dummy variables for quarterly data).

- Sample menu

  - Set range: Select a different starting and/or ending point for the current sample, within the range of data available.
  - Restore full range: self-explanatory.
  - Define, based on dummy: Given a dummy (indicator) variable with values 0 or 1, this drops from the current sample all observations for which the dummy variable has value 0.
  - Restrict, based on criterion: Similar to the item above, except that you don't need a pre-defined variable: you supply a Boolean expression (e.g. `sqft > 1400`) and the sample is restricted to observations satisfying that condition. See the entry for `genr` in the *Gretl Command Reference* for details on the Boolean operators that can be used.
  - Random sub-sample: Draw a random sample from the full dataset.
  - Drop all obs with missing values: Drop from the current sample all observations for which at least one variable has a missing value (see Section 4.6).
  - Count missing values: Give a report on observations where data values are missing. May be useful in examining a panel data set, where it's quite common to encounter missing values.
  - Set missing value code: Set a numerical value that will be interpreted as "missing" or "not available". This is intended for use with imported data, when gretl has not recognized the missing-value code used.

- Variable menu Most items under here operate on a single variable at a time. The "active" variable is set by highlighting it (clicking on its row) in the main data window. Most options will be self-explanatory. Note that you can rename a variable and can edit its descriptive label under "Edit attributes". You can also "Define a new variable" via a formula (e.g. involving some function of one or more existing variables). For the syntax of such formulae, look at the online help for "Generate variable syntax" or see the `genr` command in the *Gretl Command Reference*. One simple example:

  ```
  foo = x1 * x2
  ```

  will create a new variable `foo` as the product of the existing variables `x1` and `x2`. In these formulae, variables must be referenced by name, not number.

- Model menu For details on the various estimators offered under this menu please consult the *Gretl Command Reference*. Also see Chapter 20 regarding the estimation of nonlinear models.

- Help menu Please use this as needed! It gives details on the syntax required in various dialog entries.

2.4 Keyboard shortcuts

When working in the main gretl window, some common operations may be performed using the keyboard, as shown in the table below.

`Return`	Opens a window displaying the values of the currently selected variables: it is the same as selecting "Data, Display Values".
`Delete`	Pressing this key has the effect of deleting the selected variables. A confirmation is required, to prevent accidental deletions.
`e`	Has the same effect as selecting "Edit attributes" from the "Variable" menu.
`F2`	Same as "e". Included for compatibility with other programs.
`g`	Has the same effect as selecting "Define new variable" from the "Variable" menu (which maps onto the `genr` command).
`h`	Opens a help window for gretl commands.
`F1`	Same as "h". Included for compatibility with other programs.
`r`	Refreshes the variable list in the main window.
`t`	Graphs the selected variable; a line graph is used for time-series datasets, whereas a distribution plot is used for cross-sectional data.

2.5 The gretl toolbar

At the bottom left of the main window sits the toolbar.

The icons have the following functions, reading from left to right:

1. Launch a calculator program. A convenience function in case you want quick access to a calculator when you're working in gretl. The default program is `calc.exe` under MS Windows, or `xcalc` under the X window system. You can change the program under the "Tools, Preferences, General" menu, "Programs" tab.
2. Start a new script. Opens an editor window in which you can type a series of commands to be sent to the program as a batch.
3. Open the gretl console. A shortcut to the "Gretl console" menu item (Section 2.3 above).
4. Open the session icon window.
5. Open a window displaying available gretl function packages.
6. Open this manual in PDF format.
7. Open the help item for script commands syntax (i.e. a listing with details of all available commands).
8. Open the dialog box for defining a graph.
9. Open the dialog box for estimating a model using ordinary least squares.
10. Open a window listing the sample datasets supplied with gretl, and any other data file collections that have been installed.

Chapter 3

Modes of working

3.1 Command scripts

As you execute commands in gretl, using the GUI and filling in dialog entries, those commands are recorded in the form of a "script" or batch file. Such scripts can be edited and re-run, using either gretl or the command-line client, gretlcli.

To view the current state of the script at any point in a gretl session, choose "Command log" under the Tools menu. This log file is called `session.inp` and it is overwritten whenever you start a new session. To preserve it, save the script under a different name. Script files will be found most easily, using the GUI file selector, if you name them with the extension "`.inp`".

To open a script you have written independently, use the "File, Script files" menu item; to create a script from scratch use the "File, Script files, New script" item or the "new script" toolbar button. In either case a script window will open (see Figure 3.1).

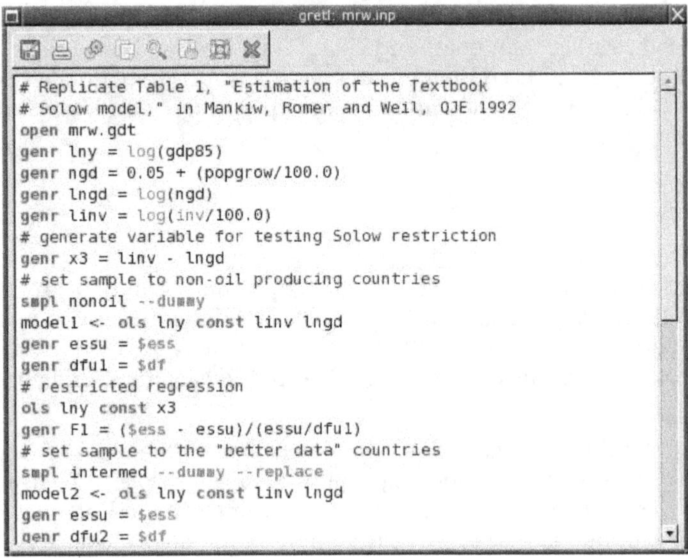

Figure 3.1: Script window, editing a command file

The toolbar at the top of the script window offers the following functions (left to right): (1) Save the file; (2) Save the file under a specified name; (3) Print the file (this option is not available on all platforms); (4) Execute the commands in the file; (5) Copy selected text; (6) Paste the selected text; (7) Find and replace text; (8) Undo the last Paste or Replace action; (9) Help (if you place the cursor in a command word and press the question mark you will get help on that command); (10) Close the window.

When you execute the script, by clicking on the Execute icon or by pressing Ctrl-r, all output is directed to a single window, where it can be edited, saved or copied to the clipboard. To learn more about the possibilities of scripting, take a look at the gretl Help item "Command reference," or start up the command-line program gretlcli and consult its help, or consult the *Gretl Command Reference*.

If you run the script when part of it is highlighted, gretl will only run that portion. Moreover, if you want to run just the current line, you can do so by pressing Ctrl-Enter.[1]

Clicking the right mouse button in the script editor window produces a pop-up menu. This gives you the option of executing either the line on which the cursor is located, or the selected region of the script if there's a selection in place. If the script is editable, this menu also gives the option of adding or removing comment markers from the start of the line or lines.

The gretl package includes over 70 "practice" scripts. Most of these relate to Ramanathan (2002), but they may also be used as a free-standing introduction to scripting in gretl and to various points of econometric theory. You can explore the practice files under "File, Script files, Practice file" There you will find a listing of the files along with a brief description of the points they illustrate and the data they employ. Open any file and run it to see the output. Note that long commands in a script can be broken over two or more lines, using backslash as a continuation character.

You can, if you wish, use the GUI controls and the scripting approach in tandem, exploiting each method where it offers greater convenience. Here are two suggestions.

- Open a data file in the GUI. Explore the data—generate graphs, run regressions, perform tests. Then open the Command log, edit out any redundant commands, and save it under a specific name. Run the script to generate a single file containing a concise record of your work.
- Start by establishing a new script file. Type in any commands that may be required to set up transformations of the data (see the `genr` command in the *Gretl Command Reference*). Typically this sort of thing can be accomplished more efficiently via commands assembled with forethought rather than point-and-click. Then save and run the script: the GUI data window will be updated accordingly. Now you can carry out further exploration of the data via the GUI. To revisit the data at a later point, open and rerun the "preparatory" script first.

Scripts and data files

One common way of doing econometric research with gretl is as follows: compose a script; execute the script; inspect the output; modify the script; run it again—with the last three steps repeated as many times as necessary. In this context, note that when you open a data file this clears out most of gretl's internal state. It's therefore probably a good idea to have your script start with an `open` command: the data file will be re-opened each time, and you can be confident you're getting "fresh" results.

One further point should be noted. When you go to open a new data file via the graphical interface, you are always prompted: opening a new data file will lose any unsaved work, do you really want to do this? When you execute a script that opens a data file, however, you are *not* prompted. The assumption is that in this case you're not going to lose any work, because the work is embodied in the script itself (and it would be annoying to be prompted at each iteration of the work cycle described above).

This means you should be careful if you've done work using the graphical interface and then decide to run a script: the current data file will be replaced without any questions asked, and it's your responsibility to save any changes to your data first.

3.2 Saving script objects

When you estimate a model using point-and-click, the model results are displayed in a separate window, offering menus which let you perform tests, draw graphs, save data from the model, and so on. Ordinarily, when you estimate a model using a script you just get a non-interactive printout of the results. You can, however, arrange for models estimated in a script

[1]This feature is not unique to gretl; other econometric packages offer the same facility. However, experience shows that while this can be remarkably useful, it can also lead to writing dinosaur scripts that are never meant to be executed all at once, but rather used as a chaotic repository to cherry-pick snippets from. Since gretl allows you to have several script windows open at the same time, you may want to keep your scripts tidy and reasonably small.

to be "captured", so that you can examine them interactively when the script is finished. Here is an example of the syntax for achieving this effect:

```
Model1 <- ols Ct 0 Yt
```

That is, you type a name for the model to be saved under, then a back-pointing "assignment arrow", then the model command. The assignment arrow is composed of the less-than sign followed by a dash; it must be separated by spaces from both the preceding name and the following command. The name for a saved object may include spaces, but in that case it must be wrapped in double quotes:

```
"Model 1" <- ols Ct 0 Yt
```

Models saved in this way will appear as icons in the gretl icon view window (see Section 3.4) after the script is executed. In addition, you can arrange to have a named model displayed (in its own window) automatically as follows:

```
Model1.show
```

Again, if the name contains spaces it must be quoted:

```
"Model 1".show
```

The same facility can be used for graphs. For example the following will create a plot of Ct against Yt, save it under the name "CrossPlot" (it will appear under this name in the icon view window), and have it displayed:

```
CrossPlot <- gnuplot Ct Yt
CrossPlot.show
```

You can also save the output from selected commands as named pieces of text (again, these will appear in the session icon window, from where you can open them later). For example this command sends the output from an augmented Dickey-Fuller test to a "text object" named ADF1 and displays it in a window:

```
ADF1 <- adf 2 x1
ADF1.show
```

Objects saved in this way (whether models, graphs or pieces of text output) can be destroyed using the command `.free` appended to the name of the object, as in `ADF1.free`.

3.3 The gretl console

A further option is available for your computing convenience. Under gretl's "Tools" menu you will find the item "Gretl console" (there is also an "open gretl console" button on the toolbar in the main window). This opens up a window in which you can type commands and execute them one by one (by pressing the Enter key) interactively. This is essentially the same as gretlcli's mode of operation, except that the GUI is updated based on commands executed from the console, enabling you to work back and forth as you wish.

In the console, you have "command history"; that is, you can use the up and down arrow keys to navigate the list of command you have entered to date. You can retrieve, edit and then re-enter a previous command.

In console mode, you can create, display and free objects (models, graphs or text) aa described above for script mode.

3.4 The Session concept

Gretl offers the idea of a "session" as a way of keeping track of your work and revisiting it later. The basic idea is to provide an iconic space containing various objects pertaining to your current working session (see Figure 3.2). You can add objects (represented by icons) to this space as you go along. If you save the session, these added objects should be available again if you re-open the session later.

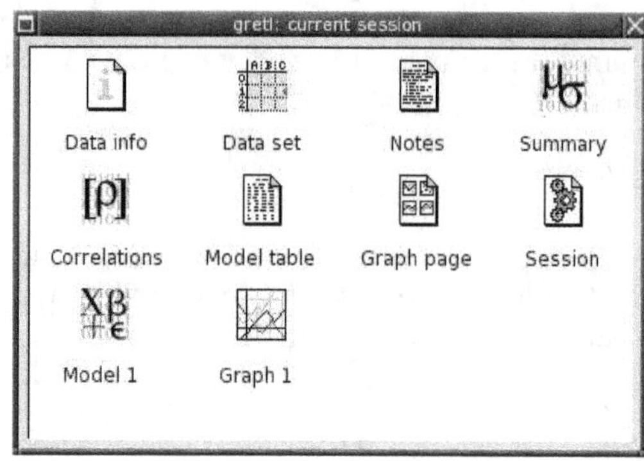

Figure 3.2: Icon view: one model and one graph have been added to the default icons

If you start gretl and open a data set, then select "Icon view" from the View menu, you should see the basic default set of icons: these give you quick access to information on the data set (if any), correlation matrix ("Correlations") and descriptive summary statistics ("Summary"). All of these are activated by double-clicking the relevant icon. The "Data set" icon is a little more complex: double-clicking opens up the data in the built-in spreadsheet, but you can also right-click on the icon for a menu of other actions.

To add a model to the Icon view, first estimate it using the Model menu. Then pull down the File menu in the model window and select "Save to session as icon. . . " or "Save as icon and close". Simply hitting the S key over the model window is a shortcut to the latter action.

To add a graph, first create it (under the View menu, "Graph specified vars", or via one of gretl's other graph-generating commands). Click on the graph window to bring up the graph menu, and select "Save to session as icon".

Once a model or graph is added its icon will appear in the Icon view window. Double-clicking on the icon redisplays the object, while right-clicking brings up a menu which lets you display or delete the object. This popup menu also gives you the option of editing graphs.

The model table

In econometric research it is common to estimate several models with a common dependent variable—the models differing in respect of which independent variables are included, or perhaps in respect of the estimator used. In this situation it is convenient to present the regression results in the form of a table, where each column contains the results (coefficient estimates and standard errors) for a given model, and each row contains the estimates for a given variable across the models. Note that some estimation methods are not compatible with the straightforward model table format, therefore gretl will not let those models be added to the model table. These methods include non-linear least squares (`nls`), generic maximum-likelihood estimators (`mle`), generic GMM (`gmm`), dynamic panel models (`dpanel` or its predecessor `arbond`), interval regressions (`intreg`), bivariate probit models (`biprobit`), AR(I)MA models (`arima` or `arma`), and (G)ARCH models (`garch` and `arch`).

In the Icon view window gretl provides a means of constructing such a table (and copying it in plain text, LaTeX or Rich Text Format). The procedure is outlined below. (The model table can also be built non-interactively, in script mode—see the entry for `modeltab` in the *Gretl Command Reference*.)

1. Estimate a model which you wish to include in the table, and in the model display window, under the File menu, select "Save to session as icon" or "Save as icon and close".
2. Repeat step 1 for the other models to be included in the table (up to a total of six models).
3. When you are done estimating the models, open the icon view of your gretl session, by selecting "Icon view" under the View menu in the main gretl window, or by clicking the "session icon view" icon on the gretl toolbar.
4. In the Icon view, there is an icon labeled "Model table". Decide which model you wish to appear in the left-most column of the model table and add it to the table, either by dragging its icon onto the Model table icon, or by right-clicking on the model icon and selecting "Add to model table" from the pop-up menu.
5. Repeat step 4 for the other models you wish to include in the table. The second model selected will appear in the second column from the left, and so on.
6. When you are finished composing the model table, display it by double-clicking on its icon. Under the Edit menu in the window which appears, you have the option of copying the table to the clipboard in various formats.
7. If the ordering of the models in the table is not what you wanted, right-click on the model table icon and select "Clear table". Then go back to step 4 above and try again.

A simple instance of gretl's model table is shown in Figure 3.3.

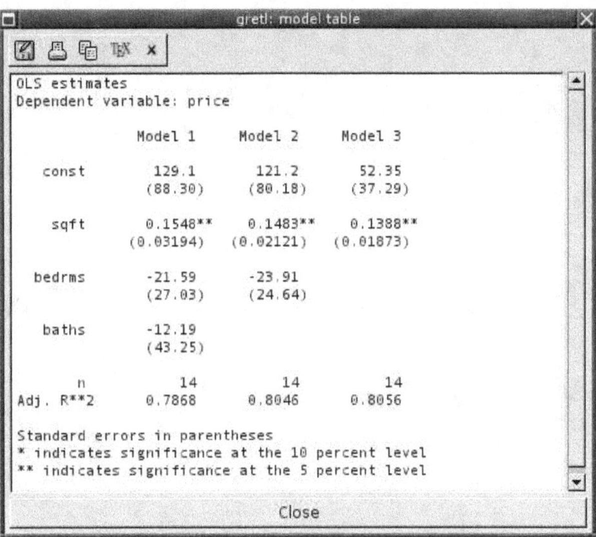

OLS estimates
Dependent variable: price

	Model 1	Model 2	Model 3
const	129.1	121.2	52.35
	(88.30)	(80.18)	(37.29)
sqft	0.1548**	0.1483**	0.1388**
	(0.03194)	(0.02121)	(0.01873)
bedrms	-21.59	-23.91	
	(27.03)	(24.64)	
baths	-12.19		
	(43.25)		
n	14	14	14
Adj. R**2	0.7868	0.8046	0.8056

Standard errors in parentheses
* indicates significance at the 10 percent level
** indicates significance at the 5 percent level

Figure 3.3: Example of model table

The graph page

The "graph page" icon in the session window offers a means of putting together several graphs for printing on a single page. This facility will work only if you have the LaTeX typesetting system installed, and are able to generate and view either PDF or PostScript output. The output format is controlled by your choice of program for compiling TeX files, which can be found under the "Programs" tab in the Preferences dialog box (under the "Tools" menu in the main window). Usually this should be pdflatex for PDF output or latex for PostScript. In the latter case you must have a working set-up for handling PostScript, which will usually include dvips, ghostscript and a viewer such as gv, ggv or kghostview.

In the Icon view window, you can drag up to eight graphs onto the graph page icon. When you double-click on the icon (or right-click and select "Display"), a page containing the selected graphs (in PDF or EPS format) will be composed and opened in your viewer. From there you should be able to print the page.

To clear the graph page, right-click on its icon and select "Clear".

As with the model table, it is also possible to manipulate the graph page via commands in script or console mode—see the entry for the `graphpg` command in the *Gretl Command Reference.*

Saving and re-opening sessions

If you create models or graphs that you think you may wish to re-examine later, then before quitting gretl select "Session files, Save session" from the File menu and give a name under which to save the session. To re-open the session later, either

- Start gretl then re-open the session file by going to the "File, Session files, Open session", or
- From the command line, type `gretl -r` *sessionfile*, where *sessionfile* is the name under which the session was saved, or
- Drag the icon representing a session file onto gretl.

Chapter 4

Data files

4.1 Data file formats

Gretl has its own native format for data files. Most users will probably not want to read or write such files outside of gretl itself, but occasionally this may be useful and details on the file formats are given in Appendix A. The program can also import data from a variety of other formats. In the GUI program this can be done via the "File, Open Data, User file" menu—note the drop-down list of acceptable file types. In script mode, simply use the open command. The supported import formats are as follows.

- Plain text files (comma-separated or "CSV" being the most common type). For details on what gretl expects of such files, see Section 4.3.
- Spreadsheets: MS Excel, Gnumeric and Open Document (ODS). The requirements for such files are given in Section 4.3.
- Stata data files (`.dta`).
- SPSS data files (`.sav`).
- SAS "xport" files (`.xpt`).
- Eviews workfiles (`.wf1`).[1]
- JMulTi data files.

When you import data from a plain text format, gretl opens a "diagnostic" window, reporting on its progress in reading the data. If you encounter a problem with ill-formatted data, the messages in this window should give you a handle on fixing the problem.

Note that gretl has a facility for writing out data in the native formats of GNU R, Octave, JMulTi and PcGive (see Appendix E). In the GUI client this option is found under the "File, Export data" menu; in the command-line client use the `store` command with the appropriate option flag.

4.2 Databases

For working with large amounts of data gretl is supplied with a database-handling routine. A *database*, as opposed to a *data file*, is not read directly into the program's workspace. A database can contain series of mixed frequencies and sample ranges. You open the database and select series to import into the working dataset. You can then save those series in a native format data file if you wish. Databases can be accessed via the menu item "File, Databases".

For details on the format of gretl databases, see Appendix A.

Online access to databases

Several gretl databases are available from Wake Forest University. Your computer must be connected to the internet for this option to work. Please see the description of the "data" command under the Help menu.

☞ Visit the gretl data page for details and updates on available data.

[1] See http://ricardo.ecn.wfu.edu/~cottrell/eviews_format/.

Foreign database formats

Thanks to Thomas Doan of *Estima*, who made available the specification of the database format used by RATS 4 (Regression Analysis of Time Series), gretl can handle such databases—or at least, a subset of same, namely time-series databases containing monthly and quarterly series.

Gretl can also import data from PcGive databases. These take the form of a pair of files, one containing the actual data (with suffix `.bn7`) and one containing supplementary information (`.in7`).

In addition, gretl offers ODBC connectivity. Be warned: this feature is meant for somewhat advanced users; there is currently no graphical interface. Interested readers will find more info in appendix B.

4.3 Creating a dataset from scratch

There are several ways of doing this:

1. Find, or create using a text editor, a plain text data file and open it via "Import".
2. Use your favorite spreadsheet to establish the data file, save it in comma-separated format if necessary (this may not be necessary if the spreadsheet format is MS Excel, Gnumeric or Open Document), then use one of the "Import" options.
3. Use gretl's built-in spreadsheet.
4. Select data series from a suitable database.
5. Use your favorite text editor or other software tools to a create data file in gretl format independently.

Here are a few comments and details on these methods.

Common points on imported data

Options (1) and (2) involve using gretl's "import" mechanism. For the program to read such data successfully, certain general conditions must be satisfied:

- The first row must contain valid variable names. A valid variable name is of 31 characters maximum; starts with a letter; and contains nothing but letters, numbers and the underscore character, _. (Longer variable names will be truncated to 31 characters.) Qualifications to the above: First, in the case of an plain text import, if the file contains no row with variable names the program will automatically add names, `v1`, `v2` and so on. Second, by "the first row" is meant the first *relevant* row. In the case of plain text imports, blank rows and rows beginning with a hash mark, #, are ignored. In the case of Excel, Gnumeric and ODS imports, you are presented with a dialog box where you can select an offset into the spreadsheet, so that gretl will ignore a specified number of rows and/or columns.
- Data values: these should constitute a rectangular block, with one variable per column (and one observation per row). The number of variables (data columns) must match the number of variable names given. See also section 4.6. Numeric data are expected, but in the case of importing from plain text, the program offers limited handling of character (string) data: if a given column contains character data only, consecutive numeric codes are substituted for the strings, and once the import is complete a table is printed showing the correspondence between the strings and the codes.
- Dates (or observation labels): Optionally, the *first* column may contain strings such as dates, or labels for cross-sectional observations. Such strings have a maximum of 15 characters (as with variable names, longer strings will be truncated). A column of this sort should be headed with the string `obs` or `date`, or the first row entry may be left blank.

For dates to be recognized as such, the date strings should adhere to one or other of a set of specific formats, as follows. For *annual* data: 4-digit years. For *quarterly* data: a 4-digit year, followed by a separator (either a period, a colon, or the letter Q), followed by a 1-digit quarter. Examples: `1997.1`, `2002:3`, `1947Q1`. For *monthly* data: a 4-digit year, followed by a period or a colon, followed by a two-digit month. Examples: `1997.01`, `2002:10`.

Plain text ("CSV") files can use comma, space, tab or semicolon as the column separator. When you open such a file via the GUI you are given the option of specifying the separator, though in most cases it should be detected automatically.

If you use a spreadsheet to prepare your data you are able to carry out various transformations of the "raw" data with ease (adding things up, taking percentages or whatever): note, however, that you can also do this sort of thing easily—perhaps more easily—within gretl, by using the tools under the "Add" menu.

Appending imported data

You may wish to establish a dataset piece by piece, by incremental importation of data from other sources. This is supported via the "File, Append data" menu items: gretl will check the new data for conformability with the existing dataset and, if everything seems OK, will merge the data. You can add new variables in this way, provided the data frequency matches that of the existing dataset. Or you can append new observations for data series that are already present; in this case the variable names must match up correctly. Note that by default (that is, if you choose "Open data" rather than "Append data"), opening a new data file closes the current one.

Using the built-in spreadsheet

Under the "File, New data set" menu you can choose the sort of dataset you want to establish (e.g. quarterly time series, cross-sectional). You will then be prompted for starting and ending dates (or observation numbers) and the name of the first variable to add to the dataset. After supplying this information you will be faced with a simple spreadsheet into which you can type data values. In the spreadsheet window, clicking the right mouse button will invoke a popup menu which enables you to add a new variable (column), to add an observation (append a row at the foot of the sheet), or to insert an observation at the selected point (move the data down and insert a blank row.)

Once you have entered data into the spreadsheet you import these into gretl's workspace using the spreadsheet's "Apply changes" button.

Please note that gretl's spreadsheet is quite basic and has no support for functions or formulas. Data transformations are done via the "Add" or "Variable" menus in the main window.

Selecting from a database

Another alternative is to establish your dataset by selecting variables from a database.

Begin with the "File, Databases" menu item. This has four forks: "Gretl native", "RATS 4", "PcGive" and "On database server". You should be able to find the file `fedstl.bin` in the file selector that opens if you choose the "Gretl native" option since this file, which contains a large collection of US macroeconomic time series, is supplied with the distribution.

You won't find anything under "RATS 4" unless you have purchased RATS data.[2] If you do possess RATS data you should go into the "Tools, Preferences, General" dialog, select the Databases tab, and fill in the correct path to your RATS files.

If your computer is connected to the internet you should find several databases (at Wake Forest University) under "On database server". You can browse these remotely; you also have the option of installing them onto your own computer. The initial remote databases window has an

[2] See www.estima.com

item showing, for each file, whether it is already installed locally (and if so, if the local version is up to date with the version at Wake Forest).

Assuming you have managed to open a database you can import selected series into gretl's workspace by using the "Series, Import" menu item in the database window, or via the popup menu that appears if you click the right mouse button, or by dragging the series into the program's main window.

Creating a gretl data file independently

It is possible to create a data file in one or other of gretl's own formats using a text editor or software tools such as awk, sed or perl. This may be a good choice if you have large amounts of data already in machine readable form. You will, of course, need to study these data formats (XML-based or "traditional") as described in Appendix A.

4.4 Structuring a dataset

Once your data are read by gretl, it may be necessary to supply some information on the nature of the data. We distinguish between three kinds of datasets:

1. Cross section
2. Time series
3. Panel data

The primary tool for doing this is the "Data, Dataset structure" menu entry in the graphical interface, or the `setobs` command for scripts and the command-line interface.

Cross sectional data

By a cross section we mean observations on a set of "units" (which may be firms, countries, individuals, or whatever) at a common point in time. This is the default interpretation for a data file: if there is insufficient information to interpret data as time-series or panel data, they are automatically interpreted as a cross section. In the unlikely event that cross-sectional data are wrongly interpreted as time series, you can correct this by selecting the "Data, Dataset structure" menu item. Click the "cross-sectional" radio button in the dialog box that appears, then click "Forward". Click "OK" to confirm your selection.

Time series data

When you import data from a spreadsheet or plain text file, gretl will make fairly strenuous efforts to glean time-series information from the first column of the data, if it looks at all plausible that such information may be present. If time-series structure is present but not recognized, again you can use the "Data, Dataset structure" menu item. Select "Time series" and click "Forward"; select the appropriate data frequency and click "Forward" again; then select or enter the starting observation and click "Forward" once more. Finally, click "OK" to confirm the time-series interpretation if it is correct (or click "Back" to make adjustments if need be).

Besides the basic business of getting a data set interpreted as time series, further issues may arise relating to the frequency of time-series data. In a gretl time-series data set, all the series must have the same frequency. Suppose you wish to make a combined dataset using series that, in their original state, are not all of the same frequency. For example, some series are monthly and some are quarterly.

Your first step is to formulate a strategy: Do you want to end up with a quarterly or a monthly data set? A basic point to note here is that "compacting" data from a higher frequency (e.g. monthly) to a lower frequency (e.g. quarterly) is usually unproblematic. You lose information in doing so, but in general it is perfectly legitimate to take (say) the average of three monthly observations to create a quarterly observation. On the other hand, "expanding" data from a lower to a higher frequency is not, in general, a valid operation.

In most cases, then, the best strategy is to start by creating a data set of the *lower* frequency, and then to compact the higher frequency data to match. When you import higher-frequency data from a database into the current data set, you are given a choice of compaction method (average, sum, start of period, or end of period). In most instances "average" is likely to be appropriate.

You *can* also import lower-frequency data into a high-frequency data set, but this is generally not recommended. What gretl does in this case is simply replicate the values of the lower-frequency series as many times as required. For example, suppose we have a quarterly series with the value 35.5 in 1990:1, the first quarter of 1990. On expansion to monthly, the value 35.5 will be assigned to the observations for January, February and March of 1990. The expanded variable is therefore useless for fine-grained time-series analysis, outside of the special case where you know that the variable in question does in fact remain constant over the sub-periods.

When the current data frequency is appropriate, gretl offers both "Compact data" and "Expand data" options under the "Data" menu. These options operate on the whole data set, compacting or exanding all series. They should be considered "expert" options and should be used with caution.

Panel data

Panel data are inherently three dimensional—the dimensions being variable, cross-sectional unit, and time-period. For example, a particular number in a panel data set might be identified as the observation on capital stock for General Motors in 1980. (A note on terminology: we use the terms "cross-sectional unit", "unit" and "group" interchangeably below to refer to the entities that compose the cross-sectional dimension of the panel. These might, for instance, be firms, countries or persons.)

For representation in a textual computer file (and also for gretl's internal calculations) the three dimensions must somehow be flattened into two. This "flattening" involves taking layers of the data that would naturally stack in a third dimension, and stacking them in the vertical dimension.

gretl always expects data to be arranged "by observation", that is, such that each row represents an observation (and each variable occupies one and only one column). In this context the flattening of a panel data set can be done in either of two ways:

- Stacked time series: the successive vertical blocks each comprise a time series for a given unit.
- Stacked cross sections: the successive vertical blocks each comprise a cross-section for a given period.

You may input data in whichever arrangement is more convenient. Internally, however, gretl always stores panel data in the form of stacked time series.

4.5 Panel data specifics

When you import panel data into gretl from a spreadsheet or comma separated format, the panel nature of the data will not be recognized automatically (most likely the data will be treated as "undated"). A panel interpretation can be imposed on the data using the graphical interface or via the `setobs` command.

In the graphical interface, use the menu item "Data, Dataset structure". In the first dialog box that appears, select "Panel". In the next dialog you have a three-way choice. The first two options, "Stacked time series" and "Stacked cross sections" are applicable if the data set is already organized in one of these two ways. If you select either of these options, the next step is to specify the number of cross-sectional units in the data set. The third option, "Use index variables", is applicable if the data set contains two variables that index the units and the time periods respectively; the next step is then to select those variables. For example, a data file might contain a country code variable and a variable representing the year of the observation. In that case gretl can reconstruct the panel structure of the data regardless of how the observation rows are organized.

The `setobs` command has options that parallel those in the graphical interface. If suitable index variables are available you can do, for example

```
setobs unitvar timevar --panel-vars
```

where `unitvar` is a variable that indexes the units and `timevar` is a variable indexing the periods. Alternatively you can use the form `setobs` *freq* `1:1` *structure*, where *freq* is replaced by the "block size" of the data (that is, the number of periods in the case of stacked time series, or the number of units in the case of stacked cross-sections) and structure is either `--stacked-time-series` or `--stacked-cross-section`. Two examples are given below: the first is suitable for a panel in the form of stacked time series with observations from 20 periods; the second for stacked cross sections with 5 units.

```
setobs 20 1:1 --stacked-time-series
setobs 5 1:1 --stacked-cross-section
```

Panel data arranged by variable

Publicly available panel data sometimes come arranged "by variable." Suppose we have data on two variables, `x1` and `x2`, for each of 50 states in each of 5 years (giving a total of 250 observations per variable). One textual representation of such a data set would start with a block for `x1`, with 50 rows corresponding to the states and 5 columns corresponding to the years. This would be followed, vertically, by a block with the same structure for variable `x2`. A fragment of such a data file is shown below, with quinquennial observations 1965-1985. Imagine the table continued for 48 more states, followed by another 50 rows for variable `x2`.

x1					
	1965	1970	1975	1980	1985
AR	100.0	110.5	118.7	131.2	160.4
AZ	100.0	104.3	113.8	120.9	140.6

If a datafile with this sort of structure is read into gretl,[3] the program will interpret the columns as distinct variables, so the data will not be usable "as is." But there is a mechanism for correcting the situation, namely the `stack` function within the `genr` command.

Consider the first data column in the fragment above: the first 50 rows of this column constitute a cross-section for the variable `x1` in the year 1965. If we could create a new variable by stacking the first 50 entries in the second column underneath the first 50 entries in the first, we would be on the way to making a data set "by observation" (in the first of the two forms mentioned above, stacked cross-sections). That is, we'd have a column comprising a cross-section for `x1` in 1965, followed by a cross-section for the same variable in 1970.

The following gretl script illustrates how we can accomplish the stacking, for both `x1` and `x2`. We assume that the original data file is called `panel.txt`, and that in this file the columns are headed with "variable names" `v1`, `v2`, ..., `v5`. (The columns are not really variables, but in the first instance we "pretend" that they are.)

```
open panel.txt
genr x1 = stack(v1..v5) --length=50
genr x2 = stack(v1..v5) --offset=50 --length=50
```

[3]Note that you will have to modify such a datafile slightly before it can be read at all. The line containing the variable name (in this example `x1`) will have to be removed, and so will the initial row containing the years, otherwise they will be taken as numerical data.

```
setobs 50 1:1 --stacked-cross-section
store panel.gdt x1 x2
```

The second line illustrates the syntax of the `stack` function. The double dots within the parentheses indicate a range of variables to be stacked: here we want to stack all 5 columns (for all 5 years).[4] The full data set contains 100 rows; in the stacking of variable `x1` we wish to read only the first 50 rows from each column: we achieve this by adding `--length=50`. Note that if you want to stack a non-contiguous set of columns you can give a comma-separated list of variable names, as in

```
genr x = stack(v1,v3,v5)
```

or you can provide within the parentheses the name of a previously created list (see chapter 14).

On line 3 we do the stacking for variable `x2`. Again we want a `length` of 50 for the components of the stacked series, but this time we want gretl to start reading from the 50th row of the original data, and we specify `--offset=50`. Line 4 imposes a panel interpretation on the data; finally, we save the data in gretl format, with the panel interpretation, discarding the original "variables" `v1` through `v5`.

The illustrative script above is appropriate when the number of variable to be processed is small. When then are many variables in the data set it will be more efficient to use a command loop to accomplish the stacking, as shown in the following script. The setup is presumed to be the same as in the previous section (50 units, 5 periods), but with 20 variables rather than 2.

```
open panel.txt
loop i=1..20
  genr k = ($i - 1) * 50
  genr x$i = stack(v1..v5) --offset=k --length=50
endloop
setobs 50 1.01 --stacked-cross-section
store panel.gdt x1 x2 x3 x4 x5 x6 x7 x8 x9 x10 \
  x11 x12 x13 x14 x15 x16 x17 x18 x19 x20
```

Panel data marker strings

It can be helpful with panel data to have the observations identified by mnemonic markers. A special function in the `genr` command is available for this purpose.

In the example above, suppose all the states are identified by two-letter codes in the left-most column of the original datafile. When the stacking operation is performed, these codes will be stacked along with the data values. If the first row is marked AR for Arkansas, then the marker AR will end up being shown on each row containing an observation for Arkansas. That's all very well, but these markers don't tell us anything about the date of the observation. To rectify this we could do:

```
genr time
genr year = 1960 + (5 * time)
genr markers = "%s:%d", marker, year
```

The first line generates a 1-based index representing the period of each observation, and the second line uses the `time` variable to generate a variable representing the year of the observation. The third line contains this special feature: if (and only if) the name of the new "variable" to generate is `markers`, the portion of the command following the equals sign is taken as a C-style format string (which must be wrapped in double quotes), followed by a comma-separated list of arguments. The arguments will be printed according to the given format to create a new set of observation markers. Valid arguments are either the names of variables in the dataset, or the string `marker` which denotes the pre-existing observation marker. The format specifiers

[4] You can also specify a list of series using the wildcard '*'; for example `stack(p*)` would stack all series whose names begin with 'v'.

which are likely to be useful in this context are %s for a string and %d for an integer. Strings can be truncated: for example %.3s will use just the first three characters of the string. To chop initial characters off an existing observation marker when constructing a new one, you can use the syntax marker + n, where n is a positive integer: in the case the first n characters will be skipped.

After the commands above are processed, then, the observation markers will look like, for example, AR:1965, where the two-letter state code and the year of the observation are spliced together with a colon.

Panel dummy variables

In a panel study you may wish to construct dummy variables of one or both of the following sorts: (a) dummies as unique identifiers for the units or groups, and (b) dummies as unique identifiers for the time periods. The former may be used to allow the intercept of the regression to differ across the units, the latter to allow the intercept to differ across periods.

Two special functions are available to create such dummies. These are found under the "Add" menu in the GUI, or under the genr command in script mode or gretlcli.

1. "unit dummies" (script command genr unitdum). This command creates a set of dummy variables identifying the cross-sectional units. The variable du_1 will have value 1 in each row corresponding to a unit 1 observation, 0 otherwise; du_2 will have value 1 in each row corresponding to a unit 2 observation, 0 otherwise; and so on.
2. "time dummies" (script command genr timedum). This command creates a set of dummy variables identifying the periods. The variable dt_1 will have value 1 in each row corresponding to a period 1 observation, 0 otherwise; dt_2 will have value 1 in each row corresponding to a period 2 observation, 0 otherwise; and so on.

If a panel data set has the YEAR of the observation entered as one of the variables you can create a periodic dummy to pick out a particular year, e.g. genr dum = (YEAR=1960). You can also create periodic dummy variables using the modulus operator, %. For instance, to create a dummy with value 1 for the first observation and every thirtieth observation thereafter, 0 otherwise, do

```
genr index
genr dum = ((index-1) % 30) = 0
```

Lags, differences, trends

If the time periods are evenly spaced you may want to use lagged values of variables in a panel regression (but see also chapter 19); you may also wish to construct first differences of variables of interest.

Once a dataset is identified as a panel, gretl will handle the generation of such variables correctly. For example the command genr x1_1 = x1(-1) will create a variable that contains the first lag of x1 where available, and the missing value code where the lag is not available (e.g. at the start of the time series for each group). When you run a regression using such variables, the program will automatically skip the missing observations.

When a panel data set has a fairly substantial time dimension, you may wish to include a trend in the analysis. The command genr time creates a variable named time which runs from 1 to T for each unit, where T is the length of the time-series dimension of the panel. If you want to create an index that runs consecutively from 1 to $m \times T$, where m is the number of units in the panel, use genr index.

Basic statistics by unit

gretl contains functions which can be used to generate basic descriptive statistics for a given variable, on a per-unit basis; these are pnobs() (number of valid cases), pmin() and pmax() (minimum and maximum) and pmean() and psd() (mean and standard deviation).

As a brief illustration, suppose we have a panel data set comprising 8 time-series observations on each of N units or groups. Then the command

```
genr pmx = pmean(x)
```

creates a series of this form: the first 8 values (corresponding to unit 1) contain the mean of x for unit 1, the next 8 values contain the mean for unit 2, and so on. The psd() function works in a similar manner. The sample standard deviation for group i is computed as

$$s_i = \sqrt{\frac{\sum(x - \bar{x}_i)^2}{T_i - 1}}$$

where T_i denotes the number of valid observations on x for the given unit, $\bar{x}_i$ denotes the group mean, and the summation is across valid observations for the group. If $T_i < 2$, however, the standard deviation is recorded as 0.

One particular use of psd() may be worth noting. If you want to form a sub-sample of a panel that contains only those units for which the variable x is time-varying, you can either use

```
smpl (pmin(x) < pmax(x)) --restrict
```

or

```
smpl (psd(x) > 0) --restrict
```

4.6 Missing data values

Representation and handling

Missing values are represented internally as DBL_MAX, the largest floating-point number that can be represented on the system (which is likely to be at least 10 to the power 300, and so should not be confused with legitimate data values). In a native-format data file they should be represented as NA. When importing CSV data gretl accepts several common representations of missing values including −999, the string NA (in upper or lower case), a single dot, or simply a blank cell. Blank cells should, of course, be properly delimited, e.g. 120.6,,5.38, in which the middle value is presumed missing.

As for handling of missing values in the course of statistical analysis, gretl does the following:

- In calculating descriptive statistics (mean, standard deviation, etc.) under the summary command, missing values are simply skipped and the sample size adjusted appropriately.
- In running regressions gretl first adjusts the beginning and end of the sample range, truncating the sample if need be. Missing values at the beginning of the sample are common in time series work due to the inclusion of lags, first differences and so on; missing values at the end of the range are not uncommon due to differential updating of series and possibly the inclusion of leads.

If gretl detects any missing values "inside" the (possibly truncated) sample range for a regression, the result depends on the character of the dataset and the estimator chosen. In many cases, the program will automatically skip the missing observations when calculating the regression results. In this situation a message is printed stating how many observations were dropped. On the other hand, the skipping of missing observations is not supported for all procedures: exceptions include all autoregressive estimators, system estimators such as SUR, and nonlinear least squares. In the case of panel data, the skipping of missing observations is supported only if their omission leaves a balanced panel. If missing observations are found in cases where they are not supported, gretl gives an error message and refuses to produce estimates.

Manipulating missing values

Some special functions are available for the handling of missing values. The boolean function missing() takes the name of a variable as its single argument; it returns a series with value 1 for each observation at which the given variable has a missing value, and value 0 otherwise (that is, if the given variable has a valid value at that observation). The function ok() is complementary to missing; it is just a shorthand for !missing (where ! is the boolean NOT operator). For example, one can count the missing values for variable x using

```
scalar nmiss_x = sum(missing(x))
```

The function zeromiss(), which again takes a single series as its argument, returns a series where all zero values are set to the missing code. This should be used with caution—one does not want to confuse missing values and zeros—but it can be useful in some contexts. For example, one can determine the first valid observation for a variable x using

```
genr time
scalar x0 = min(zeromiss(time * ok(x)))
```

The function misszero() does the opposite of zeromiss, that is, it converts all missing values to zero.

It may be worth commenting on the propagation of missing values within genr formulae. The general rule is that in arithmetical operations involving two variables, if either of the variables has a missing value at observation t then the resulting series will also have a missing value at t. The one exception to this rule is multiplication by zero: zero times a missing value produces zero (since this is mathematically valid regardless of the unknown value).

4.7 Maximum size of data sets

Basically, the size of data sets (both the number of variables and the number of observations per variable) is limited only by the characteristics of your computer. gretl allocates memory dynamically, and will ask the operating system for as much memory as your data require. Obviously, then, you are ultimately limited by the size of RAM.

Aside from the multiple-precision OLS option, gretl uses double-precision floating-point numbers throughout. The size of such numbers in bytes depends on the computer platform, but is typically eight. To give a rough notion of magnitudes, suppose we have a data set with 10,000 observations on 500 variables. That's 5 million floating-point numbers or 40 million bytes. If we define the megabyte (MB) as 1024×1024 bytes, as is standard in talking about RAM, it's slightly over 38 MB. The program needs additional memory for workspace, but even so, handling a data set of this size should be quite feasible on a current PC, which at the time of writing is likely to have at least 256 MB of RAM.

If RAM is not an issue, there is one further limitation on data size (though it's very unlikely to be a binding constraint). That is, variables and observations are indexed by signed integers, and on a typical PC these will be 32-bit values, capable of representing a maximum positive value of $2^{31} - 1 = 2,147,483,647$.

The limits mentioned above apply to gretl's "native" functionality. There are tighter limits with regard to two third-party programs that are available as add-ons to gretl for certain sorts of time-series analysis including seasonal adjustment, namely TRAMO/SEATS and X-12-ARIMA. These programs employ a fixed-size memory allocation, and can't handle series of more than 600 observations.

4.8 Data file collections

If you're using gretl in a teaching context you may be interested in adding a collection of data files and/or scripts that relate specifically to your course, in such a way that students can browse and access them easily.

There are three ways to access such collections of files:

- For data files: select the menu item "File, Open data, Sample file", or click on the folder icon on the gretl toolbar.
- For script files: select the menu item "File, Script files, Practice file".

When a user selects one of the items:

- The data or script files included in the gretl distribution are automatically shown (this includes files relating to Ramanathan's *Introductory Econometrics* and Greene's *Econometric Analysis*).
- The program looks for certain known collections of data files available as optional extras, for instance the datafiles from various econometrics textbooks (Davidson and MacKinnon, Gujarati, Stock and Watson, Verbeek, Wooldridge) and the Penn World Table (PWT 5.6). (See the data page at the gretl website for information on these collections.) If the additional files are found, they are added to the selection windows.
- The program then searches for valid file collections (not necessarily known in advance) in these places: the "system" data directory, the system script directory, the user directory, and all first-level subdirectories of these. For reference, typical values for these directories are shown in Table 4.1. (Note that PERSONAL is a placeholder that is expanded by Windows, corresponding to "My Documents" on English-language systems.)

	Linux	*MS Windows*
system data dir	/usr/share/gretl/data	c:\Program Files\gretl\data
system script dir	/usr/share/gretl/scripts	c:\Program Files\gretl\scripts
user dir	$HOME/gretl	PERSONAL\gretl

Table 4.1: Typical locations for file collections

Any valid collections will be added to the selection windows. So what constitutes a valid file collection? This comprises either a set of data files in gretl XML format (with the `.gdt` suffix) or a set of script files containing gretl commands (with `.inp` suffix), in each case accompanied by a "master file" or catalog. The gretl distribution contains several example catalog files, for instance the file `descriptions` in the `misc` sub-directory of the gretl data directory and `ps_descriptions` in the `misc` sub-directory of the scripts directory.

If you are adding your own collection, data catalogs should be named `descriptions` and script catalogs should be be named `ps_descriptions`. In each case the catalog should be placed (along with the associated data or script files) in its own specific sub-directory (e.g. `/usr/share/gretl/data/mydata` or `c:\userdata\gretl\data\mydata`).

The catalog files are plain text; if they contain non-ASCII characters they must be encoded as UTF-8. The syntax of such files is straightforward. Here, for example, are the first few lines of gretl's "misc" data catalog:

```
# Gretl: various illustrative datafiles
"arma","artificial data for ARMA script example"
"ects_nls","Nonlinear least squares example"
"hamilton","Prices and exchange rate, U.S. and Italy"
```

The first line, which must start with a hash mark, contains a short name, here "Gretl", which will appear as the label for this collection's tab in the data browser window, followed by a colon, followed by an optional short description of the collection.

Subsequent lines contain two elements, separated by a comma and wrapped in double quotation marks. The first is a datafile name (leave off the `.gdt` suffix here) and the second is a short description of the content of that datafile. There should be one such line for each datafile in the collection.

A script catalog file looks very similar, except that there are three fields in the file lines: a filename (without its `.inp` suffix), a brief description of the econometric point illustrated in the script, and a brief indication of the nature of the data used. Again, here are the first few lines of the supplied "misc" script catalog:

```
# Gretl: various sample scripts
"arma","ARMA modeling","artificial data"
"ects_nls","Nonlinear least squares (Davidson)","artificial data"
"leverage","Influential observations","artificial data"
"longley","Multicollinearity","US employment"
```

If you want to make your own data collection available to users, these are the steps:

1. Assemble the data, in whatever format is convenient.
2. Convert the data to gretl format and save as `gdt` files. It is probably easiest to convert the data by importing them into the program from plain text, CSV, or a spreadsheet format (MS Excel or Gnumeric) then saving them. You may wish to add descriptions of the individual variables (the "Variable, Edit attributes" menu item), and add information on the source of the data (the "Data, Edit info" menu item).
3. Write a descriptions file for the collection using a text editor.
4. Put the datafiles plus the descriptions file in a subdirectory of the gretl data directory (or user directory).
5. If the collection is to be distributed to other people, package the data files and catalog in some suitable manner, e.g. as a zipfile.

If you assemble such a collection, and the data are not proprietary, we would encourage you to submit the collection for packaging as a gretl optional extra.

4.9 Assembling data from multiple sources

In many contexts researchers need to bring together data from multiple source files, and in some cases these sources are not organized such that the data can simply be "stuck together" by appending rows or columns to a base dataset. In gretl, the `join` command can be used for this purpose; this command is discussed in detail in chapter 7.

Chapter 5

Sub-sampling a dataset

5.1 Introduction

Some subtle issues can arise here; this chapter attempts to explain the issues.

A sub-sample may be defined in relation to a full dataset in two different ways: we will refer to these as "setting" the sample and "restricting" the sample; these methods are discussed in sections 5.2 and 5.3 respectively. In addition section 5.4 discusses some special issues relating to panel data, and section 5.5 covers resampling with replacement, which is useful in the context of bootstrapping test statistics.

The following discussion focuses on the command-line approach. But you can also invoke the methods outlined here via the items under the Sample menu in the GUI program.

5.2 Setting the sample

By "setting" the sample we mean defining a sub-sample simply by means of adjusting the starting and/or ending point of the current sample range. This is likely to be most relevant for time-series data. For example, one has quarterly data from 1960:1 to 2003:4, and one wants to run a regression using only data from the 1970s. A suitable command is then

```
smpl 1970:1 1979:4
```

Or one wishes to set aside a block of observations at the end of the data period for out-of-sample forecasting. In that case one might do

```
smpl ; 2000:4
```

where the semicolon is shorthand for "leave the starting observation unchanged". (The semicolon may also be used in place of the second parameter, to mean that the ending observation should be unchanged.) By "unchanged" here, we mean unchanged relative to the last `smpl` setting, or relative to the full dataset if no sub-sample has been defined up to this point. For example, after

```
smpl 1970:1 2003:4
smpl ; 2000:4
```

the sample range will be 1970:1 to 2000:4.

An incremental or relative form of setting the sample range is also supported. In this case a relative offset should be given, in the form of a signed integer (or a semicolon to indicate no change), for both the starting and ending point. For example

```
smpl +1 ;
```

will advance the starting observation by one while preserving the ending observation, and

```
smpl +2 -1
```

will both advance the starting observation by two and retard the ending observation by one.

An important feature of "setting" the sample as described above is that it necessarily results in the selection of a subset of observations that are contiguous in the full dataset. The structure of the dataset is therefore unaffected (for example, if it is a quarterly time series before setting the sample, it remains a quarterly time series afterwards).

5.3 Restricting the sample

By "restricting" the sample we mean selecting observations on the basis of some Boolean (logical) criterion, or by means of a random number generator. This is likely to be most relevant for cross-sectional or panel data.

Suppose we have data on a cross-section of individuals, recording their gender, income and other characteristics. We wish to select for analysis only the women. If we have a `gender` dummy variable with value 1 for men and 0 for women we could do

```
smpl gender==0 --restrict
```

to this effect. Or suppose we want to restrict the sample to respondents with incomes over $50,000. Then we could use

```
smpl income>50000 --restrict
```

A question arises: if we issue the two commands above in sequence, what do we end up with in our sub-sample: all cases with income over 50000, or just women with income over 50000? By default, the answer is the latter: women with income over 50000. The second restriction augments the first, or in other words the final restriction is the logical product of the new restriction and any restriction that is already in place. If you want a new restriction to replace any existing restrictions you can first recreate the full dataset using

```
smpl --full
```

Alternatively, you can add the `replace` option to the `smpl` command:

```
smpl income>50000 --restrict --replace
```

This option has the effect of automatically re-establishing the full dataset before applying the new restriction.

Unlike a simple "setting" of the sample, "restricting" the sample may result in selection of non-contiguous observations from the full data set. It may therefore change the structure of the data set.

This can be seen in the case of panel data. Say we have a panel of five firms (indexed by the variable `firm`) observed in each of several years (identified by the variable `year`). Then the restriction

```
smpl year==1995 --restrict
```

produces a dataset that is not a panel, but a cross-section for the year 1995. Similarly

```
smpl firm==3 --restrict
```

produces a time-series dataset for firm number 3.

For these reasons (possible non-contiguity in the observations, possible change in the structure of the data), gretl acts differently when you "restrict" the sample as opposed to simply "setting" it. In the case of setting, the program merely records the starting and ending observations and uses these as parameters to the various commands calling for the estimation of models, the computation of statistics, and so on. In the case of restriction, the program makes a reduced copy of the dataset and by default treats this reduced copy as a simple, undated cross-section—but see the further discussion of panel data in section 5.4.

If you wish to re-impose a time-series interpretation of the reduced dataset you can do so using the setobs command, or the GUI menu item "Data, Dataset structure".

The fact that "restricting" the sample results in the creation of a reduced copy of the original dataset may raise an issue when the dataset is very large. With such a dataset in memory, the creation of a copy may lead to a situation where the computer runs low on memory for calculating regression results. You can work around this as follows:

1. Open the full data set, and impose the sample restriction.
2. Save a copy of the reduced data set to disk.
3. Close the full dataset and open the reduced one.
4. Proceed with your analysis.

Random sub-sampling

Besides restricting the sample on some deterministic criterion, it may sometimes be useful (when working with very large datasets, or perhaps to study the properties of an estimator) to draw a random sub-sample from the full dataset. This can be done using, for example,

```
smpl 100 --random
```

to select 100 cases. If you want the sample to be reproducible, you should set the seed for the random number generator first, using the set command. This sort of sampling falls under the "restriction" category: a reduced copy of the dataset is made.

5.4 Panel data

Consider for concreteness the Arellano–Bond dataset supplied with gretl (abdata.gdt). This comprises data on 140 firms ($n = 140$) observed over the years 1976-1984 ($T = 9$). The dataset is "nominally balanced" in the sense that that the time-series length is the same for all countries (this being a requirement for a dataset to count as a panel in gretl), but in fact there are many missing values (NAs).

You may want to sub-sample such a dataset in either the cross-sectional dimension (limit the sample to a subset of firms) or the time dimension (e.g. use data from the 1980s only). The simplest (but limited) way to sub-sample on firms keys off the notation used by gretl for panel observations. The full data range is printed as 1:1 (firm 1, period 1) to 140:9 (firm 140, period 9). The effect of

```
smpl 1:1 80:9
```

is to limit the sample to the first 80 firms. Note that if you instead tried smpl 1:1 80:4, gretl would insist on preserving the balance of the panel and would truncate the range to "complete" firms, as if you had typed smpl 1:1 79:9.

The firms in the Arellano–Bond dataset are anonymous, but suppose you had a panel with five named countries. With such a panel you can inform gretl of the names of the groups using the setobs command. For example, given

```
string cstr = "Portugal Italy Ireland Greece Spain"
setobs country cstr --panel-groups
```

gretl creates a string-valued series named country with group names taken from the variable cstr. Then, to include only Italy and Spain you could do

```
smpl country=="Italy" || country=="Spain" --restrict
```

or to exclude one country,

```
smpl country!="Ireland" --restrict
```

To sub-sample in the time dimension, use of `--restrict` is required. For example, the Arellano-Bond dataset contains a variable named YEAR that records the year of the observations and if one wanted to omit the first two years of data one could do

```
smpl YEAR >= 1978 --restrict
```

If a dataset does not already incude a suitable variable for this purpose one can use the command `genr time` to create a simple 1-based time index.

Note that if you apply a sample restriction that just selects certain units (firms, countries or whatever), or selects certain contiguous time-periods—such that $n > 1$, $T > 1$ and the time-series length is still the same across all included units—your sub-sample will still be interpreted by gretl as a panel.

Unbalancing restrictions

In some cases one wants to sub-sample according to a criterion that "cuts across the grain" of a panel dataset. For instance, suppose you have a micro dataset with thousands of individuals observed over several years and you want to restrict the sample to observations on employed women.

If we simply extracted from the total nT rows of the dataset those that pertain to women who were employed at time t ($t = 1, \ldots, T$) we would likely end up with a dataset that doesn't count as a panel in gretl (because the specific time-series length, T_i, would differ across individuals). In some contexts it might be OK that gretl doesn't take your sub-sample to be a panel, but if you want to apply panel-specific methods this is a problem. You can solve it by giving the `--balanced` option with `smpl`. For example, supposing your dataset contained dummy variables `gender` (with the value 1 coding for women) and `employed`, you could do

```
smpl gender==1 && employed==1 --restrict --balanced
```

What exactly does this do? Well, let's say the years of your data are 2000, 2005 and 2010, and that some women were employed in all of those years, giving a maximum T_i value of 3. But individual 526 is a women who was employed only in the year 2000 ($T_i = 1$). The effect of the `--balanced` option is then to insert "padding rows" of NAs for the years 2005 and 2010 for individual 526, and similarly for all individuals with $0 < T_i < 3$. Your sub-sample then qualifies as a panel.

5.5 Resampling and bootstrapping

Given an original data series x, the command

```
series xr = resample(x)
```

creates a new series each of whose elements is drawn at random from the elements of x. If the original series has 100 observations, each element of x is selected with probability 1/100 at each drawing. Thus the effect is to "shuffle" the elements of x, with the twist that each element of x may appear more than once, or not at all, in `xr`.

The primary use of this function is in the construction of bootstrap confidence intervals or p-values. Here is a simple example. Suppose we estimate a simple regression of y on x via OLS and find that the slope coefficient has a reported t-ratio of 2.5 with 40 degrees of freedom. The two-tailed p-value for the null hypothesis that the slope parameter equals zero is then 0.0166, using the $t(40)$ distribution. Depending on the context, however, we may doubt whether the ratio of coefficient to standard error truly follows the $t(40)$ distribution. In that case we could derive a bootstrap p-value as shown in Example 5.1.

Under the null hypothesis that the slope with respect to x is zero, y is simply equal to its mean plus an error term. We simulate y by resampling the residuals from the initial OLS and re-estimate the model. We repeat this procedure a large number of times, and record the number of cases where the absolute value of the t-ratio is greater than 2.5: the proportion of such cases is our bootstrap p-value. For a good discussion of simulation-based tests and bootstrapping, see Davidson and MacKinnon (2004, chapter 4); Davidson and Flachaire (2001) is also instructive.

Example 5.1: Calculation of bootstrap p-value

```
ols y 0 x
# save the residuals
genr ui = $uhat
scalar ybar = mean(y)
# number of replications for bootstrap
scalar replics = 10000
scalar tcount = 0
series ysim
loop replics --quiet
  # generate simulated y by resampling
  ysim = ybar + resample(ui)
  ols ysim 0 x
  scalar tsim = abs($coeff(x) / $stderr(x))
  tcount += (tsim > 2.5)
endloop
printf "proportion of cases with |t| > 2.5 = %g\n", tcount / replics
```

Chapter 6

Graphs and plots

6.1 Gnuplot graphs

A separate program, gnuplot, is called to generate graphs. Gnuplot is a very full-featured graphing program with myriad options. It is available from www.gnuplot.info (but note that a suitable copy of gnuplot is bundled with the packaged versions of gretl for MS Windows and Mac OS X). Gretl gives you direct access, via a graphical interface, to a subset of gnuplot's options and it tries to choose sensible values for you; it also allows you to take complete control over graph details if you wish.

With a graph displayed, you can click on the graph window for a pop-up menu with the following options.

- Save as PNG: Save the graph in Portable Network Graphics format (the same format that you see on screen).
- Save as postscript: Save in encapsulated postscript (EPS) format.
- Save as Windows metafile: Save in Enhanced Metafile (EMF) format.
- Save to session as icon: The graph will appear in iconic form when you select "Icon view" from the View menu.
- Zoom: Lets you select an area within the graph for closer inspection (not available for all graphs).
- Print: (Current GTK or MS Windows only) lets you print the graph directly.
- Copy to clipboard: MS Windows only, lets you paste the graph into Windows applications such as MS Word.
- Edit: Opens a controller for the plot which lets you adjust many aspects of its appearance.
- Close: Closes the graph window.

Displaying data labels

For simple X-Y scatter plots, some further options are available if the dataset includes "case markers" (that is, labels identifying each observation).[1] With a scatter plot displayed, when you move the mouse pointer over a data point its label is shown on the graph. By default these labels are transient: they do not appear in the printed or copied version of the graph. They can be removed by selecting "Clear data labels" from the graph pop-up menu. If you want the labels to be affixed permanently (so they will show up when the graph is printed or copied), select the option "Freeze data labels" from the pop-up menu; "Clear data labels" cancels this operation. The other label-related option, "All data labels", requests that case markers be shown for all observations. At present the display of case markers is disabled for graphs containing more than 250 data points.

[1]For an example of such a dataset, see the Ramanathan file data4-10: this contains data on private school enrollment for the 50 states of the USA plus Washington, DC; the case markers are the two-letter codes for the states.

GUI plot editor

Selecting the Edit option in the graph popup menu opens an editing dialog box, shown in Figure 6.1. Notice that there are several tabs, allowing you to adjust many aspects of a graph's appearance: font, title, axis scaling, line colors and types, and so on. You can also add lines or descriptive labels to a graph (under the Lines and Labels tabs). The "Apply" button applies your changes without closing the editor; "OK" applies the changes and closes the dialog.

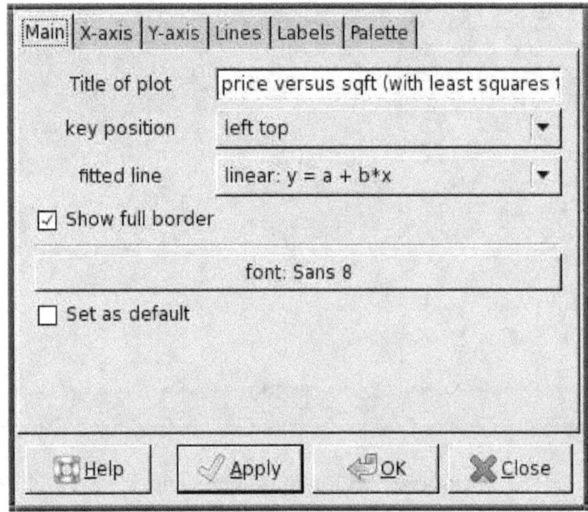

Figure 6.1: gretl's gnuplot controller

Publication-quality graphics: advanced options

The GUI plot editor has two limitations. First, it cannot represent all the myriad options that gnuplot offers. Users who are sufficiently familiar with gnuplot to know what they're missing in the plot editor presumably don't need much help from gretl, so long as they can get hold of the gnuplot command file that gretl has put together. Second, even if the plot editor meets your needs, in terms of fine-tuning the graph you see on screen, a few details may need further work in order to get optimal results for publication.

Either way, the first step in advanced tweaking of a graph is to get access to the graph command file.

- In the graph display window, right-click and choose "Save to session as icon".
- If it's not already open, open the icon view window—either via the menu item View/Icon view, or by clicking the "session icon view" button on the main-window toolbar.
- Right-click on the icon representing the newly added graph and select "Edit plot commands" from the pop-up menu.
- You get a window displaying the plot file (Figure 6.2).

Here are the basic things you can do in this window. Obviously, you can edit the file you just opened. You can also send it for processing by gnuplot, by clicking the "Execute" (cogwheel) icon in the toolbar. Or you can use the "Save as" button to save a copy for editing and processing as you wish.

Unless you're a gnuplot expert, most likely you'll only need to edit a couple of lines at the top of the file, specifying a driver (plus options) and an output file. We offer here a brief summary of some points that may be useful.

First, gnuplot's output mode is set via the command `set term` followed by the name of a supported driver ("terminal" in gnuplot parlance) plus various possible options. (The top line

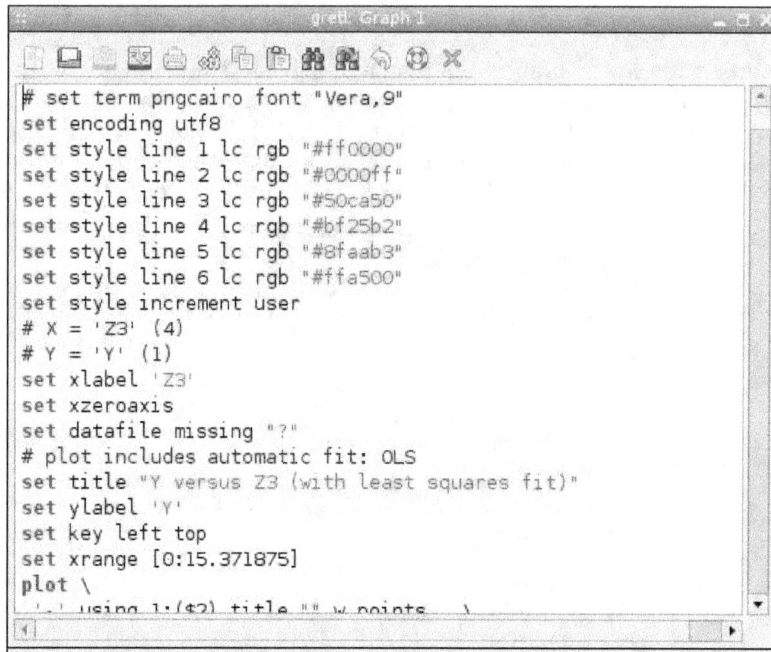

Figure 6.2: Plot commands editor

in the plot commands window shows the `set term` line that gretl used to make a PNG file, commented out.) The graphic formats that are most suitable for publication are PDF and EPS. These are supported by the gnuplot `term` types `pdf`, `pdfcairo` and `postscript` (with the `eps` option). The `pdfcairo` driver has the virtue that is behaves in a very similar manner to the PNG one, the output of which you see on screen. This is provided by the version of gnuplot that is included in the gretl packages for MS Windows and Mac OS X; if you're on Linux it may or may be supported. If `pdfcairo` is not available, the `pdf` terminal may be available; the `postscript` terminal is almost certainly available.

Besides selecting a term type, if you want to get gnuplot to write the actual output file you need to append a `set output` line giving a filename. Here are a few examples of the first two lines you might type in the window editing your plot commands. We'll make these more "realistic" shortly.

```
set term pdfcairo
set output 'mygraph.pdf'

set term pdf
set output 'mygraph.pdf'

set term postscript eps
set output 'mygraph.eps'
```

There are a couple of things worth remarking here. First, you may want to adjust the size of the graph, and second you may want to change the font. The default sizes produced by the above drivers are 5 inches by 3 inches for `pdfcairo` and `pdf`, and 5 inches by 3.5 inches for `postscript eps`. In each case you can change this by giving a size specification, which takes the form XX,YY (examples below).

You may ask, why bother changing the size in the gnuplot command file? After all, PDF and EPS are both vector formats, so the graphs can be scaled at will. True, but a uniform scaling will also affect the font size, which may end looking wrong. You can get optimal results by experimenting with the `font` and `size` options to gnuplot's `set term` command. Here are some examples (comments follow below).

```
# pdfcairo, regular size, slightly amended
set term pdfcairo font "Sans,6" size 5in,3.5in
# or small size
set term pdfcairo font "Sans,5" size 3in,2in

# pdf, regular size, slightly amended
set term pdf font "Helvetica,8" size 5in,3.5in
# or small
set term pdf font "Helvetica,6" size 3in,2in

# postscript, regular
set term post eps solid font "Helvetica,16"
# or small
set term post eps solid font "Helvetica,12" size 3in,2in
```

On the first line we set a sans serif font for pdfcairo at a suitable size for a 5×3.5 inch plot (which you may find looks better than the rather "letterboxy" default of 5×3). And on the second we illustrate what you might do to get a smaller 3×2 inch plot. You can specify the plot size in centimeters if you prefer, as in

```
set term pdfcairo font "Sans,6" size 6cm,4cm
```

We then repeat the exercise for the pdf terminal. Notice that here we're specifying one of the 35 standard PostScript fonts, namely Helvetica. Unlike pdfcairo, the plain pdf driver is unlikely to be able to find fonts other than these.

In the third pair of lines we illustrate options for the postscript driver (which, as you see, can be abbreviated as post). Note that here we have added the option solid. Unlike most other drivers, this one uses dashed lines unless you specify the solid option. Also note that we've (apparently) specified a much larger font in this case. That's because the eps option in effect tells the postscript driver to work at half-size (among other things), so we need to double the font size.

Table 6.1 summarizes the basics for the three drivers we have mentioned.

Terminal	default size (inches)	suggested font
pdfcairo	5×3	Sans,6
pdf	5×3	Helvetica,8
post eps	5×3.5	Helvetica,16

Table 6.1: Drivers for publication-quality graphics

To find out more about gnuplot visit www.gnuplot.info. This site has documentation for the current version of the program in various formats.

Additional tips

To be written. Line widths, enhanced text. Show a "before and after" example.

6.2 Plotting graphs from scripts

When working with scripts, you may want to have a graph shown onto your display or saved into a file. In fact, if in your usual workflow you find yourself creating similar graphs over and over again, you might want to consider the option of writing a script which automates this process for you. Gretl gives you two main tools for doing this: one is a command called gnuplot, whose main use is to create standard plot quickly. The other one is the plot command block, which has a more elaborate syntax but offers you more control on output.

The `gnuplot` command

The `gnuplot` command is described at length in the *Gretl Command Reference* and the online help system. Here, we just summarize its main features: basically, it consists of the `gnuplot` keyword, followed by a list of items, telling the command *what* you want plotted and a list of options, telling it *how* you want it plotted.

For example, the line

```
gnuplot y1 y2 x
```

will give you a basic XY plot of the two series `y1` and `y2` on the vertical axis versus the series `x` on the horizontal axis. In general, the arguments to the `gnuplot` command is a list of series, the last of which goes on the x-axis, while all the other ones go onto the y-axis. By default, the `gnuplot` command gives you a scatterplot. If you just have one variable on the y-axis, then gretl will also draw a the OLS interpolation, if the fit is good enough.[2]

Several aspects of the behavior described above can be modified. You do this by appending options to the command. Most options can be broadly grouped in three categories:

1. Plot styles: we support points (the default choice), lines, lines and points together, and impulses (vertical lines).
2. Algorithm for the fitted line: here you can choose between linear, quadratic and cubic interpolation, but also more exotic choices, such as semi-log, inverse or loess (non-parametric). Of course, you can also turn this feature off.
3. Input and output: you can choose whether you want your graph on your computer screen (and possibly use the in-built graphical widget to further customize it — see above, page 35), or rather save it to a file. We support several graphical formats, among which PNG and PDF, to make it easy to incorporate your plots into text documents.

The following script uses the AWM dataset to exemplify some traditional plots in macroeconomics:

```
open AWM.gdt --quiet

# --- consumption and income, different styles -----------

gnuplot PCR YER
gnuplot PCR YER --output=display
gnuplot PCR YER --output=display --time-series
gnuplot PCR YER --output=display --time-series --with-lines

# --- Phillips' curve, different fitted lines ------------

gnuplot INFQ URX --output=display
gnuplot INFQ URX --suppress-fitted --output=display
gnuplot INFQ URX --inverse-fit --output=display
gnuplot INFQ URX --loess-fit --output=display
```

FIXME: comment on the above

For more detail, consult the *Gretl Command Reference.*

The `plot` command block

The `plot` environment is a way to pass information to Gnuplot in a more structured way, so that customization of basic plots becomes easier. It has the following characteristics:

[2]The technical condition for this is that the two-tailed p-value for the slope coefficient should be under 10%.

The block starts with the `plot` keyword, followed by a required parameter: the name of a list, a single series or a matrix. This parameter specifies the data to be plotted. The starting line may be prefixed with the `savename <-` apparatus to save a plot as an icon in the GUI program. The block ends with `end plot`.

Inside the block you have zero or more lines of these types, identified by an initial keyword:

`option`: specify a single option (details below)

`options`: specify multiple options on a single line; if more than one option is given on a line, the options should be separated by spaces.

`literal`: a command to be passed to gnuplot literally

`printf`: a printf statement whose result will be passed to gnuplot literally; this allows the use of string variables without having to resort to @-style string substitution.

The options available are basically those of the current `gnuplot` command, but with a few differences. For one thing you don't need the leading double-dash in an "option" (or "options") line. Besides that,

- You can't use the option `--matrix=whatever` with `plot`: that possibility is handled by providing the name of a matrix on the initial `plot` line.
- The `--input=filename` option is not supported: use `gnuplot` for the case where you're supplying the entire plot specification yourself.
- The several options pertaining to the presence and type of a fitted line, are replaced in `plot` by a single option `fit` which requires a parameter. Supported values for the parameter are: none, linear, quadratic, cubic, inverse, semilog and loess. Example:

  ```
  option fit=quadratic
  ```

As with `gnuplot`, the default is to show a linear fit in an X-Y scatter if it's significant at the 10 percent level.

Here's a simple example, the plot specification from the "bandplot" package, which shows how to achieve the same result via the `gnuplot` command and a `plot` block, respectively—the latter occupies a few more lines but is clearer

```
gnuplot 1 2 3 4 --with-lines --matrix=plotmat \
--suppress-fitted --output=display \
{ set linetype 3 lc rgb "#0000ff"; set title "@title"; \
  set nokey; set xlabel "@xname"; }

plot plotmat
  options with-lines fit=none
  literal set linetype 3 lc rgb "#0000ff"
  literal set nokey
  printf "set title \"%s\"", title
  printf "set xlabel \"%s\"", xname
end plot --output=display
```

Note that `--output=display` is appended to `end plot`; also note that if you give a matrix to `plot` it's assumed you want to plot all the columns. In addition, if you give a single series and the dataset is time series, it's assumed you want a time-series plot.

FIXME: provide an example with real data.

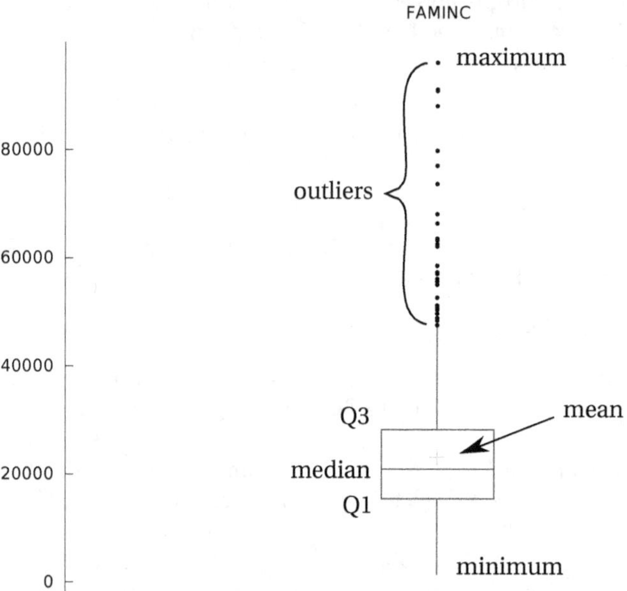

Figure 6.3: Sample boxplot

6.3 Boxplots

These plots (after Tukey and Chambers) display the distribution of a variable. The central box encloses the middle 50 percent of the data, i.e. it is bounded by the first and third quartiles. The "whiskers" extend to from each end of the box for a range equal to 1.5 times the interquartile range. Observations outside that range are considered outliers and represented via dots.[3] A line is drawn across the box at the median and a "+" sign identifies the mean—see Figure 6.3.

In the case of boxplots with confidence intervals, dotted lines show the limits of an approximate 90 percent confidence interval for the median. This is obtained by the bootstrap method, which can take a while if the data series is very long. For details on constructing boxplots, see the entry for `boxplot` in the *Gretl Command Reference* or use the Help button that appears when you select one of the boxplot items under the menu item "View, Graph specified vars" in the main gretl window.

Factorized boxplots

A nice feature which is quite useful for data visualization is the conditional, or factorized boxplot. This type of plot allows you to examine the distribution of a variable conditional on the value of some discrete factor.

As an example, we'll use one of the datasets supplied with gretl, that is `rac3d`, which contains an example taken from Cameron and Trivedi (2013) on the health conditions of 5190 people. The script below compares the unconditional (marginal) distribution of the number of illnesses in the past 2 weeks with the distribution of the same variable, conditional on age classes.

```
open rac3d.gdt
# unconditional boxplot
boxplot ILLNESS --output=display
# create a discrete variable for age class:
# 0 = below 20, 1 = between 20 and 39, etc
```

[3]To give you an intuitive idea, if a variable is normally distributed, the chances of picking an outlier by this definition are slightly below 0.7%.

```
series age_class = floor(AGE/0.2)
# conditional boxplot
boxplot ILLNESS age_class --factorized --output=display
```

After running the code above, you should see two graphs similar to Figure 6.4. By comparing the marginal plot to the factorized one, the effect of age on the mean number of illnesses is quite evident: by joining the green crosses you get what is technically known as the conditional mean function, or regression function if you prefer.

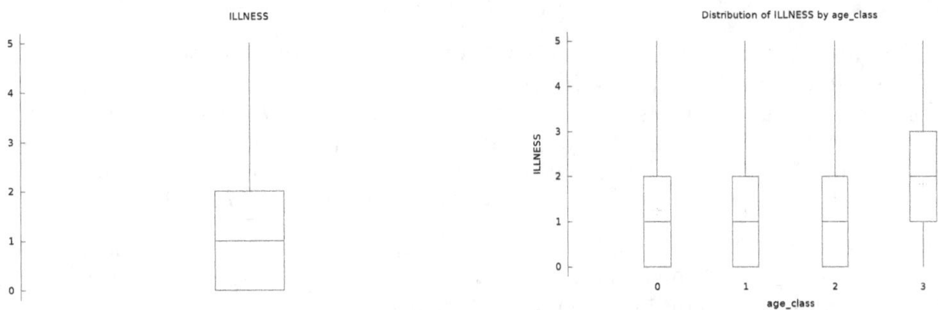

Figure 6.4: Conditional and unconditional distribution of illnesses

Chapter 7

Joining data sources

7.1 Introduction

Gretl provides two commands for adding data from file to an existing dataset in the program's workspace, namely append and join. The append command, which has been available for a long time, is relatively simple and is described in the *Gretl Command Reference*. Here we focus on the join command, which is much more flexible and sophisticated. This chapter gives an overview of the functionality of join along with a detailed account of its syntax and options. We provide several toy examples and discuss one real-world case at length.

First, a note on terminology: in the following we use the terms "left-hand" and "inner" to refer to the dataset that is already in memory, and the terms "right-hand" and "outer" to refer to the dataset in the file from which additional data are to be drawn.

Two main features of join are worth emphasizing at the outset:

- "Key" variables can be used to match specific observations (rows) in the inner and outer datasets, and this match need not be 1 to 1.
- A row filter may be applied to screen out unwanted observations in the outer dataset.

As will be explained below, these features support rather complex concatenation and manipulation of data from different sources.

A further aspect of join should be noted—one that makes this command particularly useful when dealing with very large data files. That is, when gretl executes a join operation it does not, in general, read into memory the entire content of the right-hand side dataset. Only those columns that are actually needed for the operation are read in full. This makes join faster and less demanding of computer memory than the methods available in most other software. On the other hand, gretl's asymmetrical treatment of the "inner" and "outer" datasets in join may require some getting used to, for users of other packages.

7.2 Basic syntax

The minimal invocation of join is

join *filename varname*

where *filename* is the name of a data file and *varname* is the name of a series to be imported. Only two sorts of data file are supported at present: delimited text files (where the delimiter may be comma, space, tab or semicolon) and "native" gretl data files (gdt or gdtb). A series named *varname* may already be present in the left-hand dataset, but that is not required. The series to be imported may be numerical or string-valued. For most of the discussion below we assume that just a single series is imported by each join command, but see section 7.7 for an account of multiple imports.

The effect of the minimal version of join is this: gretl looks for a data column labeled *varname* in the specified file; if such a column is found and the number of observations on the right matches the number of observations in the current sample range on the left, then the values from the right are copied into the relevant range of observations on the left. If *varname* does not already exist on the left, any observations outside of the current sample are set to NA; if it exists already then observations outside of the current sample are left unchanged.

The case where you want to rename a series on import is handled by the `--data` option. This option has one required argument, the name by which the series is known on the right. At this point we need to explain something about right-hand variable names (column headings).

Right-hand names

We accept on input arbitrary column heading strings, but if these strings do not qualify as valid gretl identifiers they are automatically converted, and in the context of `join` you must use the converted names. A gretl identifier must start with a letter, contain nothing but (ASCII) letters, digits and the underscore character, and must not exceed 31 characters. The rules used in name conversion are:

1. Skip any leading non-letters.
2. Until the 31-character is reached or the input is exhausted: transcribe "legal" characters; skip "illegal" characters apart from spaces; and replace one or more consecutive spaces with an underscore, unless the last character transcribed is an underscore in which case space is skipped.

In the unlikely event that this policy yields an empty string, we replace the original with `col`*n*, where *n* is replaced by the 1-based index of the column in question among those used in the join operation. If you are in doubt regarding the converted name of a given column, the function `fixname()` can be used as a check: it takes the original string as an argument and returns the converted name. Examples:

```
? eval fixname("valid_identifier")
valid_identifier
? eval fixname("12. Some name")
Some_name
```

Returning to the use of the `--data` option, suppose we have a column headed `"12. Some name"` on the right and wish to import it as `x`. After figuring how the right-hand name converts, we can do

```
join foo.csv x --data="Some_name"
```

No right-hand names?

Some data files have no column headings; they jump straight into the data (and you need to determine from accompanying documentation what the columns represent). Since gretl expects column headings, you have to take steps to get the importation right. It is generally a good idea to insert a suitable header row into the data file. However, if for some reason that's not practical, you should give the `--no-header` option, in which case gretl will name the columns on the right as `col1`, `col2` and so on. If you do not do either of these things you will likely lose the first row of data, since gretl will attempt to make variable names out of it, as described above.

7.3 Filtering

Rows from the outer dataset can be filtered using the `--filter` option. The required parameter for this option is a Boolean condition, that is, an expression which evaluates to non-zero (true, include the row) or zero (false, skip the row) for each of the outer rows. The filter expression may include any of the following terms: up to three "right-hand" series (under their converted names as explained above); scalar or string variables defined "on the left"; any of the operators and functions available in gretl (including user-defined functions); and numeric or string constants.

Here are a few simple examples of potentially valid filter options (assuming that the specified right-hand side columns are found):

```
# 1. relationship between two right-hand variables
--filter="x15<=x17"

# 2. comparison of right-hand variable with constant
--filter="nkids>2"

# 3. comparison of string-valued right-hand variable with string constant
--filter="SEX==\"F\""

# 4. filter on valid values of a right-hand variable
--filter=!missing(income)

# 5. compound condition
--filter="x < 100 && (x > 0 || y > 0)"
```

Note that if you are comparing against a string constant (as in example 3 above) it is necessary to put the string in "escaped" double-quotes (each double-quote preceded by a backslash) so the interpreter knows that F is not supposed to be the name of a variable.

It is safest to enclose the whole filter expression in double quotes, however this is not strictly required unless the expression contains spaces or the equals sign.

In general, an error is flagged if a missing value is encountered in a series referenced in a filter expression. This is because the condition then becomes indeterminate; taking example 2 above, if the `nkids` value is NA on any given row we are not in a position to evaluate the condition `nkids>2`. However, you can use the `missing()` function—or `ok()`, which is a shorthand for `!missing()`—if you need a filter that keys off the missing or non-missing status of a variable.

7.4 Matching with keys

Things get interesting when we come to key-matching. The purpose of this facility is perhaps best introduced by example. Suppose that (as with many survey and census-based datasets) we have a dataset that is composed of two or more related files, each having a different unit of observation; for example we have a "persons" data file and a "households" data file. Table 7.1 shows a simple, artificial case. The file `people.csv` contains a unique identifier for the individuals, `pid`. The households file, `hholds.csv`, contains the unique household identifier `hid`, which is also present in the persons file.

As a first example of `join` with keys, let's add the household-level variable `xh` to the persons dataset:

```
open people.csv --quiet
join hholds.csv xh --ikey=hid
print --byobs
```

The basic key option is named `ikey`; this indicates "inner key", that is, the key variable found in the left-hand or inner dataset. By default it is assumed that the right-hand dataset contains a column of the same name, though as we'll see below that assumption can be overridden. The `join` command above says, find a series named `xh` in the right-hand dataset and add it to the left-hand one, using the values of `hid` to match rows. Looking at the data in Table 7.1 we can see how this should work. Persons 1 and 2 are both members of household 1, so they should both get values of 1 for `xh`; persons 3 and 4 are members of household 2, so that `xh` = 4; and so on. Note that the order in which the key values occur on the right-hand side does not matter. The gretl output from the `print` command is shown in the lower panel of Table 7.1.

Note that key variables are treated conceptually as integers. If a specified key contains fractional values these are truncated.

Two extensions of the basic key mechanism are available.

- If the outer dataset contains a relevant key variable but it goes under a different name from the inner key, you can use the `--okey` option to specify the outer key. (As with

people.csv

```
pid,hid,gender,age,xp
1,1,M,50,1
2,1,F,40,2
3,2,M,30,3
4,2,F,25,2
5,3,M,40,3
6,4,F,35,4
7,4,M,70,3
8,4,F,60,3
9,5,F,20,4
10,6,M,40,4
```

hholds.csv

```
hid,country,xh
1,US,1
6,IT,12
3,UK,6
4,IT,8
2,US,4
5,IT,10
```

pid	hid	xh
1	1	1
2	1	1
3	2	4
4	2	4
5	3	6
6	4	8
7	4	8
8	4	8
9	5	10
10	6	12

Table 7.1: Two linked CSV data files, and the effect of a `join`

other right-hand names, this does not have to be a valid gretl identifier.) So, for example, if hholds.csv contained the hid information, but under the name HHOLD, the join command above could be modified as

```
join hholds.csv xh --ikey=hid --okey=HHOLD
```

- If a single key is not sufficient to generate the matches you want, you can specify a double key in the form of two series names separated by a comma; in this case the importation of data is restricted to those rows on which both keys match. The syntax here is, for example

  ```
  join foo.csv x --ikey=key1,key2
  ```

  Again, the --okey option may be used if the corresponding right-hand columns are named differently. The same number of keys must be given on the left and the right, but when a double key is used and only one of the key names differs on the right, the name that is in common may be omitted (although the comma separator must be retained). For example, the second of the following lines is acceptable shorthand for the first:

  ```
  join foo.csv x --ikey=key1,Lkey2 --okey=key1,Rkey2
  join foo.csv x --ikey=key1,Lkey2 --okey=,Rkey2
  ```

The number of key-matches

The example shown in Table 7.1 is an instance of a 1 to 1 match: applying the matching criterion produces exactly one value of the variable xh corresponding to each row of the inner dataset. Two other possibilities arise: it may be that some rows in the inner dataset have no match on the right, and/or some rows on the left have multiple matches on the right. The latter case ("1 to n matching") is addressed in detail in the next section; here we discuss the former.

The case where there's no match on the right is handled differently depending on whether the join operation is adding a new series to the inner dataset or modifying an existing one. If it's a new series, then the unmatched rows automatically get NA for the imported data. If, on the other hand, the join is pulling in values for a series that is already present on the left, only matched rows will be updated—or in other words, we do *not* overwite an existing value on the left with NA in the case where there's no match on the right.

These defaults may not produce the desired results in every case but gretl provides the means to modify the effect if need be. We will illustrate with two scenarios.

First, consider adding a new series recording "number of hours worked" when the inner dataset contains individuals and the outer file contains data on jobs. If an individual does not appear in the jobs file, we may want to take her hours worked as implicitly zero rather than NA. In this case gretl's misszero() function can be used to turn NA into 0 in the imported series.

Second, consider updating a series via join, when the outer file is presumed to contain all available updated values, such that "no match" should be taken as an implicit NA. In this case we want the (presumably out-of-date) values on any unmatched rows to be overwritten with NA. Let the series in question be called x (both on the left and the right) and let the common key be called pid. The solution is then

```
join update.csv tmpvar --data=x --ikey=pid
x = tmpvar
```

As a new variable, tmpvar will get NA for all unmatched rows; we then transcribe its values into x. In a more complicated case one might use the smpl command to limit the sample range before assigning tmpvar to x, or use the conditional assignment operator ?:.

One further point should be mentioned here. Given some missing values in an imported series you may sometimes want to know whether (a) the NAs were explicitly represented in the outer data file or (b) they arose due to "no match". You can find this out by using a method described in the following section, namely the count variant of the aggregation option: this will give you a series with 0 values for all and only unmatched rows.

7.5 Aggregation

In the case of 1 to n matching of rows ($n > 1$) the user must specify an "aggregation method"; that is, a method for mapping from n rows down to one. This is handled by the `--aggr` option which requires a single argument from the following list:

Code	*Value returned*
`count`	count of matches
`avg`	mean of matching values
`sum`	sum of matching values
`min`	minimum of matching values
`max`	maximum of matching values
`seq:`i	the ith matching value (e.g. `seq:2`)
`min(`*aux*`)`	minimum of matching values of auxiliary variable
`max(`*aux*`)`	maximum of matching values of auxiliary variable

Note that the `count` aggregation method is special, in that there is no need for a "data series" on the right; the imported series is simply a function of the specified key(s). All the other methods require that "actual data" are found on the right. Also note that when `count` is used, the value returned when no match is found is (as one might expect) zero rather than `NA`.

The basic use of the `seq` method is shown above: following the colon you give a positive integer representing the (1-based) position of the observation in the sequence of matched rows. Alternatively, a negative integer can be used to count down from the last match (`seq:-1` selects the last match, `seq:-2` the second-last match, and so on). If the specified sequence number is out of bounds for a given observation this method returns `NA`.

Referring again to the data in Table 7.1, suppose we want to import data from the persons file into a dataset established at household level. Here's an example where we use the individual `age` data from `people.csv` to add the average and minimum age of household members.

```
open hholds.csv --quiet
join people.csv avgage --ikey=hid --data=age --aggr=avg
join people.csv minage --ikey=hid --data=age --aggr=min
```

Here's a further example where we add to the household data the sum of the personal data `xp`, with the twist that we apply filters to get the sum specifically for household members under the age of 40, and for women.

```
open hholds.csv --quiet
join people.csv young_xp --ikey=hid --filter="age<40" --data=xp --aggr=sum
join people.csv female_xp --ikey=hid --filter="gender==\"F\"" --data=xp --aggr=sum
```

The possibility of using an auxiliary variable with the `min` and `max` modes of aggregation gives extra flexibility. For example, suppose we want for each household the income of its oldest member:

```
open hholds.csv --quiet
join people.csv oldest_xp --ikey=hid --data=xp --aggr=max(age)
```

7.6 String-valued key variables

The examples above use numerical variables (household and individual ID numbers) in the matching process. It is also possible to use string-valued variables, in which case a match means that the string values of the key variables compare equal (with case sensitivity). When using double keys, you can mix numerical and string keys, but naturally you cannot mix a string variable on the left (via `ikey`) with a numerical one on the right (via `okey`), or vice versa.

Here's a simple example. Suppose that alongside hholds.csv we have a file countries.csv with the following content:

```
country,GDP
UK,100
US,500
IT,150
FR,180
```

The variable country, which is also found in hholds.csv, is string-valued. We can pull the GDP of the country in which the household resides into our households dataset with

```
open hholds.csv -q
join countries.csv GDP --ikey=country
```

which gives

```
             hid      country        GDP

1              1            1        500
2              6            2        150
3              3            3        100
4              4            2        150
5              2            1        500
6              5            2        150
```

7.7 Importing multiple series

The examples given so far have been limited in one respect. While several columns in the outer data file may be referenced (as keys, or in filtering or aggregation) only one column has actually provided data—and correspondingly only one series in the inner dataset has been created or modified—per invocation of join. However, join can handle the importation of several series at once. This section gives an account of the required syntax along with certain restrictions that apply to the multiple-import case.

There are two ways to specify more than one series for importation:

1. The *varname* field in the command can take the form of a space-separated list of names rather than a single name.

2. Alternatively, you can give the name of an array of strings in place of *varname*: the elements of this array should be the names of the series to import.

Here are the limitations:

1. The --data option, which permits the renaming of a series on import, is not available. When importing multiple series you are obliged to accept their "outer" names, fixed up as described in section 7.2.

2. While the other join options are available, they necessarily apply uniformly to all the series imported via a given command. This means that if you want to import several series but using different keys, filters or aggregation methods you must use a sequence of commands.

Here are a couple of examples of multiple imports.

```
# open base datafile containing keys
open PUMSdata.gdt

# join using a list of import names
```

```
join ss13pnc.csv SCHL WAGP WKHP --ikey=SERIALNO,SPORDER

# using a strings array: may be worthwhile if the array
# will be used for more than one purpose
strings S = array(3)
S[1] = "SCHL"
S[2] = "WAGP"
S[3] = "WKHP"
join ss13pnc.csv S --ikey=SERIALNO,SPORDER
```

7.8 A real-world case

For a real use-case for `join` with cross-sectional data, we turn to the Bank of Italy's *Survey on Household Income and Wealth* (SHIW).[1] In ASCII form the 2010 survey results comprise 47 MB of data in 29 files. In this exercise we will draw on five of the SHIW files to construct a replica of the dataset used in Thomas Mroz's famous paper (Mroz, 1987) on women's labor force participation, which contains data on married women between the age of 30 and 60 along with certain characteristics of their households and husbands.

Our general strategy is as follows: we create a "core" dataset by opening the file `carcom10.csv`, which contains basic data on the individuals. After dropping unwanted individuals (all but married women), we use the resulting dataset as a base for pulling in further data via the `join` command.

The complete script to do the job is given in the Appendix to this chapter; here we walk through the script with comments interspersed. We assume that all the relevant files from the Bank of Italy survey are contained in a subdirectory called `SHIW`.

Starting with `carcom10.csv`, we use the `--cols` option to the `open` command to import specific series, namely `NQUEST` (household ID number), `NORD` (sequence number for individuals within each household), `SEX` (male = 1, female = 2), `PARENT` (status in household: 1 = head of household, 2 = spouse of head, etc.), `STACIV` (marital status: married = 1), `STUDIO` (educational level, coded from 1 to 8), `ETA` (age in years) and `ACOM4C` (size of town).

```
open SHIW/carcom10.csv --cols=1,2,3,4,9,10,29,41
```

We then restrict the sample to married women from 30 to 60 years of age, and additionally restrict the sample of women to those who are either heads of households or spouses of the head.

```
smpl SEX==2 && ETA>=30 && ETA<=60 && STACIV==1 --restrict
smpl PARENT<3 --restrict
```

For compatibility with the Mroz dataset as presented in the gretl data file `mroz87.gdt`, we rename the age and education variables as `WA` and `WE` respectively, we compute the `CIT` dummy and finally we store the reduced base dataset in gretl format.

```
rename ETA WA
rename STUDIO WE
series CIT = (ACOM4C > 2)

store mroz_rep.gdt
```

The next step will be to get data on working hours from the jobs file `allb1.csv`. There's a complication here. We need the total hours worked over the course of the year (for both the women and their husbands). This is not available as such, but the variables `ORETOT` and `MESILAV` give, respectively, average hours worked per week and the number of months worked in 2010, each on a per-job basis. If each person held at most one job over the year we could compute his or her annual hours as

[1] Details of the survey can be found at http://www.bancaditalia.it/statistiche/indcamp/bilfait/dismicro. The ASCII (CSV) data files for the 2010 survey are available at http://www.bancaditalia.it/statistiche/indcamp/bilfait/dismicro/annuale/ascii/ind10_ascii.zip.

```
HRS = ORETOT * 52 * MESILAV/12
```

However, some people had more than one job, and in this case what we want is the sum of annual hours across their jobs. We could use `join` with the `seq` aggregation method to construct this sum, but it is probably more straightforward to read the `allb1` data, compute the HRS values per job as shown above, and save the results to a temporary CSV file.

```
open SHIW/allb1.csv --cols=1,2,8,11 --quiet
series HRS = misszero(ORETOT) * 52 * misszero(MESILAV)/12
store HRS.csv NQUEST NORD HRS
```

Now we can reopen the base dataset and join the hours variable from `HRS.csv`. Note that we need a double key here: the women are uniquely identified by the combination of NQUEST and NORD. We don't need an `okey` specification since these keys go under the same names in the right-hand file. We define labor force participation, LFP, based on hours.

```
open mroz_rep.gdt
join HRS.csv WHRS --ikey=NQUEST,NORD --data=HRS --aggr=sum
WHRS = misszero(WHRS)
LFP = WHRS > 0
```

For reference, here's how we could have used `seq` to avoid writing a temporary file:

```
join SHIW/allb1.csv njobs --ikey=NQUEST,NORD --data=ORETOT --aggr=count
series WHRS = 0
loop i=1..max(njobs) -q
  join SHIW/allb1.csv htmp --ikey=NQUEST,NORD --data=ORETOT --aggr="seq:$i"
  join SHIW/allb1.csv mtmp --ikey=NQUEST,NORD --data=MESILAV --aggr="seq:$i"
  WHRS += misszero(htmp) * 52 * misszero(mtmp)/12
endloop
```

To generate the work experience variable, AX, we use the file `lavoro.csv`: this contains a variable named ETALAV which records the age at which the person first started work.

```
join SHIW/lavoro.csv ETALAV --ikey=NQUEST,NORD
series AX = misszero(WA - ETALAV)
```

We compute the woman's hourly wage, WW, as the ratio of total employment income to annual working hours. This requires drawing the series YL (payroll income) and YM (net self-employment income) from the persons file `rper10.csv`.

```
join SHIW/rper10.csv YL YM --ikey=NQUEST,NORD --aggr=sum
series WW = LFP ? (YL + YM)/WHRS : 0
```

The family's net disposable income is available as Y in the file `rfam10.csv`; we import this as FAMINC.

```
join SHIW/rfam10.csv FAMINC --ikey=NQUEST --data=Y
```

Data on number of children are now obtained by applying the `count` method. For the Mroz replication we want the number of children under the age of 6, and also the number aged 6 to 18.

```
join SHIW/carcom10.csv KIDS --ikey=NQUEST --aggr=count --filter="ETA<=18"
join SHIW/carcom10.csv KL6 --ikey=NQUEST --aggr=count --filter=ETA<6
series K618 = KIDS - KL6
```

We want to add data on the women's husbands, but how do we find them? To do this we create an additional inner key which we'll call H_ID (husband ID), by sub-sampling in turn on the observations falling into each of two classes: (a) those where the woman is recorded as head of household and (b) those where the husband has that status. In each case we want the individual ID (NORD) of the household member whose status is complementary to that of the woman in question. So for case (a) we subsample using PARENT==1 (head of household) and filter the join using PARENT==2 (spouse of head); in case (b) we do the converse. We thus construct H_ID piece-wise.

```
# for women who are household heads
smpl PARENT==1 --restrict --replace
join SHIW/carcom10.csv H_ID --ikey=NQUEST --data=NORD --filter="PARENT==2"
# for women who are not household heads
smpl PARENT==2 --restrict --replace
join SHIW/carcom10.csv H_ID --ikey=NQUEST --data=NORD --filter="PARENT==1"
smpl full
```

Now we can use our new inner key to retrieve the husbands' data, matching H_ID on the left with NORD on the right within each household.

```
join SHIW/carcom10.csv HA --ikey=NQUEST,H_ID --okey=NQUEST,NORD --data=ETA
join SHIW/carcom10.csv HE --ikey=NQUEST,H_ID --okey=NQUEST,NORD --data=STUDIO
join HRS.csv HHRS --ikey=NQUEST,H_ID --okey=NQUEST,NORD --data=HRS --aggr=sum
HHRS = misszero(HHRS)
```

The remainder of the script is straightforward and does not require discussion here: we recode the education variables for compatibility; delete some intermediate series that are not needed any more; add informative labels; and save the final product. See the Appendix for details.

To compare the results from this dataset with those from the earlier US data used by Mroz, one can copy the input file `heckit.inp` (supplied with the gretl package) and substitute `mroz_rep.gdt` for `mroz87.gdt`. It turns out that the results are qualitatively very similar.

7.9 The representation of dates

Up to this point all the data we have considered have been cross-sectional. In the following sections we discuss data that have a time dimension, and before proceeding it may be useful to say something about the representation of dates. Gretl takes the ISO 8601 standard as its reference point but provides mean of converting dates provided in other formats; it also offers a set of calendrical functions for manipulating dates (`isodate`, `isoconv`, `epochday` and others).

ISO 8601 recognizes two formats for daily dates, "extended" and "basic". In both formats dates are given as 4-digit year, 2-digit month and 2-digit day, in that order. In extended format a dash is inserted between the fields—as in `2013-10-21` or more generally YYYY-MM-DD—while in basic format the fields are run together (YYYYMMDD). Extended format is more easily parsed by human readers while basic format is more suitable for computer processing, since one can apply ordinary arithmetic to compare dates as equal, earlier or later. The standard also recognizes YYYY-MM as representing year and month, e.g. `2010-11` for November 2010,[2] as well as a plain four-digit number for year alone.

One problem for economists is that the "quarter" is not a period covered by ISO 8601. This could be presented by YYYY-Q (with only one digit following the dash) but in gretl output we in fact use a colon, as in `2013:2` for the second quarter of 2013. (For printed output of months gretl also uses a colon, as in `2013:06`. A difficulty with following ISO here is that in a statistical context a string such as `1980-10` may look more like a subtraction than a date.) Anyway, at present we are more interested in the parsing of dates on input rather than in what gretl prints. And in that context note that "excess precision" is acceptable: a month may be represented by its first day (e.g. `2005-05-01` for May, 2005), and a quarter may be represented by its first month and day (`2005-07-01` for the third quarter of 2005).

Some additional points regarding dates will be taken up as they become relevant in practical cases of joining data.

[2]The form YYYYMM is *not* recognized for year and month.

7.10 Time-series data

Suppose our left-hand dataset is recognized by gretl as time series with a supported frequency (annual, quarterly, monthly, weekly, daily or hourly). This will be the case if the original data were read from a file that contained suitable time or date information, or if a time-series interpretation has been imposed using either the `setobs` command or its GUI equivalent. Then—apart, perhaps, from some very special cases—joining additional data is bound to involve matching observations by time-period.

In this case, contrary to the cross-sectional case, the inner dataset has a natural ordering which gretl is aware of; hence, no "inner key" is required. All we need is a means of identifying the period on the right and this is why, in a time-series context, we'll refer to the outer key as the "time key". Most likely, this information will appear in a single column in the outer data file, often but not always the first column.

The `join` command provides a simple (but limited) default for extracting period information from the outer data file, plus an option that can be used if the default is not applicable, as follows.

- The default assumptions are: (1) the time key appears in the first column; (2) the heading of this column is either left blank or is one of `obs`, `date`, `year`, `period`, `observation`, or `observation_date` (on a case-insensitive comparison); and (3) the time format conforms to ISO 8601 where applicable ("extended" daily date format `YYYY-MM-DD`, monthly format `YYYY-MM`, or annual format `YYYY`).
- If dates do not appear in the first column of the outer file, or if the column heading or format is not as just described, the `--tkey` option can be used to indicate which column should be used and/or what format should be assumed.

Setting the time-key column and/or format

The `--tkey` option requires a parameter holding the name of the column in which the time key is located and/or a string specifying the format in which dates/times are written in the time-key column. This parameter should be enclosed in double-quotes. If both elements are present they should be separated by a comma; if only a format is given it should be preceded by a comma. Some examples:

```
--tkey="Period,%m/%d/%Y"
--tkey="Period"
--tkey="obsperiod"
--tkey=",%Ym%m"
```

The first of these applies if `Period` is not the first column on the right, and dates are given in the US format of month, day, year, separated by slashes. The second implies that although `Period` is not the first column, the date format is ISO 8601. The third again implies that the date format is OK; here the name is required even if `obsperiod` is the first column since this heading is not one recognized by gretl's heuristic. The last example implies that dates are in the first column (with one of the recognized headings), but are given in the non-standard format year, "`m`", month.

The date format string should be composed using the codes employed by the POSIX function `strptime`; Table 7.2 contains a list of the most relevant codes.[3]

Example: daily stock prices

We show below the first few lines of a file named `IBM.csv` containing stock-price data for IBM corporation.

[3] The `%q` code for quarter is not present in `strptime`; it is added for use with `join` since quarterly data are common in macroeconomics.

Code	Meaning
%%	The % character.
%b	The month name according to the current locale, either abbreviated or in full.
%C	The century number (0-99).
%d	The day of month (1-31).
%D	Equivalent to %m/%d/%y. (This is the American style date, very confusing to non-Americans, especially since %d/%m/%y is widely used in Europe. The ISO 8601 standard format is %Y-%m-%d.)
%H	The hour (0-23).
%j	The day number in the year (1-366).
%m	The month number (1-12).
%n	Arbitrary whitespace.
%q	The quarter (1-4).
%w	The weekday number (0-6) with Sunday = 0.
%y	The year within century (0-99). When a century is not otherwise specified, values in the range 69-99 refer to years in the twentieth century (1969-1999); values in the range 00-68 refer to years in the twenty-first century (2000-2068).
%Y	The year, including century (for example, 1991).

Table 7.2: Date format codes

```
Date,Open,High,Low,Close,Volume,Adj Close
2013-08-02,195.50,195.50,193.22,195.16,3861000,195.16
2013-08-01,196.65,197.17,195.41,195.81,2856900,195.81
2013-07-31,194.49,196.91,194.49,195.04,3810000,195.04
```

Note that the data are in reverse time-series order—that won't matter to `join`, the data can appear in any order. Also note that the first column is headed `Date` and holds daily dates as ISO 8601 extended. That means we can pull the data into gretl very easily. In the following fragment we create a suitably dimensioned empty daily dataset then rely on the default behavior of `join` with time-series data to import the closing stock price.

```
nulldata 500
setobs 5 2012-01-01
join IBM.csv Close
```

To make explicit what we're doing, we could accomplish exactly the same using the `--tkey` option:

```
join IBM.csv Close --tkey="Date,%Y-%m-%d"
```

Example: OECD quarterly data

Table 7.3 shows an excerpt from a CSV file provided by the OECD statistical site (`stat.oecd.org`) in response to a request for GDP at constant prices for several countries.[4]

This is an instance of data in what we call *atomic format*, that is, a format in which each line of the outer file contains a single data-point and extracting data mainly requires filtering the appropriate lines. The outer time key is under the `Period` heading, and has the format `Q<quarter>-<year>`. Assuming that the file in Table 7.3 has the name `oecd.csv`, the following script reconstructs the time series of Gross Domestic Product for several countries:

[4]Retrieved 2013-08-05. The OECD files in fact contain two leading columns with very long labels; these are irrelevant to the present example and can be omitted without altering the sample script.

```
Frequency,Period,Country,Value,Flags
"Quarterly","Q1-1960","France",463876.148126845,E
"Quarterly","Q1-1960","Germany",768802.119278467,E
"Quarterly","Q1-1960","Italy",414629.791450547,E
"Quarterly","Q1-1960","United Kingdom",578437.090291889,E
"Quarterly","Q2-1960","France",465618.977328614,E
"Quarterly","Q2-1960","Germany",782484.138122549,E
"Quarterly","Q2-1960","Italy",420714.910290157,E
"Quarterly","Q2-1960","United Kingdom",572853.474696578,E
"Quarterly","Q3-1960","France",469104.41925852,E
"Quarterly","Q3-1960","Germany",809532.161494483,E
"Quarterly","Q3-1960","Italy",426893.675840156,E
"Quarterly","Q3-1960","United Kingdom",581252.066618986,E
"Quarterly","Q4-1960","France",474664.327992619,E
"Quarterly","Q4-1960","Germany",817806.132384948,E
"Quarterly","Q4-1960","Italy",427221.338414114,E
...
```

Table 7.3: Example of CSV file as provided by the OECD statistical website

```
nulldata 220
setobs 4 1960:1

join oecd.csv FRA --tkey="Period,Q%q-%Y" --data=Value --filter="Country==\"France\""
join oecd.csv GER --tkey="Period,Q%q-%Y" --data=Value --filter="Country==\"Germany\""
join oecd.csv ITA --tkey="Period,Q%q-%Y" --data=Value --filter="Country==\"Italy\""
join oecd.csv  UK --tkey="Period,Q%q-%Y" --data=Value --filter="Country==\"United Kingdom\""
```

Note the use of the format codes `%q` for the quarter and `%Y` for the 4-digit year. A touch of elegance could have been added by storing the invariant options to `join` using the `setopt` command, as in

```
setopt join persist --tkey="Period,Q%q-%Y" --data=Value
join oecd.csv FRA --filter="Country==\"France\""
join oecd.csv GER --filter="Country==\"Germany\""
join oecd.csv ITA --filter="Country==\"Italy\""
join oecd.csv  UK --filter="Country==\"United Kingdom\""
setopt join clear
```

If one were importing a large number of such series it might be worth rewriting the sequence of joins as a loop, as in

```
sprintf countries "France Germany Italy \"United Kingdom\""
sprintf vnames "FRA GER ITA UK"
setopt join persist --tkey="Period,Q%q-%Y" --data=Value

loop foreach i @countries
  string vname = strsplit(vnames, i)
  join oecd.csv @vname --filter="Country==\"$i\""
endloop
setopt join clear
```

7.11 Special handling of time columns

When dealing with straight time series data the `tkey` mechanism described above should suffice in almost all cases. In some contexts, however, time enters the picture in a more complex way; examples include panel data (see section 7.12) and so-called realtime data (see chapter 8). To handle such cases `join` provides the `--tconvert` option. This can be used to select certain columns in the right-hand data file for special treatment: strings representing dates in these columns will be converted to numerical values: 8-digit numbers on the pattern YYYYMMDD (ISO basic daily format). Once dates are in this form it is easy to use them in key-matching or filtering.

By default it is assumed that the strings in the selected columns are in ISO extended format, `YYYY-MM-DD`. If that is not the case you can supply a time-format string using the `--tconv-fmt` option. The format string should be written using the codes shown in Table 7.2.

Here are some examples:

```
# select one column for treatment
--tconvert=start_date

# select two columns for treatment
--tconvert="start_date,end_date"

# specify US-style daily date format
--tconv-fmt="%m/%d/%Y"

# specify quarterly date-strings (as in 2004q1)
--tconv-fmt="%Yq%q"
```

Some points to note:

- If a specified column is not selected for a substantive role in the join operation (as data to be imported, as a key, or as an auxiliary variable for use in aggregation) the column in question is not read and so no conversion is carried out.
- If a specified column contains numerical rather than string values, no conversion is carried out.
- If a string value in a selected column fails parsing using the relevant time format (user-specified or default), the converted value is `NA`.
- On successful conversion, the output is always in daily-date form as stated above. If you specify a monthly or quarterly time format, the converted date is the first day of the month or quarter.

7.12 Panel data

In section 7.10 we gave an example of reading quarterly GDP data for several countries from an OECD file. In that context we imported each country's data as a distinct time-series variable. Now suppose we want the GDP data in panel format instead (stacked time series). How can we do this with `join`?

As a reminder, here's what the OECD data look like:

```
Frequency,Period,Country,Value,Flags
"Quarterly","Q1-1960","France",463876.148126845,E
"Quarterly","Q1-1960","Germany",768802.119278467,E
"Quarterly","Q1-1960","Italy",414629.791450547,E
"Quarterly","Q1-1960","United Kingdom",578437.090291889,E
"Quarterly","Q2-1960","France",465618.977328614,E
```

and so on. If we have four countries and quarterly observations running from 1960:1 to 2013:2 ($T = 214$ quarters) we might set up our panel workspace like this:

```
scalar N = 4
scalar T = 214
scalar NT = N*T
nulldata NT --preserve
setobs T 1.1 --stacked-time-series
```

The relevant outer keys are obvious: `Country` for the country and `Period` for the time period. Our task is now to construct matching keys in the inner dataset. This can be done via two panel-specific options to the `setobs` command. Let's work on the time dimension first:

```
setobs 4 1960:1 --panel-time
series quarter = $obsdate
```

This variant of setobs allows us to tell gretl that time in our panel is quarterly, starting in the first quarter of 1960. Having set that, the accessor $obsdate will give us a series of 8-digit dates representing the first day of each quarter—19600101, 19600401, 19600701, and so on, repeating for each country. As we explained in section 7.11, we can use the --tconvert option on the outer series Period to get exactly matching values (in this case using a format of Q%q-%Y for parsing the Period values).

Now for the country names:

```
string cstrs
sprintf cstrs "France Germany Italy \"United Kingdom\""
setobs country cstrs --panel-groups
```

Here we write into the string cstrs the names of the countries, using escaped double-quotes to handle the space in "United Kingdom", then pass this string to setobs with the --panel-groups option, preceded by the identifier country. This asks gretl to construct a string-valued series named country, in which each name will repeat T times.

We're now ready to join. Assuming the OECD file is named oecd.csv we do

```
join oecd.csv GDP --data=Value \
 --ikey=country,quarter --okey=Country,Period \
 --tconvert=Period --tconv-fmt="Q%q-%Y"
```

Other input formats

The OECD file discussed above is in the most convenient format for join, with one data-point per line. But sometimes we may want to make a panel from a data file structured like this:

```
# Real GDP
Period,France,Germany,Italy,"United Kingdom"
"Q1-1960",463863,768757,414630,578437
"Q2-1960",465605,782438,420715,572853
"Q3-1960",469091,809484,426894,581252
"Q4-1960",474651,817758,427221,584779
"Q1-1961",482285,826031,442528,594684
...
```

Call this file side_by_side.csv. Assuming the same initial set-up as above, we can panelize the data by setting the sample to each country's time series in turn and importing the relevant column. The only point to watch here is that the string "United Kingdom", being a column heading, will become United_Kingdom on importing (see section 7.2) so we'll need a slightly different set of country strings.

```
sprintf cstrs "France Germany Italy United_Kingdom"
setobs country cstrs --panel-groups
loop foreach i @cstrs --quiet
  smpl country=="$i" --restrict --replace
  join side_by_side.csv GDP --data=$i \
  --ikey=quarter --okey=Period \
  --tconvert=Period --tconv-fmt="Q%q-%Y"
endloop
smpl full
```

If our working dataset and the outer data file are dimensioned such that there are just as many time-series observations on the right as there are time slots on the left—and the observations on the right are contiguous, in chronological order, and start on the same date as the working dataset—we could dispense with the key apparatus and just use the first line of the join command shown above. However, in general it is safer to use keys to ensure that the data end up in correct registration.

7.13 Memo: `join` options

Basic syntax: `join` *filename varname(s)* [*options*]

flag	*effect*
`--data`	Give the name of the data column on the right, in case it differs from *varname* (7.2); single import only
`--filter`	Specify a condition for filtering data rows (7.3)
`--ikey`	Specify up to two keys for matching data rows (7.4)
`--okey`	Specify outer key name(s) in case they differ the inner ones (7.4)
`--aggr`	Select an aggregation method for 1 to n joins (7.5)
`--tkey`	Specify right-hand time key (7.10)
`--tconvert`	Select outer date columns for conversion to numeric form (7.11)
`--tconv-fmt`	Specify a format for use with `tconvert` (7.11)
`--no-header`	Treat the first row on the right as data (7.2)
`--verbose`	Report on progress in reading the outer data

Appendix: the full Mroz data script

```
# start with everybody; get gender, age and a few other variables
# directly while we're at it
open SHIW/carcom10.csv --cols=1,2,3,4,9,10,29,41

# subsample on married women between the ages of 30 and 60
smpl SEX==2 && ETA>=30 && ETA<=60 && STACIV==1 --restrict
# for simplicity, restrict to heads of households and their spouses
smpl PARENT<3  --restrict

# rename the age and education variables for compatibility; compute
# the "city" dummy and finally save the reduced base dataset
rename ETA WA
rename STUDIO WE
series CIT = (ACOM4C>2)
store mroz_rep.gdt

# make a temp file holding annual hours worked per job
open SHIW/allb1.csv --cols=1,2,8,11 --quiet
series HRS = misszero(ORETOT) * 52 * misszero(MESILAV)/12
store HRS.csv NQUEST NORD HRS

# reopen the base dataset and begin drawing assorted data in
open mroz_rep.gdt

# women's annual hours (summed across jobs)
join HRS.csv WHRS --ikey=NQUEST,NORD --data=HRS --aggr=sum
WHRS = misszero(WHRS)

# labor force participation
LFP = WHRS > 0

# work experience: ETALAV = age when started first job
join SHIW/lavoro.csv ETALAV --ikey=NQUEST,NORD
series AX = misszero(WA - ETALAV)

# women's hourly wages
join SHIW/rper10.csv YL YM --ikey=NQUEST,NORD --aggr=sum
series WW = LFP ? (YL + YM)/WHRS : 0

# family income (Y = net disposable income)
join SHIW/rfam10.csv FAMINC --ikey=NQUEST --data=Y

# get data on children using the "count" method
join SHIW/carcom10.csv KIDS --ikey=NQUEST --aggr=count --filter="ETA<=18"
join SHIW/carcom10.csv KL6 --ikey=NQUEST --aggr=count --filter=ETA<6
series K618 = KIDS - KL6

# data on husbands: we first construct an auxiliary inner key for
# husbands, using the little trick of subsampling the inner dataset
#
# for women who are household heads
smpl PARENT==1 --restrict --replace
join SHIW/carcom10.csv H_ID --ikey=NQUEST --data=NORD --filter="PARENT==2"
# for women who are not household heads
smpl PARENT==2 --restrict --replace
join SHIW/carcom10.csv H_ID --ikey=NQUEST --data=NORD --filter="PARENT==1"
smpl full

# add husbands' data via the newly-added secondary inner key
join SHIW/carcom10.csv HA --ikey=NQUEST,H_ID --okey=NQUEST,NORD --data=ETA
join SHIW/carcom10.csv HE --ikey=NQUEST,H_ID --okey=NQUEST,NORD --data=STUDIO
join HRS.csv HHRS --ikey=NQUEST,H_ID --okey=NQUEST,NORD --data=HRS --aggr=sum
```

```
HHRS = misszero(HHRS)

# final cleanup begins

# recode educational attainment as years of education
matrix eduyrs = {0, 5, 8, 11, 13, 16, 18, 21}
series WE = replace(WE, seq(1,8), eduyrs)
series HE = replace(HE, seq(1,8), eduyrs)

# cut some cruft
delete SEX STACIV KIDS YL YM PARENT H_ID ETALAV

# add some labels for the series
setinfo LFP -d "1 if woman worked in 2010"
setinfo WHRS -d "Wife's hours of work in 2010"
setinfo KL6 -d "Number of children less than 6 years old in household"
setinfo K618 -d "Number of children between ages 6 and 18 in household"
setinfo WA -d "Wife's age"
setinfo WE -d "Wife's educational attainment, in years"
setinfo WW -d "Wife's average hourly earnings, in 2010 euros"
setinfo HHRS -d "Husband's hours worked in 2010"
setinfo HA -d "Husband's age"
setinfo HE -d "Husband's educational attainment, in years"
setinfo FAMINC -d "Family income, in 2010 euros"
setinfo AX -d "Actual years of wife's previous labor market experience"
setinfo CIT -d "1 if live in large city"

# save the final product
store mroz_rep.gdt
```

Chapter 8

Realtime data

8.1 Introduction

As of gretl version 1.9.13 the join command (see chapter 7) has been enhanced to deal with so-called realtime datasets in a straightforward manner. Such datasets contain information on when the observations in a time series were actually published by the relevant statistical agency and how they have been revised over time. Probably the most popular sources of such data are the "Alfred" online database at the St. Louis Fed (http://alfred.stlouisfed.org/) and the OECD's StatExtracts site, http://stats.oecd.org/. The examples in this chapter deal with files downloaded from these sources, but should be easy to adapt to files with a slightly different format.

As already stated, join requires a column-oriented plain text file, where the columns may be separated by commas, tabs, spaces or semicolons. Alfred and the OECD provide the option to download realtime data in this format (tab-delimited files from Alfred, comma-delimited from the OECD). If you have a realtime dataset in a spreadsheet file you must export it to a delimited text file before using it with join.

Representing revision histories is more complex than just storing a standard time series, because for each observation period you have in general more than one published value over time, along with the information on when each of these values were valid or current. Sometimes this is represented in spreadsheets with two time axes, one for the observation period and another one for the publication date or "vintage". The filled cells then form an upper triangle (or a "guillotine blade" shape, if the publication dates do not reach back far enough to complete the triangle). This format can be useful for giving a human reader an overview of realtime data, but it is not optimal for automatic processing; for that purpose "atomic" format is best.

8.2 Atomic format for realtime data

What we are calling atomic format is exactly the format used by Alfred if you choose the option "Observations by Real-Time Period", and by the OECD if you select all editions of a series for download as plain text (CSV).[1] A file in this format contains one actual data-point per line, together with associated metadata. This is illustrated in Table 8.1, where we show the first three lines from an Alfred file and an OECD file (slightly modified).[2]

Consider the first data line in the Alfred file: in the observation_date column we find 1960-01-01, indicating that the data-point on this line, namely 112.0, is an observation or measurement (in this case, of the US index of industrial production) that refers to the period starting on January 1st 1960. The realtime_start_date value of 1960-02-16 tells us that this value was published on February 16th 1960, and the realtime_end_date value says that this vintage remained current through March 15th 1960. On the next day (as we can see from the following line) this data-point was revised slightly downward to 111.0.

Daily dates in Alfred files are given in ISO extended format, YYYY-MM-DD, but below we describe how to deal with differently formatted dates. Note that daily dates are appropriate for the last two columns, which jointly record the interval over which a given data vintage was current. Daily dates might, however, be considered overly precise for the first column, since the data period may well be the year, quarter or month (as it is in fact here). However, following Alfred's

[1] If you choose to download in Excel format from OECD you get a file in the triangular or guillotine format mentioned above.

[2] In the Alfred file we have used commas rather than tabs as the column delimiter; in the OECD example we have shortened the name in the Variable column.

Alfred: monthly US industrial production

```
observation_date,INDPRO,realtime_start_date,realtime_end_date
1960-01-01,112.0000,1960-02-16,1960-03-15
1960-01-01,111.0000,1960-03-16,1961-10-15
```

OECD: monthly UK industrial production

```
Country,Variable,Frequency,Time,Edition,Value,Flags
"United Kingdom","INDPRO","Monthly","Jan-1990","February 1999",100,
"United Kingdom","INDPRO","Monthly","Feb-1990","February 1999",99.3,
```

Table 8.1: Variant atomic formats for realtime data

practice it is acceptable to specify a daily date, indicating the first day of the period, even for non-daily data.[3]

Compare the first data line of the OECD example. There's a greater amount of leading metadata, which is left implicit in the Alfred file. Here `Time` is the equivalent of Alfred's `observation_date`, and `Edition` the equivalent of Alfred's `realtime_start_date`. So we read that in February 1999 a value of 100 was current for the UK index of industrial production for January 1990, and from the next line we see that in the same vintage month a value of 99.3 was current for industrial production in February 1990.

Besides the different names and ordering of the columns, there are a few more substantive differences between Alfred and OECD files, most of which are irrelevant for `join` but some of which are (possibly) relevant.

The first (irrelevant) difference is the ordering of the lines. It appears (though we're not sure how consistent this is) that in Alfred files the lines are sorted by observation date first and then by publication date—so that all revisions of a given observation are grouped together—while OECD files are sorted first by revision date (`Edition`) and then by observation date (`Time`). If we want the next revision of UK industrial production for January 1990 in the OECD file we have to scan down several lines until we find

```
"United Kingdom","INDPRO","Monthly","Jan-1990","March 1999",100,
```

This difference is basically irrelevant because `join` can handle the case where the lines appear in random order, although some operations can be coded more conveniently if we're able to assume chronological ordering (either on the Alfred or the OECD pattern, it doesn't matter).

The second (also irrelevant) difference is that the OECD seems to include periodic "Edition" lines even when there is no change from the previous value (as illustrated above, where the UK industrial production index for January 1990 is reported as 100 as of March 1999, the same value that we saw to be current in February 1999), while Alfred reports a new value only when it differs from what was previously current.

A third difference lies in the dating of the revisions or editions. As we have seen, Alfred gives a specific daily date while (in the UK industrial production file at any rate), the OECD just dates each edition to a month. This is not necessarily relevant for `join`, but it does raise the question of whether the OECD might date revisions to a finer granularity in some of their files, in which case one would have to be on the lookout for a different date format.

The final difference is that Alfred supplies an "end date" for each data vintage while the OECD supplies only a starting date. But there is less to this difference than meets the eye: according to the Alfred webmaster, "by design, a new vintage must start immediately following (the day after) the lapse of the old vintage"—so the end date conveys no independent information.[4]

---

[3]Notice that this implies that in the Alfred example it is not clear without further information whether the observation period is the first quarter of 1960, the month January 1960, or the day January 1st 1960. However, we assume that this information is always available in context.

[4]Email received from Travis May of the Federal Reserve Bank of St. Louis, 2013-10-17. This closes off the possibility that a given vintage could lapse or expire some time before the next vintage becomes available, hence giving rise to a "hole" in an Alfred realtime file.

8.3 More on time-related options

Before we get properly started it is worth saying a little more about the --tkey and --tconvert options to join (first introduced in section 7.11), as they apply in the case of realtime data.

When you're working with regular time series data tkey is likely to be useful while tconvert is unlikely to be applicable (see section 7.10). On the other hand, when you're working with panel data tkey is definitely not applicable but tconvert may well be helpful (section 7.12). When working with realtime data, however, depending on the task in hand both options may be useful. You will likely need tkey; you may well wish to select at least one column for tconvert treatment; and in fact you may want to name a given column in both contexts—that is, include the tkey variable among the tconvert columns.

Why might this make sense? Well, think of the --tconvert option as a "preprocessing" directive: it asks gretl to convert date strings to numerical values (8-digit ISO basic dates) "at source", as they are read from the outer datafile. The --tkey option, on the other hand, singles out a column as the one to use for matching rows with the inner dataset. So you would want to name a column in both roles if (a) it should be used for matching periods and also (b) it is desirable to have the values from this column in numerical form, most likely for use in filtering.

As we have seen, you can supply specific formats in connection with both tkey and tconvert (in the latter case via the companion option --tconv-fmt) to handle the case where the date strings on the right are not ISO-friendly at source. This raises the question of how the format specifications work if a given column is named under both options. Here are the rules that gretl applies:

1. If a format is given with the --tkey option it always applies to the tkey column alone; and for that column it overrides any format given via the --tconv-fmt option.
2. If a format is given via tconv-fmt it is assumed to apply to all the tconvert columns, unless this assumption is overriden by rule 1.

8.4 Getting a certain data vintage

The most common application of realtime data is to "travel back in time" and retrieve the data that were current as of a certain date in the past. This would enable you to replicate a forecast or other statistical result that could have been produced at that date.

For example, suppose we are interested in a variable of monthly frequency named INDPRO, realtime data on which is stored in an Alfred file named INDPRO.txt, and we want to check the status quo as of June 15th 2011.

If we don't already have a suitable dataset into which to import the INDPRO data, our first steps will be to create an appropriately dimensioned empty dataset using the nulldata command and then specify its time-series character via setobs, as in

```
nulldata 132
setobs 12 2004:01
```

For convenience we can put the name of our realtime file into a string variable. On Windows this might look like

```
string fname = "C:/Users/yourname/Downloads/INDPRO.txt"
```

We can then import the data vintage 2011-06-15 using join, arbitrarily choosing the self-explanatory identifier ip_asof_20110615.

```
join @fname ip_asof_20110615 --tkey=observation_date --data=INDPRO \
--tconvert="realtime_start_date" \
--filter="realtime_start_date<=20110615" --aggr=max(realtime_start_date)
```

Here some detailed explanations of the various options are warranted:

- The `--tkey` option specifies the column which should be treated as holding the observation period identifiers to be matched against the periods in the current gretl dataset.[5] The more general form of this option is `--tkey="colname,format"` (note the double quotes here), so if the dates do not come in standard format, we can tell gretl how to parse them by using the appropriate conversion specifiers as shown in Table 7.2. For example, here we could have written `--tkey="observation_date,%Y-%m-%d"`.
- Next, `--data=INDPRO` tells gretl that we want to retrieve the entries stored in the column named `INDPRO`.
- As explained in section 7.11 the `--tconvert` option selects certain columns in the right-hand data file for conversion from date strings to 8-digit numbers on the pattern YYYYMMDD. We'll need this for the next step, filtering, since the transformation to numerical values makes it possible to perform basic arithmetic on dates. Note that since date strings in Alfred files conform to gretl's default assumption it is not necessary to use the `--tconv-fmt` option here.
- The `--filter` option specification in combination with the subsequent `--aggr` aggregation treatment is the central piece of our data retrieval; notice how we use the date constant 20110615 in ISO basic form to do numerical comparisons, and how we perform the numerical `max` operation on the converted column `realtime_start_date`. It would also have been possible to predefine a scalar variable, as in

  ```
  vintage = 20110615
  ```

  and then use `vintage` in the `join` command instead. Here we tell `join` that we only want to extract those publications that (1) already appeared before (and including) June 15th 2011, and (2) were not yet obsoleted by a newer release.[6]

As a result, your dataset will now contain a time series named `ip_asof_20110615` with the values that a researcher would have had available on June 15th 2011. Of course, all values for the observations after June 2011 will be missing (and probably a few before that, too), because they only have become available later on.

8.5 Getting the n-th release for each observation period

For some purposes it may be useful to retrieve the n-th published value of each observation, where n is a fixed positive integer, irrespective of *when* each of these n-th releases was published. Suppose we are interested in the third release, then the relevant `join` command becomes:

```
join @fname ip_3rdpub --tkey=observation_date --data=INDPRO --aggr="seq:3"
```

Since we do not need the `realtime_start_date` information for this retrieval, we have dropped the `--tconvert` option here. Note that this formulation assumes that the source file is ordered chronologically, otherwise using the option `--aggr="seq:3"`, which retrieves the third value from each sequence of matches, could have yielded a result different from the one intended. However, this assumption holds for Alfred files and is probably rather safe in general.

The values of the variable imported as `ip_3rdpub` in this way were published at different dates, so the variable is effectively a mix of different vintages. Depending on the type of variable, this may also imply drastic jumps in the values; for example, index numbers are regularly re-based to different base periods. This problem also carries over to inflation-adjusted economic variables, where the base period of the price index changes over time. Mixing vintages in general also means mixing different scales in the output, with which you would have to deal appropriately.[7]

---

[5] Strictly speaking, using `--tkey` is unnecessary in this example because we could just have relied on the default, which is to use the first column in the source file for the periods. However, being explicit is often a good idea.

[6] By implementing the second condition through the `max` aggregation on the `realtime_start_date` column alone, without using the `realtime_end_date` column, we make use of the fact that Alfred files cannot have "holes" as explained before.

[7] Some user-contributed functions may be available that address this issue, but it is beyond our scope here.

8.6 Getting the values at a fixed lag after the observation period

New data releases may take place on any day of the month, and as we have seen the specific day of each release is recorded in realtime files from Alfred. However, if you are working with, say, monthly or quarterly data you may sometimes want to adjust the granularity of your realtime axis to a monthly or quarterly frequency. For example, in order to analyse the data revision process for monthly industrial production you might be interested in the extent of revisions between the data available two and three months after each observation period.

This is a relatively complicated task and there is more than one way of accomplishing it. Either you have to make several passes through the outer dataset or you need a sophisticated filter, written as a hansl function. Either way you will want to make use of some of gretl's built-in calendrical functions.

We'll assume that a suitably dimensioned workspace has been set up as described above. Given that, the key ingredients of the join are a filtering function which we'll call `rel_ok` (for "release is OK") and the `join` command which calls it. Here's the function:

```
function series rel_ok (series obsdate, series reldate, int p)
  series y_obs, m_obs, y_rel, m_rel
  # get year and month from observation date
  isoconv(obsdate, &y_obs, &m_obs)
  # get year and month from release date
  isoconv(reldate, &y_rel, &m_rel)
  # find the delta in months
  series dm = (12*y_rel + m_rel) - (12*y_obs + m_obs)
  # and implement the filter
  return dm <= p
end function
```

And here's the command:

```
scalar lag = 3  # choose your fixed lag here
join @fname ip_plus3 --data=INDPRO --tkey=observation_date \
--tconvert="observation_date,realtime_start_date" \
--filter="rel_ok(observation_date, realtime_start_date, lag)" \
--aggr=max(realtime_start_date)
```

Note that we use `--tconvert` to convert both the observation date and the realtime start date (or release date) to 8-digit numerical values. Both of these series are passed to the filter, which uses the built-in function `isoconv` to extract year and month. We can then calculate `dm`, the "delta months" since the observation date, for each release. The filter condition is that this delta should be no greater than the specified lag, p.[8]

This filter condition may be satisfied by more than one release, but only the latest of those will actually be the vintage that was current at the end of the n-th month after the observation period, so we add the option `--aggr=max(realtime_start_date)`. If instead you want to target the release at the *beginning* of the n-th month you would have to use a slightly more complicated filter function.

An illustration

Figure 8.1 shows four time series for the monthly index of US industrial production from October 2005 to June 2009: the value as of first publication plus the values current 3, 6 and 12

Another even more complicated issue in the realtime context is that of "benchmark revisions" applied by statistical agencies, where the underlying definition or composition of a variable changes on some date, which goes beyond a mere rescaling. However, this type of structural change is not, in principle, a feature of realtime data alone, but applies to any time-series data.

[8] The filter is written on the assumption that the lag is expressed in months; on that understanding it could be used with annual or quarterly data as well as monthly. The idea could be generalized to cover weekly or daily data without much difficulty.

months out from the observation date.[9] From visual inspection it would seem that over much of this period the Federal reserve was fairly consistently overestimating industrial production at first release and shortly thereafter, relative to the figure they arrived at with a lag of a year.

The script that produced this Figure is shown in full in Example 8.1. Note that in this script we are using a somewhat more efficient version of the `rel_ok` function shown above, where we pass the required series arguments in "pointer" form to avoid having to copy them (see chapter 13).

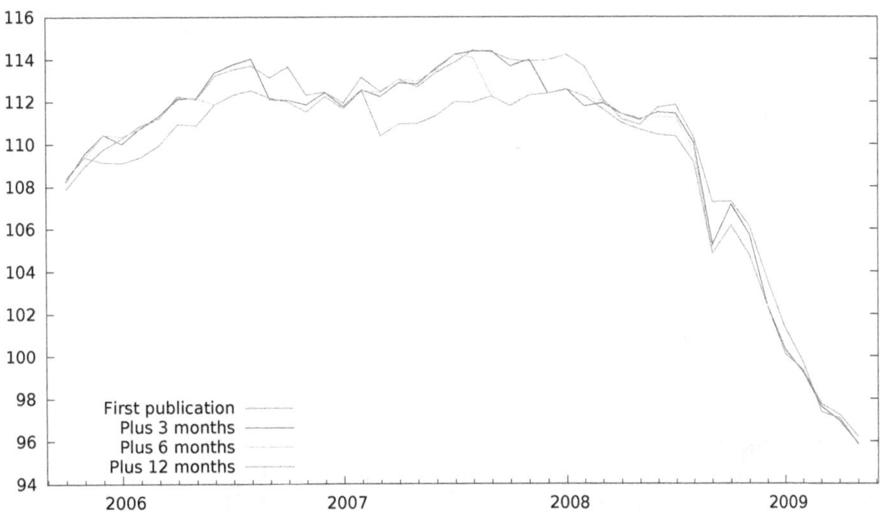

Figure 8.1: Successive revisions to US industrial production

8.7 Getting the revision history for an observation

For our final example we show how to retrieve the revision history for a given observation (again using Alfred data on US industrial production). In this exercise we are switching the time axis: the observation period is a fixed point and time is "vintage time".

A suitable script is shown in Example 8.2. We first select an observation to track (January 1970). We start the clock in the following month, when a data-point for this period was first published, and let it run to the end of the vintage history (in this file, March 2013). Our outer time key is the realtime start date and we filter on observation date; we name the imported `INDPRO` values as `ip_jan70`. Since it sometimes happens that more than one revision occurs in a given month we need to select an aggregation method: here we choose to take the last revision in the month.

Recall from section 8.2 that Alfred records a new revision only when the data-point in question actually changes. This means that our imported series will contain missing values for all months when no real revision took place. However, we can apply a simple autoregressive procedure to fill in the data: each missing value equals the prior non-missing value.

Figure 8.2 displays the revision history. Over this sample period the periodic re-basing of the index overshadows amendments due to accrual of new information.

[9]Why not a longer series? Because if we try to extend it in either direction we immediately run into the index re-basing problem mentioned in section 8.5, with big (staggered) leaps downward in all the series.

Example 8.1: Retrieving successive realtime lags of US industrial production

```
function series rel_ok (series *obsdate, series *reldate, int p)
  series y_obs, m_obs, y_rel, m_rel
  isoconv(obsdate, &y_obs, &m_obs)
  isoconv(reldate, &y_rel, &m_rel)
  series dm = (12*y_rel + m_rel) - (12*y_obs + m_obs)
  return dm <= p
end function

nulldata 45
setobs 12 2005:10

string fname = "INDPRO.txt"

# initial published values
join @fname firstpub --data=INDPRO --tkey=observation_date \
--tconvert=realtime_start_date --aggr=min(realtime_start_date)

# plus 3 months
join @fname plus3 --data=INDPRO --tkey=observation_date \
--tconvert="observation_date,realtime_start_date" \
--filter="rel_ok(&observation_date, &realtime_start_date, 3)" \
--aggr=max(realtime_start_date)

# plus 6 months
join @fname plus6 --data=INDPRO --tkey=observation_date \
--tconvert="observation_date,realtime_start_date" \
--filter="rel_ok(&observation_date, &realtime_start_date, 6)" \
--aggr=max(realtime_start_date)

# plus 12 months
join @fname plus12 --data=INDPRO --tkey=observation_date \
--tconvert="observation_date,realtime_start_date" \
--filter="rel_ok(&observation_date, &realtime_start_date, 12)" \
--aggr=max(realtime_start_date)

setinfo firstpub --graph-name="First publication"
setinfo plus3 --graph-name="Plus 3 months"
setinfo plus6 --graph-name="Plus 6 months"
setinfo plus12 --graph-name="Plus 12 months"

gnuplot firstpub plus3 plus6 plus12 --time --with-lines \
 --output=realtime.pdf { set key left bottom; }
```

Example 8.2: Retrieving a revision history

```
# choose the observation to track here (YYYYMMDD)
scalar target = 19700101

nulldata 518 --preserve
setobs 12 1970:02

join INDPRO.txt ip_jan70 --data=INDPRO --tkey=realtime_start_date \
--tconvert=observation_date \
--filter="observation_date==target" --aggr=seq:-1

ip_jan70 = ok(ip_jan70) ? ip_jan70 : ip_jan70(-1)
```

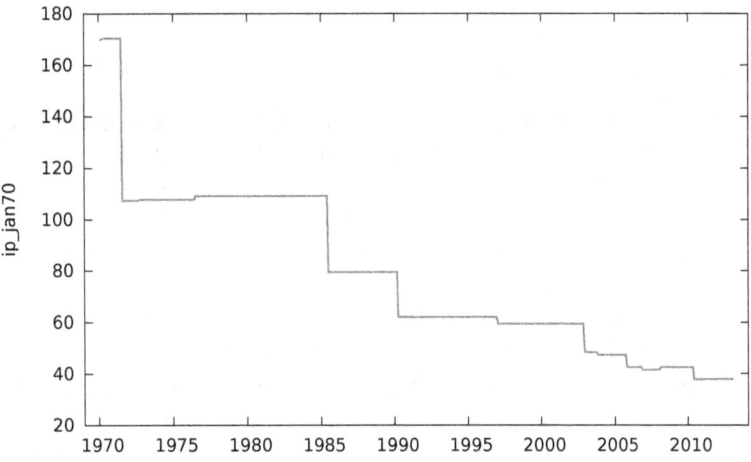

Figure 8.2: Vintages of the index of US industrial production for January 1970

Chapter 9

Special functions in genr

9.1 Introduction

The `genr` command provides a flexible means of defining new variables. It is documented in the *Gretl Command Reference*. This chapter offers a more expansive discussion of some of the special functions available via `genr` and some of the finer points of the command.

9.2 Long-run variance

As is well known, the variance of the average of T random variables $x_1, x_2, \ldots, x_T$ with equal variance σ^2 equals σ^2/T if the data are uncorrelated. In this case, the sample variance of x_t over the sample size provides a consistent estimator.

If, however, there is serial correlation among the x_ts, the variance of $\bar{X} = T^{-1}\sum_{t=1}^{T} x_t$ must be estimated differently. One of the most widely used statistics for this purpose is a nonparametric kernel estimator with the Bartlett kernel defined as

$$\hat{\omega}^2(k) = T^{-1}\sum_{t=k}^{T-k}\left[\sum_{i=-k}^{k} w_i(x_t - \bar{X})(x_{t-i} - \bar{X})\right], \tag{9.1}$$

where the integer k is known as the window size and the w_i terms are the so-called *Bartlett weights*, defined as $w_i = 1 - \frac{|i|}{k+1}$. It can be shown that, for k large enough, $\hat{\omega}^2(k)/T$ yields a consistent estimator of the variance of $\bar{X}$.

Gretl implements this estimator by means of the function `lrvar()`, which takes two arguments: the series whose long-run variance must be estimated and the scalar k. If k is negative, the popular choice $T^{1/3}$ is used.

9.3 Cumulative densities and p-values

The two functions `cdf` and `pvalue` provide complementary means of examining values from several probability distributions: the standard normal, Student's t, χ^2, F, gamma, and binomial. The syntax of these functions is set out in the *Gretl Command Reference*; here we expand on some subtleties.

The cumulative density function or CDF for a random variable is the integral of the variable's density from its lower limit (typically either $-\infty$ or 0) to any specified value x. The p-value (at least the one-tailed, right-hand p-value as returned by the `pvalue` function) is the complementary probability, the integral from x to the upper limit of the distribution, typically $+\infty$.

In principle, therefore, there is no need for two distinct functions: given a CDF value p_0 you could easily find the corresponding p-value as $1 - p_0$ (or vice versa). In practice, with finite-precision computer arithmetic, the two functions are not redundant. This requires a little explanation. In gretl, as in most statistical programs, floating point numbers are represented as "doubles" — double-precision values that typically have a storage size of eight bytes or 64 bits. Since there are only so many bits available, only so many floating-point numbers can be represented: *doubles do not model the real line.* Typically doubles can represent numbers over the range (roughly) $\pm 1.7977 \times 10^{308}$, but only to about 15 digits of precision.

Suppose you're interested in the left tail of the χ^2 distribution with 50 degrees of freedom: you'd like to know the CDF value for $x = 0.9$. Take a look at the following interactive session:

```
? genr p1 = cdf(X, 50, 0.9)
Generated scalar p1 (ID 2) = 8.94977e-35
? genr p2 = pvalue(X, 50, 0.9)
Generated scalar p2 (ID 3) = 1
? genr test = 1 - p2
Generated scalar test (ID 4) = 0
```

The `cdf` function has produced an accurate value, but the `pvalue` function gives an answer of 1, from which it is not possible to retrieve the answer to the CDF question. This may seem surprising at first, but consider: if the value of `p1` above is correct, then the correct value for `p2` is $1 - 8.94977 \times 10^{-35}$. But there's no way that value can be represented as a double: that would require over 30 digits of precision.

Of course this is an extreme example. If the x in question is not too far off into one or other tail of the distribution, the `cdf` and `pvalue` functions will in fact produce complementary answers, as shown below:

```
? genr p1 = cdf(X, 50, 30)
Generated scalar p1 (ID 2) = 0.0111648
? genr p2 = pvalue(X, 50, 30)
Generated scalar p2 (ID 3) = 0.988835
? genr test = 1 - p2
Generated scalar test (ID 4) = 0.0111648
```

But the moral is that if you want to examine extreme values you should be careful in selecting the function you need, in the knowledge that values very close to zero can be represented as doubles while values very close to 1 cannot.

9.4 Retrieving internal variables

The `genr` command provides a means of retrieving various values calculated by the program in the course of estimating models or testing hypotheses. The variables that can be retrieved in this way are listed in the *Gretl Command Reference*; here we say a bit more about the special variables `$test` and `$pvalue`.

These variables hold, respectively, the value of the last test statistic calculated using an explicit testing command and the p-value for that test statistic. If no such test has been performed at the time when these variables are referenced, they will produce the missing value code. The "explicit testing commands" that work in this way are as follows: `add` (joint test for the significance of variables added to a model); `adf` (Augmented Dickey–Fuller test, see below); `arch` (test for ARCH); `chow` (Chow test for a structural break); `coeffsum` (test for the sum of specified coefficients); `cusum` (the Harvey–Collier t-statistic); `kpss` (KPSS stationarity test, no p-value available); `lmtest` (see below); `meantest` (test for difference of means); `omit` (joint test for the significance of variables omitted from a model); `reset` (Ramsey's RESET); `restrict` (general linear restriction); `runs` (runs test for randomness); `testuhat` (test for normality of residual); and `vartest` (test for difference of variances). In most cases both a `$test` and a `$pvalue` are stored; the exception is the KPSS test, for which a p-value is not currently available.

An important point to notice about this mechanism is that the internal variables `$test` and `$pvalue` are over-written each time one of the tests listed above is performed. If you want to reference these values, you must do so at the correct point in the sequence of gretl commands.

A related point is that some of the test commands generate, by default, more than one test statistic and p-value; in these cases only the last values are stored. To get proper control over the retrieval of values via `$test` and `$pvalue` you should formulate the test command in such a way that the result is unambiguous. This comment applies in particular to the `adf` and `lmtest` commands.

- By default, the `adf` command generates three variants of the Dickey–Fuller test: one based on a regression including a constant, one using a constant and linear trend, and one using a constant and a quadratic trend. When you wish to reference `$test` or `$pvalue` in connection with this command, you can control the variant that is recorded by using one of the flags `--nc`, `--c`, `--ct` or `--ctt` with `adf`.

- By default, the `lmtest` command (which must follow an OLS regression) performs several diagnostic tests on the regression in question. To control what is recorded in `$test` and `$pvalue` you should limit the test using one of the flags `--logs`, `--autocorr`, `--squares` or `--white`.

As an aid in working with values retrieved using `$test` and `$pvalue`, the nature of the test to which these values relate is written into the descriptive label for the generated variable. You can read the label for the variable using the `label` command (with just one argument, the name of the variable), to check that you have retrieved the right value. The following interactive session illustrates this point.

```
? adf 4 x1 --c
Augmented Dickey-Fuller tests, order 4, for x1
sample size 59
unit-root null hypothesis: a = 1
  test with constant
  model: (1 - L)y = b0 + (a-1)*y(-1) + ... + e
  estimated value of (a - 1): -0.216889
  test statistic: t = -1.83491
  asymptotic p-value 0.3638
P-values based on MacKinnon (JAE, 1996)
? genr pv = $pvalue
Generated scalar pv (ID 13) = 0.363844
? label pv
  pv=Dickey-Fuller pvalue (scalar)
```

9.5 The discrete Fourier transform

The discrete Fourier transform can be best thought of as a linear, invertible transform of a complex vector. Hence, if $\mathbf{x}$ is an n-dimensional vector whose k-th element is $x_k = a_k + ib_k$, then the output of the discrete Fourier transform is a vector $\mathbf{f} = \mathcal{F}(\mathbf{x})$ whose k-th element is

$$f_k = \sum_{j=0}^{n-1} e^{-i\omega(j,k)} x_j$$

where $\omega(j,k) = 2\pi i \frac{jk}{n}$. Since the transformation is invertible, the vector $\mathbf{x}$ can be recovered from $\mathbf{f}$ via the so-called inverse transform

$$x_k = \frac{1}{n} \sum_{j=0}^{n-1} e^{i\omega(j,k)} f_j.$$

The Fourier transform is used in many diverse situations on account of this key property: the convolution of two vectors can be performed efficiently by multiplying the elements of their Fourier transforms and inverting the result. If

$$z_k = \sum_{j=1}^{n} x_j y_{k-j},$$

then

$$\mathcal{F}(\mathbf{z}) = \mathcal{F}(\mathbf{x}) \odot \mathcal{F}(\mathbf{y}).$$

That is, $\mathcal{F}(\mathbf{z})_k = \mathcal{F}(\mathbf{x})_k \mathcal{F}(\mathbf{y})_k$.

For computing the Fourier transform, gretl uses the external library `fftw3`: see Frigo and Johnson (2005). This guarantees extreme speed and accuracy. In fact, the CPU time needed to perform the transform is $O(n \log n)$ for any n. This is why the array of numerical techniques employed in fftw3 is commonly known as the *Fast* Fourier Transform.

Gretl provides two matrix functions[1] for performing the Fourier transform and its inverse: fft and ffti. In fact, gretl's implementation of the Fourier transform is somewhat more specialized: the input to the fft function is understood to be real. Conversely, ffti takes a complex argument and delivers a real result. For example:

```
x1 = { 1 ; 2 ; 3 }
# perform the transform
f = fft(a)
# perform the inverse transform
x2 = ffti(f)
```

yields

$$x_1 = \begin{bmatrix} 1 \\ 2 \\ 3 \end{bmatrix} \qquad f = \begin{bmatrix} 6 & 0 \\ -1.5 & 0.866 \\ -1.5 & -0.866 \end{bmatrix} \qquad x_2 = \begin{bmatrix} 1 \\ 2 \\ 3 \end{bmatrix}$$

where the first column of f holds the real part and the second holds the complex part. In general, if the input to fft has n columns, the output has $2n$ columns, where the real parts are stored in the odd columns and the complex parts in the even ones. Should it be necessary to compute the Fourier transform on several vectors with the same number of elements, it is numerically more efficient to group them into a matrix rather than invoking fft for each vector separately.

As an example, consider the multiplication of two polynomials:

$$\begin{aligned} a(x) &= 1 + 0.5x \\ b(x) &= 1 + 0.3x - 0.8x^2 \\ c(x) = a(x) \cdot b(x) &= 1 + 0.8x - 0.65x^2 - 0.4x^3 \end{aligned}$$

The coefficients of the polynomial $c(x)$ are the convolution of the coefficients of $a(x)$ and $b(x)$; the following gretl code fragment illustrates how to compute the coefficients of $c(x)$:

```
# define the two polynomials
a = { 1, 0.5, 0, 0 }'
b = { 1, 0.3, -0.8, 0 }'
# perform the transforms
fa = fft(a)
fb = fft(b)
# complex-multiply the two transforms
fc = cmult(fa, fb)
# compute the coefficients of c via the inverse transform
c = ffti(fc)
```

Maximum efficiency would have been achieved by grouping a and b into a matrix. The computational advantage is so little in this case that the exercise is a bit silly, but the following alternative may be preferable for a large number of rows/columns:

```
# define the two polynomials
a = { 1 ; 0.5; 0 ; 0 }
b = { 1 ; 0.3 ; -0.8 ; 0 }
# perform the transforms jointly
f = fft(a ~ b)
# complex-multiply the two transforms
fc = cmult(f[,1:2], f[,3:4])
# compute the coefficients of c via the inverse transform
c = ffti(fc)
```

[1]See chapter 15.

Traditionally, the Fourier transform in econometrics has been mostly used in time-series analysis, the periodogram being the best known example. Example script 9.1 shows how to compute the periodogram of a time series via the fft function.

Example 9.1: Periodogram via the Fourier transform

```
nulldata 50
# generate an AR(1) process
series e = normal()
series x = 0
x = 0.9*x(-1) + e
# compute the periodogram
scale = 2*pi*$nobs
X = { x }
F = fft(X)
S = sumr(F.^2)
S = S[2:($nobs/2)+1]/scale
omega = seq(1,($nobs/2))' .* (2*pi/$nobs)
omega = omega ~ S
# compare the built-in command
pergm x
print omega
```

Chapter 10

Gretl data types

10.1 Introduction

Gretl offers the following data types:

`scalar`	holds a single numerical value
`series`	holds n numerical values, where n is the number of observations in the current dataset
`matrix`	holds a rectangular array of numerical values, of any dimensions
`list`	holds the ID numbers of a set of series
`string`	holds an array of characters
`bundle`	holds a variable number of objects of various types
`array`	holds a variable number of objects of a given type

The "numerical values" mentioned above are all double-precision floating point numbers.

In this chapter we give a run-down of the basic characteristics of each of these types and also explain their "life cycle" (creation, modification and destruction). The list and matrix types, whose uses are relatively complex, are discussed at greater length in the following two chapters.

10.2 Series

We begin with the `series` type, which is the oldest and in a sense the most basic type in gretl. When you open a data file in the gretl GUI, what you see in the main window are the ID numbers, names (and descriptions, if available) of the series read from the file. All the series existing at any point in a gretl session are of the same length, although some may have missing values. The variables that can be added via the items under the Add menu in the main window (logs, squares and so on) are also series.

For a gretl session to contain any series, a common series length must be established. This is usually achieved by opening a data file, or importing a series from a database, in which case the length is set by the first import. But one can also use the `nulldata` command, which takes as it single argument the desired length, a positive integer.

Each series has these basic attributes: an ID number, a name, and of course n numerical values. In addition a series may have a description (which is shown in the main window and is also accessible via the `labels` command), a "display name" for use in graphs, a record of the compaction method used in reducing the variable's frequency (for time-series data only) and a flag marking the variable as discrete. These attributes can be edited in the GUI by choosing Edit Attributes (either under the Variable menu or via right-click), or by means of the `setinfo` command.

In the context of most commands you are able to reference series by name or by ID number as you wish. The main exception is the definition or modification of variables via a formula; here you must use names since ID numbers would get confused with numerical constants.

Note that series ID numbers are always consecutive, and the ID number for a given series will change if you delete a lower-numbered series. In some contexts, where gretl is liable to get confused by such changes, deletion of low-numbered series is disallowed.

Discrete series

It is possible to mark variables of the series type as *discrete*. The meaning and uses of this facility are explained in chapter 11.

10.3 Scalars

The scalar type is relatively simple: just a convenient named holder for a single numerical value. Scalars have none of the additional attributes pertaining to series, do not have ID numbers, and must be referenced by name. A common use of scalar variables is to record information made available by gretl commands for further processing, as in `scalar s2 = $sigma^2` to record the square of the standard error of the regression following an estimation command such as `ols`.

You can define and work with scalars in gretl without having any dataset in place.

In the gretl GUI, scalar variables can be inspected and their values edited via the "Icon view" (see the `View` menu in the main window).

10.4 Matrices

Matrices in gretl work much as in other mathematical software (e.g. MATLAB, Octave). Like scalars they have no ID numbers and must be referenced by name, and they can be used without any dataset in place. Matrix indexing is 1-based: the top-left element of matrix `A` is `A[1,1]`. Matrices are discussed at length in chapter 15; advanced users of gretl will want to study this chapter in detail.

Matrices have two optional attribute beyond their numerical content: they may have column and/or row names attached; these are displayed when the matrix is printed. See the `colnames` and `rownames` functions for details.

In the gretl GUI, matrices can be inspected, analysed and edited via the `Icon view` item under the `View` menu in the main window: each currently defined matrix is represented by an icon.

10.5 Lists

As with matrices, lists merit an explication of their own (see chapter 14). Briefly, named lists can (and should!) be used to make command scripts less verbose and repetitious, and more easily modifiable. Since lists are in fact lists of series ID numbers they can be used only when a dataset is in place.

In the gretl GUI, named lists can be inspected and edited under the Data menu in the main window, via the item Define or edit list.

10.6 Strings

String variables may be used for labeling, or for constructing commands. They are discussed in chapter 14. They must be referenced by name; they can be defined in the absence of a dataset.

Such variables can be created and modified via the command-line in the gretl console or via script; there is no means of editing them via the gretl GUI.

10.7 Bundles

A `bundle` is a container or wrapper for various sorts of objects—specifically, scalars, series, matrices, strings, arrays and bundles. (Yes, a bundle can contain other bundles). A bundle takes the form of a hash table or associative array: each item placed in the bundle is associated with a key which can used to retrieve it subsequently. We begin by explaining the mechanics of bundles then offer some thoughts on what they are good for.

To use a bundle you first either "declare" it, as in

```
bundle foo
```

or define an empty bundle using the null keyword:

```
bundle foo = null
```

These two formulations are basically equivalent, in that they both create an empty bundle. The difference is that the second variant may be reused—if a bundle named foo already exists the effect is to empty it—while the first may only be used once in a given gretl session; it is an error to declare a variable that already exists.

To add an object to a bundle you assign to a compound left-hand value: the name of the bundle followed by the key. Two forms of syntax are acceptable in this context. The currently recommended syntax is *bundlename.key*; that is, the name of the bundle followed by a dot, then the key. Both the bundle name and the key must be valid gretl identifiers.[1] For example, the statement

```
foo.matrix1 = m
```

adds an object called m (presumably a matrix) to bundle foo under the key matrix1. The original syntax (and the only form supported prior to version 1.9.12 of gretl) requires that the key be given as a quoted string literal enclosed in square brackets, as in

```
foo["matrix1"] = m
```

While the old syntax is still acceptable, we suggest that it be used only if you wish to preserve compatibility with gretl version 1.9.11 (December, 2012) or earlier.

To get an item out of a bundle, again use the name of the bundle followed by the key, as in

```
matrix bm = foo.matrix1
# or using the old notation
matrix bm = foo["matrix1"]
```

Note that the key identifying an object within a given bundle is necessarily unique. If you reuse an existing key in a new assignment, the effect is to replace the object which was previously stored under the given key. It is not required that the type of the replacement object is the same as that of the original.

Also note that when you add an object to a bundle, what in fact happens is that the bundle acquires a copy of the object. The external object retains its own identity and is unaffected if the bundled object is replaced by another. Consider the following script fragment:

```
bundle foo
matrix m = I(3)
foo.mykey = m
scalar x = 20
foo.mykey = x
```

After the above commands are completed bundle foo does not contain a matrix under mykey, but the original matrix m is still in good health.

To delete an object from a bundle use the delete command, with the bundle/key combination, as in

```
delete foo.mykey
```

[1]As a reminder: 31 characters maximum, starting with a letter and composed of just letters, numbers or underscore.

This destroys the object associated with the key and removes the key from the hash table.

To determine whether a bundle contains an object associated with a given key, use the `inbundle()` function. This takes two arguments: the name of the bundle and the key string. The value returned by this function is an integer which codes for the type of the object (0 for no match, 1 for scalar, 2 for series, 3 for matrix, 4 for string, 5 for bundle and 6 for array). The function `typestr()` may be used to get the string corresponding to this code. For example:

```
scalar type = inbundle(foo, x)
if type == 0
  print "x: no such object"
else
  printf "x is of type %s\n", typestr(type)
endif
```

Besides adding, accessing, replacing and deleting individual items, the other operations that are supported for bundles are union, printing and deletion. As regards union, if bundles `b1` and `b2` are defined you can say

```
bundle b3 = b1 + b2
```

to create a new bundle that is the union of the two others. The algorithm is: create a new bundle that is a copy of `b1`, then add any items from `b2` whose keys are not already present in the new bundle. (This means that bundle union is not commutative if the bundles have one or more key strings in common.)

If `b` is a bundle and you say `print b`, you get a listing of the bundle's keys along with the types of the corresponding objects, as in

```
? print b
bundle b:
 x (scalar)
 mat (matrix)
 inside (bundle)
```

What are bundles good for?

Bundles are unlikely to be of interest in the context of standalone gretl scripts, but they can be very useful in the context of complex function packages where a good deal of information has to be passed around between the component functions. Instead of using a lengthy list of individual arguments, function *A* can bundle up the required data and pass it to functions *B* and *C*, where relevant information can be extracted via a mnemonic key.

In this context bundles should be passed in "pointer" form (see chapter 13) as illustrated in the following trivial example, where a bundle is created at one level then filled out by a separate function.

```
# modification of bundle (pointer) by user function

function void fill_out_bundle (bundle *b)
  b.mat =  I(3)
  b.str = "foo"
  b.x = 32
end function

bundle my_bundle
fill_out_bundle(&my_bundle)
```

The bundle type can also be used to advantage as the *return value* from a packaged function, in cases where a package writer wants to give the user the option of accessing various results. In the gretl GUI, function packages that return a bundle are treated specially: the output window that displays the printed results acquires a menu showing the bundled items (their names and types), from which the user can save items of interest. For example, a function package that estimates a model might return a bundle containing a vector of parameter estimates, a residual series and a covariance matrix for the parameter estimates, among other possibilities.

As a refinement to support the use of bundles as a function return type, the `setnote` function can be used to add a brief explanatory note to a bundled item—such notes will then be shown in the GUI menu. This function takes three arguments: the name of a bundle, a key string, and the note. For example

```
setnote(b, "vcv", "covariance matrix")
```

After this, the object under the key `vcv` in bundle `b` will be shown as "covariance matrix" in a GUI menu.

10.8 Arrays

The gretl array type was developed in summer 2014 and is first documented in gretl 1.10. Arrays are intended for scripting use only: they have no GUI representation and they're unlikely ever to acquire one.[2] Since this type is a relative newcomer there are at present relatively few built-in functions that take arrays as arguments or that return arrays; further development can be expected.

A gretl array is, as you might expect, a container which can hold zero or more objects of a certain type, indexed by consecutive integers starting at 1. It is one-dimensional. This type is implemented by a quite "generic" back-end. The types of object that can be put into arrays are strings, matrices, bundles and lists; a given array can hold only one of these types.

Of gretl's "primary" types, then, neither scalars nor series are supported by the array mechanism. There would be little point in supporting arrays of scalars as such since the matrix type already plays that role, and more flexibly. As for series, they have a special status as elements of a dataset (which is in a sense an "array of series" already) and in addition we have the list type which already functions as a sort of array for subsets of the series in a dataset.

Creating an array

An array can be brought into existence in any of four ways: bare declaration, assignment from `null`, or using one of the functions `array()` or `defarray()`. In each case one of the specific type-words `strings`, `matrices`, `bundles` or `lists` must be used. Here are some examples:

```
# make an empty array of strings
strings S
# make an empty array of matrices
matrices M = null
# make an array with space for four bundles
bundles B = array(4)
# make an array with three specified strings
strings P = defarray("foo", "bar", "baz")
```

The "bare declaration" form and the "`= null`" form have the same effect of creating an empty array, but the second can be used in contexts where bare declaration is not allowed (and it can also be used to destroy the content of an existing array and reduce it to size zero). The `array()` function expects a positive integer argument and can be used to create an array of pre-given size; in this case the elements are initialized appropriately as empty strings, null matrices, or empty bundles or lists. The `defarray()` function takes a variable number of arguments (one or more), each of which may be the name of a variable of the appropriate type or an expression which evaluates to an object of the appropriate type.

[2]However, it's possible to save arrays "invisibly" in the context of a GUI session, by virtue of the fact that they can be packed into bundles (see below), and bundles can be saved as part of a "session".

Setting and getting elements

There are two ways to set the value of an array element: you can set a particular element using the array index, or you can append an element using the += operator:

```
# first case
strings S = array(3)
S[2] = "string the second"
# alternative
matrices M = null
M += mnormal(T,k)
```

In the first method the index must (of course) be within bounds; that is, greater than zero and no greater than the current length of the array. When the second method is used it automatically extends the length of the array by 1.

To get hold of an element, the array index must be used:

```
# for S an array of strings
string s = S[5]
# for M an array of matrices
printf "\n%#12.5\n", M[1]
```

Operations on whole arrays

At present only one operation is available for arrays as a whole, namely appending. You can do, for example

```
# for S1 and S2 both arrays of strings
strings BigS = S1 + S2
# or
S1 += S2
```

In each case the result is an array of strings whose length is the sum of the lengths of S1 and S2—and similarly for the other supported types.

Arrays as function arguments

One can write hansl functions that take as arguments any of the array types; in addition arrays can be passed to function in "pointerized" form.[3] In addition hansl functions may return any of the array types. Here is a trivial example for strings:

```
function void printstrings (strings *S)
  loop i=1..nelem(S) -q
    printf "element %d: '%s'\n", i, S[i]
  endloop
end function

function strings mkstrs (int n)
  strings S = array(n)
  loop i=1..n -q
    S[i] = sprintf("member %d", i)
  endloop
  return S
end function

strings Foo = mkstrs(5)
print Foo
printstrings(&Foo)
```

[3]With the exception of an array of lists. Our thinking on how exactly to handle arrays of lists as function arguments is not yet very far advanced.

A couple of points are worth noting here. First, the `nelem()` function works to give the number of elements in any of the "container" types (lists, arrays, bundles, matrices). Second, if you do "`print Foo`" for Foo an array, you'll see something like:

```
? print Foo
Array of strings, length 5
```

Arrays and bundles

As mentioned, the `bundle` type is supported by the array mechanism. In addition, arrays (of whatever type) can be put into bundles:

```
matrices M = array(8)
# set values of M[i] here...
bundle b
b.M = M
```

The mutual "packability" of bundles and arrays means that it's possible to go quite far down the rabbit-hole... users are advised not to get carried away.

10.9 The life cycle of gretl objects

Creation

The most basic way to create a new variable of any type is by *declaration*, where one states the type followed by the name of the variable to create, as in

```
scalar x
series y
matrix A
```

and so forth. In that case the object in question is given a default initialization, as follows: a new scalar has value NA (missing); a new series is filled with NAs; a new matrix is null (zero rows and columns); a new string is empty; a new list has no members, and a new bundle is empty.

Declaration can be supplemented by a definite initialization, as in

```
scalar x = pi
series y = log(x)
matrix A = zeros(10,4)
```

The traditional way of creating a new variable in gretl was via the `genr` command (which is still supported), as in

```
genr x = y/100
```

Here the type of `x` is left implicit and will be determined automatically depending on the context: if `y` is a scalar, a series or a matrix `x` will inherit `y`'s type (otherwise an error will be generated, since division is applicable to these types only). Moreover, the type of a new variable can be left implicit *without* use of `genr`:[4]

```
x = y/100
```

In "modern" gretl scripting we recommend that you state the type of a new variable explicitly. This makes the intent clearer to a reader of the script and also guards against errors that might otherwise be difficult to understand (i.e. a certain variable turns out to be of the wrong type for some subsequent calculation, but you don't notice at first because you didn't say what type you needed). An exception to this rule might reasonably be granted for clear and simple cases where there's little possibility of confusion.

[4]Apart from the bundle type: that must always be specified.

Modification

Typically, the values of variables of all types are modified by assignment, using the = operator with the name of the variable on the left and a suitable value or formula on the right:

```
z = normal()
x = 100 * log(y) - log(y(-1))
M = qform(a, X)
```

By a "suitable" value we mean one that is conformable for the type in question. A gretl variable acquires its type when it is first created and this cannot be changed via assignment; for example, if you have a matrix A and later want a string A, you will have to delete the matrix first.

☞ One point to watch out for in gretl scripting is type conflicts having to do with the names of series brought in from a data file. For example, in setting up a command loop (see chapter 12) it is very common to call the loop index i. Now a loop index is a scalar (typically incremented each time round the loop). If you open a data file that happens to contain a series named i you will get a type error ("Types not conformable for operation") when you try to use i as a loop index.

Although the type of an existing variable cannot be changed on the fly, gretl nonetheless tries to be as "understanding" as possible. For example if x is a series and you say

```
x = 100
```

gretl will give the series a constant value of 100 rather than complaining that you are trying to assign a scalar to a series. This issue is particularly relevant for the matrix type—see chapter 15 for details.

Besides using the regular assignment operator you also have the option of using an "inflected" equals sign, as in the C programming language. This is shorthand for the case where the new value of the variable is a function of the old value. For example,

```
x += 100 # in longhand: x = x + 100
x *= 100 # in longhand: x = x * 100
```

For scalar variables you can use a more condensed shorthand for simple increment or decrement by 1, namely trailing ++ or -- respectively:

```
x = 100
x--      # x now equals 99
x++      # x now equals 100
```

In the case of objects holding more than one value—series, matrices and bundles—you can modify particular values within the object using an expression within square brackets to identify the elements to access. We have discussed this above for the bundle type and chapter 15 goes into details for matrices. As for series, there are two ways to specify particular values for modification: you can use a simple 1-based index, or if the dataset is a time series or panel (or if it has marker strings that identify the observations) you can use an appropriate observation string. Such strings are displayed by gretl when you print data with the --byobs flag. Examples:

```
x[13]      = 100  # simple index: the 13th observation
x[1995:4]  = 100  # date: quarterly time series
x[2003:08] = 100  # date: monthly time series
x["AZ"]    = 100  # the observation with marker string "AZ"
x[3:15]    = 100  # panel: the 15th observation for the 3rd unit
```

Note that with quarterly or monthly time series there is no ambiguity between a simple index number and a date, since dates always contain a colon. With annual time-series data, however, such ambiguity exists and it is resolved by the rule that a number in brackets is always read as a simple index: x[1905] means the nineteen-hundred and fifth observation, *not* the observation for the year 1905. You can specify a year by quotation, as in x["1905"].

Destruction

Objects of the types discussed above, *with the important exception of named lists*, are all destroyed using the `delete` command: `delete` *objectname*.

Lists are an exception for this reason: in the context of gretl commands, a named list expands to the ID numbers of the member series, so if you say

```
delete L
```

for `L` a list, the effect is to delete all the series in `L`; the list itself is not destroyed, but ends up empty. To delete the list itself (without deleting the member series) you must invert the command and use the `list` keyword:

```
list L delete
```

Chapter 11

Discrete variables

When a variable can take only a finite, typically small, number of values, then it is said to be *discrete*. In gretl, variables of the series type (only) can be marked as discrete. (When we speak of "variables" below this should be understood as referring to series.) Some gretl commands act in a slightly different way when applied to discrete variables; moreover, gretl provides a few commands that only apply to discrete variables. Specifically, the `dummify` and `xtab` commands (see below) are available only for discrete variables, while the `freq` (frequency distribution) command produces different output for discrete variables.

11.1 Declaring variables as discrete

Gretl uses a simple heuristic to judge whether a given variable should be treated as discrete, but you also have the option of explicitly marking a variable as discrete, in which case the heuristic check is bypassed.

The heuristic is as follows: First, are all the values of the variable "reasonably round", where this is taken to mean that they are all integer multiples of 0.25? If this criterion is met, we then ask whether the variable takes on a "fairly small" set of distinct values, where "fairly small" is defined as less than or equal to 8. If both conditions are satisfied, the variable is automatically considered discrete.

To mark a variable as discrete you have two options.

1. From the graphical interface, select "Variable, Edit Attributes" from the menu. A dialog box will appear and, if the variable seems suitable, you will see a tick box labeled "Treat this variable as discrete". This dialog box can also be invoked via the context menu (right-click on a variable) or by pressing the F2 key.

2. From the command-line interface, via the `discrete` command. The command takes one or more arguments, which can be either variables or list of variables. For example:

   ```
   list xlist = x1 x2 x3
   discrete z1 xlist z2
   ```

   This syntax makes it possible to declare as discrete many variables at once, which cannot presently be done via the graphical interface. The switch `--reverse` reverses the declaration of a variable as discrete, or in other words marks it as continuous. For example:

   ```
   discrete foo
   # now foo is discrete
   discrete foo --reverse
   # now foo is continuous
   ```

The command-line variant is more powerful, in that you can mark a variable as discrete even if it does not seem to be suitable for this treatment.

Note that marking a variable as discrete does not affect its content. It is the user's responsibility to make sure that marking a variable as discrete is a sensible thing to do. Note that if you want to recode a continuous variable into classes, you can use gretl's arithmetical functionality, as in the following example:

```
nulldata 100
# generate a series with mean 2 and variance 1
```

```
series x = normal() + 2
# split into 4 classes
series z = (x>0) + (x>2) + (x>4)
# now declare z as discrete
discrete z
```

Once a variable is marked as discrete, this setting is remembered when you save the data file.

11.2 Commands for discrete variables

The dummify command

The dummify command takes as argument a series x and creates dummy variables for each distinct value present in x, which must have already been declared as discrete. Example:

```
open greene22_2
discrete Z5 # mark Z5 as discrete
dummify Z5
```

The effect of the above command is to generate 5 new dummy variables, labeled DZ5_1 through DZ5_5, which correspond to the different values in Z5. Hence, the variable DZ5_4 is 1 if Z5 equals 4 and 0 otherwise. This functionality is also available through the graphical interface by selecting the menu item "Add, Dummies for selected discrete variables".

The dummify command can also be used with the following syntax:

```
list dlist = dummify(x)
```

This not only creates the dummy variables, but also a named list (see section 14.1) that can be used afterwards. The following example computes summary statistics for the variable Y for each value of Z5:

```
open greene22_2
discrete Z5 # mark Z5 as discrete
list foo = dummify(Z5)
loop foreach i foo
  smpl $i --restrict --replace
  summary Y
endloop
smpl --full
```

Since dummify generates a list, it can be used directly in commands that call for a list as input, such as ols. For example:

```
open greene22_2
discrete Z5 # mark Z5 as discrete
ols Y 0 dummify(Z5)
```

The freq command

The freq command displays absolute and relative frequencies for a given variable. The way frequencies are counted depends on whether the variable is continuous or discrete. This command is also available via the graphical interface by selecting the "Variable, Frequency distribution" menu entry.

For discrete variables, frequencies are counted for each distinct value that the variable takes. For continuous variables, values are grouped into "bins" and then the frequencies are counted for each bin. The number of bins, by default, is computed as a function of the number of valid observations in the currently selected sample via the rule shown in Table 11.1. However, when the command is invoked through the menu item "Variable, Frequency Plot", this default can be overridden by the user.

For example, the following code

Observations	Bins
$8 \leq n < 16$	5
$16 \leq n < 50$	7
$50 \leq n \leq 850$	$\lceil \sqrt{n} \rceil$
$n > 850$	29

Table 11.1: Number of bins for various sample sizes

```
open greene19_1
freq TUCE
discrete TUCE # mark TUCE as discrete
freq TUCE
```

yields

```
Read datafile /usr/local/share/gretl/data/greene/greene19_1.gdt
periodicity: 1, maxobs: 32,
observations range: 1-32

Listing 5 variables:
  0) const     1) GPA       2) TUCE      3) PSI       4) GRADE

? freq TUCE

Frequency distribution for TUCE, obs 1-32
number of bins = 7, mean = 21.9375, sd = 3.90151

       interval          midpt   frequency    rel.     cum.

          <  13.417     12.000        1      3.12%    3.12% *
    13.417 - 16.250     14.833        1      3.12%    6.25% *
    16.250 - 19.083     17.667        6     18.75%   25.00% ******
    19.083 - 21.917     20.500        6     18.75%   43.75% ******
    21.917 - 24.750     23.333        9     28.12%   71.88% **********
    24.750 - 27.583     26.167        7     21.88%   93.75% *******
          >= 27.583     29.000        2      6.25%  100.00% **

Test for null hypothesis of normal distribution:
Chi-square(2) = 1.872 with p-value 0.39211
? discrete TUCE # mark TUCE as discrete
? freq TUCE

Frequency distribution for TUCE, obs 1-32

         frequency    rel.     cum.

  12           1      3.12%    3.12% *
  14           1      3.12%    6.25% *
  17           3      9.38%   15.62% ***
  19           3      9.38%   25.00% ***
  20           2      6.25%   31.25% **
  21           4     12.50%   43.75% ****
  22           2      6.25%   50.00% **
  23           4     12.50%   62.50% ****
  24           3      9.38%   71.88% ***
  25           4     12.50%   84.38% ****
  26           2      6.25%   90.62% **
  27           1      3.12%   93.75% *
  28           1      3.12%   96.88% *
  29           1      3.12%  100.00% *
```

```
Test for null hypothesis of normal distribution:
Chi-square(2) = 1.872 with p-value 0.39211
```

As can be seen from the sample output, a Doornik-Hansen test for normality is computed automatically. This test is suppressed for discrete variables where the number of distinct values is less than 10.

This command accepts two options: `--quiet`, to avoid generation of the histogram when invoked from the command line and `--gamma`, for replacing the normality test with Locke's nonparametric test, whose null hypothesis is that the data follow a Gamma distribution.

If the distinct values of a discrete variable need to be saved, the `values()` matrix construct can be used (see chapter 15).

The xtab command

The `xtab` command cab be invoked in either of the following ways. First,

```
xtab ylist ; xlist
```

where `ylist` and `xlist` are lists of discrete variables. This produces cross-tabulations (two-way frequencies) of each of the variables in `ylist` (by row) against each of the variables in `xlist` (by column). Or second,

```
xtab xlist
```

In the second case a full set of cross-tabulations is generated; that is, each variable in `xlist` is tabulated against each other variable in the list. In the graphical interface, this command is represented by the "Cross Tabulation" item under the View menu, which is active if at least two variables are selected.

Here is an example of use:

```
open greene22_2
discrete Z* # mark Z1-Z8 as discrete
xtab Z1 Z4 ; Z5 Z6
```

which produces

```
Cross-tabulation of Z1 (rows) against Z5 (columns)

       [   1][   2][   3][   4][   5]  TOT.

[   0]    20    91    75    93    36    315
[   1]    28    73    54    97    34    286

TOTAL     48   164   129   190    70    601

Pearson chi-square test = 5.48233 (4 df, p-value = 0.241287)

Cross-tabulation of Z1 (rows) against Z6 (columns)

       [   9][  12][  14][  16][  17][  18][  20]  TOT.

[   0]     4    36   106    70    52    45     2    315
[   1]     3     8    48    45    37    67    78    286

TOTAL      7    44   154   115    89   112    80    601

Pearson chi-square test = 123.177 (6 df, p-value = 3.50375e-24)

Cross-tabulation of Z4 (rows) against Z5 (columns)
```

```
        [   1][   2][   3][   4][   5]  TOT.

[   0]     17    60    35    45    14    171
[   1]     31   104    94   145    56    430

TOTAL      48   164   129   190    70    601

Pearson chi-square test = 11.1615 (4 df, p-value = 0.0248074)

Cross-tabulation of Z4 (rows) against Z6 (columns)

        [   9][  12][  14][  16][  17][  18][  20]  TOT.

[   0]      1     8    39    47    30    32    14    171
[   1]      6    36   115    68    59    80    66    430

TOTAL       7    44   154   115    89   112    80    601

Pearson chi-square test = 18.3426 (6 df, p-value = 0.0054306)
```

Pearson's χ^2 test for independence is automatically displayed, provided that all cells have expected frequencies under independence greater than 10^{-7}. However, a common rule of thumb states that this statistic is valid only if the expected frequency is 5 or greater for at least 80 percent of the cells. If this condition is not met a warning is printed.

Additionally, the `--row` or `--column` options can be given: in this case, the output displays row or column percentages, respectively.

If you want to cut and paste the output of `xtab` to some other program, e.g. a spreadsheet, you may want to use the `--zeros` option; this option causes cells with zero frequency to display the number 0 instead of being empty.

Chapter 12

Loop constructs

12.1 Introduction

The command `loop` opens a special mode in which gretl accepts a block of commands to be repeated zero or more times. This feature may be useful for, among other things, Monte Carlo simulations, bootstrapping of test statistics and iterative estimation procedures. The general form of a loop is:

```
loop control-expression [ --progressive | --verbose | --quiet ]
   loop body
endloop
```

Five forms of control-expression are available, as explained in section 12.2.

Not all gretl commands are available within loops. The commands that are not presently accepted in this context are shown in Table 12.1.

Table 12.1: Commands not usable in loops

corrgm	cusum	function	hurst	include	leverage	nulldata	rename
rmplot	run	setmiss	vif				

By default, the `genr` command operates quietly in the context of a loop (without printing information on the variable generated). To force the printing of feedback from `genr` you may specify the `--verbose` option to `loop`. The `--quiet` option suppresses the usual printout of the number of iterations performed, which may be desirable when loops are nested.

The `--progressive` option to `loop` modifies the behavior of the commands `print` and `store`, and certain estimation commands, in a manner that may be useful with Monte Carlo analyses (see Section 12.3).

The following sections explain the various forms of the loop control expression and provide some examples of use of loops.

☞ If you are carrying out a substantial Monte Carlo analysis with many thousands of repetitions, memory capacity and processing time may be an issue. To minimize the use of computer resources, run your script using the command-line program, gretlcli, with output redirected to a file.

12.2 Loop control variants

Count loop

The simplest form of loop control is a direct specification of the number of times the loop should be repeated. We refer to this as a "count loop". The number of repetitions may be a numerical constant, as in `loop 1000`, or may be read from a scalar variable, as in `loop replics`.

In the case where the loop count is given by a variable, say `replics`, in concept `replics` is an integer; if the value is not integral, it is converted to an integer by truncation. Note that `replics` is evaluated only once, when the loop is initially compiled.

While loop

A second sort of control expression takes the form of the keyword `while` followed by a boolean expression. For example,

```
loop while essdiff > .00001
```

Execution of the commands within the loop will continue so long as (a) the specified condition evaluates as true and (b) the number of iterations does not exceed the value of the internal variable `loop_maxiter`. By default this equals 100000, but you can specify a different value (or remove the limit) via the `set` command (see the *Gretl Command Reference*).

Index loop

A third form of loop control uses an index variable, for example `i`.[1] In this case you specify starting and ending values for the index, which is incremented by one each time round the loop. The syntax looks like this: `loop i=1..20`.

The index variable may be a pre-existing scalar; if this is not the case, the variable is created automatically and is destroyed on exit from the loop.

The index may be used within the loop body in either of two ways: you can access the integer value of `i` or you can use its string representation, `$i`.

The starting and ending values for the index can be given in numerical form, by reference to predefined scalar variables, or as expressions that evaluate to scalars. In the latter two cases the variables are evaluated once, at the start of the loop. In addition, with time series data you can give the starting and ending values in the form of dates, as in `loop i=1950:1..1999:4`.

This form of loop control is intended to be quick and easy, and as such it is subject to certain limitations. In particular, the index variable is always incremented by one at each iteration. If, for example, you have

```
loop i=m..n
```

where `m` and `n` are scalar variables with values `m` > `n` at the time of execution, the index will not be decremented; rather, the loop will simply be bypassed.

If you need more complex loop control, see the "`for`" form below.

The index loop is particularly useful in conjunction with the `values()` matrix function when some operation must be carried out for each value of some discrete variable (see chapter 11). Consider the following example:

```
open greene22_2
discrete Z8
v8 = values(Z8)
loop i=1..rows(v8)
  scalar xi = v8[i]
  smpl (Z8=xi) --restrict --replace
  printf "mean(Y | Z8 = %g) = %8.5f, sd(Y | Z8 = %g) = %g\n", \
    xi, mean(Y), xi, sd(Y)
endloop
```

In this case, we evaluate the conditional mean and standard deviation of the variable `Y` for each value of `Z8`.

[1] It is common programming practice to use simple, one-character names for such variables. However, you may use any name that is acceptable by gretl: up to 31 characters, starting with a letter, and containing nothing but letters, numerals and the underscore character.

Foreach loop

The fourth form of loop control also uses an index variable, in this case to index a specified list of strings. The loop is executed once for each string in the list. This can be useful for performing repetitive operations on a list of variables. Here is an example of the syntax:

```
loop foreach i peach pear plum
   print "$i"
endloop
```

This loop will execute three times, printing out "peach", "pear" and "plum" on the respective iterations. The numerical value of the index starts at 1 and is incremented by 1 at each iteration.

If you wish to loop across a list of variables that are contiguous in the dataset, you can give the names of the first and last variables in the list, separated by "`..`", rather than having to type all the names. For example, say we have 50 variables `AK`, `AL`, ..., `WY`, containing income levels for the states of the US. To run a regression of income on time for each of the states we could do:

```
genr time
loop foreach i AL..WY
   ols $i const time
endloop
```

This loop variant can also be used for looping across the elements in a *named list* (see chapter 14). For example:

```
list ylist = y1 y2 y3
loop foreach i ylist
   ols $i const x1 x2
endloop
```

Note that if you use this idiom inside a function (see chapter 13), looping across a list that has been supplied to the function as an argument, it is necessary to use the syntax *listname*.`$i` to reference the list-member variables. In the context of the example above, this would mean replacing the third line with

```
   ols ylist.$i const x1 x2
```

For loop

The final form of loop control emulates the `for` statement in the C programming language. The sytax is `loop for`, followed by three component expressions, separated by semicolons and surrounded by parentheses. The three components are as follows:

1. Initialization: This is evaluated only once, at the start of the loop. Common example: setting a scalar control variable to some starting value.

2. Continuation condition: this is evaluated at the top of each iteration (including the first). If the expression evaluates as true (non-zero), iteration continues, otherwise it stops. Common example: an inequality expressing a bound on a control variable.

3. Modifier: an expression which modifies the value of some variable. This is evaluated prior to checking the continuation condition, on each iteration after the first. Common example: a control variable is incremented or decremented.

Here's a simple example:

```
loop for (r=0.01; r<.991; r+=.01)
```

In this example the variable `r` will take on the values 0.01, 0.02, ..., 0.99 across the 99 iterations. Note that due to the finite precision of floating point arithmetic on computers it may be necessary to use a continuation condition such as the above, `r<.991`, rather than the more "natural" `r<=.99`. (Using double-precision numbers on an x86 processor, at the point where you would expect `r` to equal 0.99 it may in fact have value 0.990000000000001.)

Any or all of the three expressions governing a `for` loop may be omitted—the minimal form is `(;;)`. If the continuation test is omitted it is implicitly true, so you have an infinite loop unless you arrange for some other way out, such as a `break` statement.

If the initialization expression in a `for` loop takes the common form of setting a scalar variable to a given value, the string representation of that scalar's value is made available within the loop via the accessor `$varname`.

12.3 Progressive mode

If the `--progressive` option is given for a command loop, special behavior is invoked for certain commands, namely, `print`, `store` and simple estimation commands. By "simple" here we mean commands which (a) estimate a single equation (as opposed to a system of equations) and (b) do so by means of a single command statement (as opposed to a block of statements, as with `nls` and `mle`). The paradigm is `ols`; other possibilities include `tsls`, `wls`, `logit` and so on.

The special behavior is as follows.

Estimators: The results from each individual iteration of the estimator are not printed. Instead, after the loop is completed you get a printout of (a) the mean value of each estimated coefficient across all the repetitions, (b) the standard deviation of those coefficient estimates, (c) the mean value of the estimated standard error for each coefficient, and (d) the standard deviation of the estimated standard errors. This makes sense only if there is some random input at each step.

`print`: When this command is used to print the value of a variable, you do not get a print each time round the loop. Instead, when the loop is terminated you get a printout of the mean and standard deviation of the variable, across the repetitions of the loop. This mode is intended for use with variables that have a scalar value at each iteration, for example the error sum of squares from a regression. Data series cannot be printed in this way, and neither can matrices.

`store`: This command writes out the values of the specified scalars, from each time round the loop, to a specified file. Thus it keeps a complete record of their values across the iterations. For example, coefficient estimates could be saved in this way so as to permit subsequent examination of their frequency distribution. Only one such `store` can be used in a given loop.

12.4 Loop examples

Monte Carlo example

A simple example of a Monte Carlo loop in "progressive" mode is shown in Example 12.1.

This loop will print out summary statistics for the 'a' and 'b' estimates and R^2 across the 100 repetitions. After running the loop, `coeffs.gdt`, which contains the individual coefficient estimates from all the runs, can be opened in gretl to examine the frequency distribution of the estimates in detail.

The command `nulldata` is useful for Monte Carlo work. Instead of opening a "real" data set, `nulldata 50` (for instance) opens a dummy data set, containing just a constant and an index variable, with a series length of 50. Constructed variables can then be added. See the `set` command for information on generating repeatable pseudo-random series.

Iterated least squares

Example 12.2 uses a "while" loop to replicate the estimation of a nonlinear consumption function of the form

Example 12.1: Simple Monte Carlo loop

```
nulldata 50
set seed 547
series x = 100 * uniform()
# open a "progressive" loop, to be repeated 100 times
loop 100 --progressive
   series u = 10 * normal()
   # construct the dependent variable
   series y = 10*x + u
   # run OLS regression
   ols y const x
   # grab the coefficient estimates and R-squared
   scalar a = $coeff(const)
   scalar b = $coeff(x)
   scalar r2 = $rsq
   # arrange for printing of stats on these
   print a b r2
   # and save the coefficients to file
   store coeffs.gdt a b
endloop
```

$$C = \alpha + \beta Y^{\gamma} + \epsilon$$

as presented in Greene (2000), Example 11.3. This script is included in the gretl distribution under the name `greene11_3.inp`; you can find it in gretl under the menu item "File, Script files, Practice file, Greene...".

The option `--print-final` for the `ols` command arranges matters so that the regression results will not be printed each time round the loop, but the results from the regression on the last iteration will be printed when the loop terminates.

Example 12.3 shows how a loop can be used to estimate an ARMA model, exploiting the "outer product of the gradient" (OPG) regression discussed by Davidson and MacKinnon (1993).

Further examples of `loop` usage that may be of interest can be found in chapter 16.

Example 12.2: Nonlinear consumption function

```
open greene11_3.gdt
# run initial OLS
ols C 0 Y
scalar essbak = $ess
scalar essdiff = 1
scalar beta = $coeff(Y)
scalar gamma = 1

# iterate OLS till the error sum of squares converges
loop while essdiff > .00001
   # form the linearized variables
   series C0 = C + gamma * beta * Y^gamma * log(Y)
   series x1 = Y^gamma
   series x2 = beta * Y^gamma * log(Y)
   # run OLS
   ols C0 0 x1 x2 --print-final --no-df-corr --vcv
   beta = $coeff[2]
   gamma = $coeff[3]
   ess = $ess
   essdiff = abs(ess - essbak)/essbak
   essbak = ess
endloop

# print parameter estimates using their "proper names"
printf "alpha = %g\n", $coeff[1]
printf "beta  = %g\n", beta
printf "gamma = %g\n", gamma
```

Example 12.3: ARMA 1, 1

```
# Estimation of an ARMA(1,1) model "manually", using a loop

open arma.gdt

scalar c = 0
scalar a = 0.1
scalar m = 0.1

series e = 0.0
series de_c = e
series de_a = e
series de_m = e

scalar crit = 1

loop while crit > 1.0e-9 --quiet
   # one-step forecast errors
   e = y - c - a*y(-1) - m*e(-1)

   # log-likelihood
   scalar loglik = -0.5 * sum(e^2)
   print loglik

   # partials of e with respect to c, a, and m
   de_c = -1 - m * de_c(-1)
   de_a = -y(-1) -m * de_a(-1)
   de_m = -e(-1) -m * de_m(-1)

   # partials of l with respect to c, a and m
   series sc_c = -de_c * e
   series sc_a = -de_a * e
   series sc_m = -de_m * e

   # OPG regression
   ols const sc_c sc_a sc_m --print-final --no-df-corr --vcv

   # Update the parameters
   c += $coeff[1]
   a += $coeff[2]
   m += $coeff[3]

   # show progress
   printf "  constant        = %.8g (gradient %#.6g)\n", c, $coeff[1]
   printf "  ar1 coefficient = %.8g (gradient %#.6g)\n", a, $coeff[2]
   printf "  ma1 coefficient = %.8g (gradient %#.6g)\n", m, $coeff[3]

   crit = $T - $ess
   print crit
endloop

scalar se_c = $stderr[1]
scalar se_a = $stderr[2]
scalar se_m = $stderr[3]

printf "\n"
printf "constant        = %.8g (se = %#.6g, t = %.4f)\n", c, se_c, c/se_c
printf "ar1 coefficient = %.8g (se = %#.6g, t = %.4f)\n", a, se_a, a/se_a
printf "ma1 coefficient = %.8g (se = %#.6g, t = %.4f)\n", m, se_m, m/se_m
```

Chapter 13

User-defined functions

13.1 Defining a function

Gretl offers a mechanism for defining functions, which may be called via the command line, in the context of a script, or (if packaged appropriately, see section 13.5) via the program's graphical interface.

The syntax for defining a function looks like this:[1]

`function` *return-type function-name (parameters)*
  function body
`end function`

The opening line of a function definition contains these elements, in strict order:

1. The keyword `function`.
2. *return-type*, which states the type of value returned by the function, if any. This must be one of `void` (if the function does not return anything), `scalar`, `series`, `matrix`, `list`, `string` or `bundle`.
3. *function-name*, the unique identifier for the function. Function names have a maximum length of 31 characters; they must start with a letter and can contain only letters, numerals and the underscore character. You will get an error if you try to define a function having the same name as an existing gretl command.
4. The function's *parameters*, in the form of a comma-separated list enclosed in parentheses. This may be run into the function name, or separated by white space as shown. In case the function takes no arguments (unusual, but acceptable) this should be indicated by placing the keyword `void` between the parameter-list parentheses.

Function parameters can be of any of the types shown below.[2]

Type	Description
`bool`	scalar variable acting as a Boolean switch
`int`	scalar variable acting as an integer
`scalar`	scalar variable
`series`	data series
`list`	named list of series
`matrix`	matrix or vector
`string`	string variable or string literal
`bundle`	all-purpose container (see section 10.7)

Each element in the listing of parameters must include two terms: a type specifier, and the name by which the parameter shall be known within the function. An example follows:

[1] A somewhat different syntax was in force prior to gretl version 1.8.4, and remained acceptable up to version 1.9.90. The old syntax is no longer supported; see section 13.6 for details on updating.

[2] An additional parameter type is available for GUI use, namely obs; this is equivalent to `int` except for the way it is represented in the graphical interface for calling a function.

```
function scalar myfunc (series y, list xvars, bool verbose)
```

Each of the type-specifiers, with the exception of `list` and `string`, may be modified by prepending an asterisk to the associated parameter name, as in

```
function scalar myfunc (series *y, scalar *b)
```

The meaning of this modification is explained below (see section 13.4); it is related to the use of pointer arguments in the C programming language.

Function parameters: optional refinements

Besides the required elements mentioned above, the specification of a function parameter may include some additional fields, as follows:

- The `const` modifier.
- For `scalar` or `int` parameters: minimum, maximum and/or default values; or for `bool` parameters, just a default value.
- For optional arguments in pointer form, and additionally for string or list arguments in standard form (see section 13.4), the special default value `null`.
- For all parameters, a descriptive string.
- For `int` parameters with minimum and maximum values specified, a set of strings to associate with the allowed numerical values (value labels).

The first two of these options may be useful in many contexts; the last two may be helpful if a function is to be packaged for use in the gretl GUI (but probably not otherwise). We now expand on each of the options.

- The `const` modifier: must be given as a prefix to the basic parameter specification, as in

  ```
  const matrix M
  ```

  This constitutes a promise that the corresponding argument will not be modified within the function; gretl will flag an error if the function attempts to modify the argument.

- Minimum, maximum and default values for `scalar` or `int` types: These values should directly follow the name of the parameter, enclosed in square brackets and with the individual elements separated by colons. For example, suppose we have an integer parameter `order` for which we wish to specify a minimum of 1, a maximum of 12, and a default of 4. We can write

  ```
  int order[1:12:4]
  ```

  If you wish to omit any of the three specifiers, leave the corresponding field empty. For example `[1::4]` would specify a minimum of 1 and a default of 4 while leaving the maximum unlimited. However, as a special case, it is acceptable to give just one value, with no colons: in that case the value is interpreted as a default. So for example

  ```
  int k[0]
  ```

  designates a default value of 0 for the parameter `k`, with no minimum or maximum specified. If you wished to specify a minimum of zero with no maximum or default you would have to write

  ```
  int k[0::]
  ```

  For a parameter of type `bool` (whose values are just zero or non-zero), you can specify a default of 1 (true) or 0 (false), as in

```
bool verbose[0]
```

- Descriptive string: This will show up as an aid to the user if the function is packaged (see section 13.5 below) and called via gretl's graphical interface. The string should be enclosed in double quotes and separated from the preceding elements of the parameter specification with a space, as in

```
series y "dependent variable"
```

- Value labels: These may be used only with `int` parameters for which minimum and maximum values have been specified, so there is a fixed number of admissible values, and the number of labels must match the number of values. They will show up in the graphical interface in the form of a drop-down list, making the function writer's intent clearer when an integer argument represents a categorical selection. A set of value labels must be enclosed in braces, and the individual labels must be enclosed in double quotes and separated by commas or spaces. For example:

```
int case[1:3:1] {"Fixed effects", "Between model", "Random effects"}
```

If two or more of the trailing optional fields are given in a parameter specification, they must be given in the order shown above: min-max-default, description, value labels. Note that there is no facility for "escaping" characters within descriptive strings or value labels; these may contain spaces but they cannot contain the double-quote character.

Here is an example of a well-formed function specification using all the elements mentioned above:

```
function matrix myfunc (series y "dependent variable",
                        list X "regressors",
                        int p[0::1] "lag order",
                        int c[1:2:1] "criterion" {"AIC", "BIC"},
                        bool quiet[0])
```

One advantage of specifying default values for parameters, where applicable, is that in script or command-line mode users may omit trailing arguments that have defaults. For example, `myfunc` above could be invoked with just two arguments, corresponding to `y` and `X`; implicitly `p` = 1, `c` = 1 and `quiet` is false.

Functions taking no parameters

You may define a function that has no parameters (these are called "routines" in some programming languages). In this case, use the keyword `void` in place of the listing of parameters:

```
function matrix myfunc2 (void)
```

The function body

The *function body* is composed of gretl commands, or calls to user-defined functions (that is, function calls may be nested). A function may call itself (that is, functions may be recursive). While the function body may contain function calls, it may not contain function definitions. That is, you cannot define a function inside another function. For further details, see section 13.4.

13.2 Calling a function

A user function is called by typing its name followed by zero or more arguments enclosed in parentheses. If there are two or more arguments these should be separated by commas.

There are automatic checks in place to ensure that the number of arguments given in a function call matches the number of parameters, and that the types of the given arguments match the types specified in the definition of the function. An error is flagged if either of these conditions is violated. One qualification: allowance is made for omitting arguments at the end of the list, provided that default values are specified in the function definition. To be precise, the check is that the number of arguments is at least equal to the number of *required* parameters, and is no greater than the total number of parameters.

A scalar, series or matrix argument to a function may be given either as the name of a pre-existing variable or as an expression which evaluates to a variable of the appropriate type. Scalar arguments may also be given as numerical values. List arguments must be specified by name.

The following trivial example illustrates a function call that correctly matches the function definition.

```
# function definition
function scalar ols_ess(series y, list xvars)
  ols y 0 xvars --quiet
  scalar myess = $ess
  printf "ESS = %g\n", myess
  return myess
end function
# main script
open data4-1
list xlist = 2 3 4
# function call (the return value is ignored here)
ols_ess(price, xlist)
```

The function call gives two arguments: the first is a data series specified by name and the second is a named list of regressors. Note that while the function offers the variable myess as a return value, it is ignored by the caller in this instance. (As a side note here, if you want a function to calculate some value having to do with a regression, but are not interested in the full results of the regression, you may wish to use the --quiet flag with the estimation command as shown above.)

A second example shows how to write a function call that assigns a return value to a variable in the caller:

```
# function definition
function series get_uhat(series y, list xvars)
  ols y 0 xvars --quiet
  series uh = $uhat
  return uh
end function
# main script
open data4-1
list xlist = 2 3 4
# function call
series resid = get_uhat(price, xlist)
```

13.3 Deleting a function

If you have defined a function and subsequently wish to clear it out of memory, you can do so using the keywords delete or clear, as in

```
function myfunc delete
function get_uhat clear
```

Note, however, that if myfunc is already a defined function, providing a new definition automatically overwrites the previous one, so it should rarely be necessary to delete functions explicitly.

13.4 Function programming details

Variables versus pointers

Series, scalar, matrix, bundle, as well as array arguments to functions can be passed in two ways: "as they are", or as pointers. For example, consider the following:

```
function series triple1(series x)
  return 3*x
end function

function series triple2(series *x)
  return 3*x
end function
```

These two functions are nearly identical (and yield the same result); the only difference is that you need to feed a series into `triple1`, as in `triple1(myseries)`, while `triple2` must be supplied a *pointer* to a series, as in `triple2(&myseries)`.

Why make the distinction? There are two main reasons for doing so: modularity and performance.

By modularity we mean the insulation of a function from the rest of the script which calls it. One of the many benefits of this approach is that your functions are easily reusable in other contexts. To achieve modularity, *variables created within a function are local to that function, and are destroyed when the function exits*, unless they are made available as return values and these values are "picked up" or assigned by the caller.

In addition, functions do not have access to variables in "outer scope" (that is, variables that exist in the script from which the function is called) except insofar as these are explicitly passed to the function as arguments.

By default, when a variable is passed to a function as an argument, what the function actually "gets" is a *copy* of the outer variable, which means that the value of the outer variable is not modified by anything that goes on inside the function. But the use of pointers allows a function and its caller to cooperate such that an outer variable can be modified by the function. In effect, this allows a function to "return" more than one value (although only one variable can be returned directly—see below). The parameter in question is marked with a prefix of * in the function definition, and the corresponding argument is marked with the complementary prefix & in the caller. For example,

```
function series get_uhat_and_ess(series y, list xvars, scalar *ess)
  ols y 0 xvars --quiet
  ess = $ess
  series uh = $uhat
  return uh
end function
# main script
open data4-1
list xlist = 2 3 4
# function call
scalar SSR
series resid = get_uhat_and_ess(price, xlist, &SSR)
```

In the above, we may say that the function is given the *address* of the scalar variable `SSR`, and it assigns a value to that variable (under the local name `ess`). (For anyone used to programming in C: note that it is not necessary, or even possible, to "dereference" the variable in question within the function using the * operator. Unadorned use of the name of the variable is sufficient to access the variable in outer scope.)

An "address" parameter of this sort can be used as a means of offering optional information to the caller. (That is, the corresponding argument is not strictly needed, but will be used if present). In that case the parameter should be given a default value of `null` and the the function should test to see if the caller supplied a corresponding argument or not, using the built-in function `isnull()`. For example, here is the simple function shown above, modified to make the filling out of the `ess` value optional.

```
function series get_uhat_and_ess(series y, list xvars, scalar *ess[null])
  ols y 0 xvars --quiet
  if !isnull(ess)
     ess = $ess
  endif
  return $uhat
end function
```

If the caller does not care to get the `ess` value, it can use `null` in place of a real argument:

```
series resid = get_uhat_and_ess(price, xlist, null)
```

Alternatively, trailing function arguments that have default values may be omitted, so the following would also be a valid call:

```
series resid = get_uhat_and_ess(price, xlist)
```

Pointer arguments may also be useful for optimizing performance: even if a variable is not modified inside the function, it may be a good idea to pass it as a pointer if it occupies a lot of memory. Otherwise, the time gretl spends transcribing the value of the variable to the local copy may be non-negligible, compared to the time the function spends doing the job it was written for.

Example 13.1 takes this to the extreme. We define two functions which return the number of rows of a matrix (a pretty fast operation). Function `a()` gets a matrix as argument; function `b()` gets a pointer to a matrix. The functions are evaluated 500 times on a matrix with 2000 rows and 2000 columns; on a typical system floating-point numbers take 8 bytes of memory, so the total size of the matrix is roughly 32 megabytes.

Running the code in example 13.1 will produce output similar to the following (the actual numbers of course depend on the machine you're using):

```
Elapsed time:
        a: without pointers (copy) = 3.3274 seconds,
        b: with pointers (no copy) = 0.00463796 seconds,
```

If a pointer argument is used for this sort of purpose—and the object to which the pointer points is not modified (is treated as read-only) by the function—one can signal this to the user by adding the `const` qualifier, as shown for function `b()` in Example 13.1. When a pointer argument is qualified in this way, any attempt to modify the object within the function will generate an error.

However, combining the `const` flag with the pointer mechanism is technically redundant for the following reason: if you mark a matrix argument as `const` then gretl will in fact pass it in pointer mode internally (since it can't be modified within the function there's no downside to simply making it available to the function rather than copying it). So in the Example we could give the signature of the `b()` function as simply

```
function scalar b(const matrix X)
```

and call it via `r = b(X)`, for the same speed-up relative to function `a()`. Nonetheless, the version of `b()` shown in the example has the virtue of making clearer what is going on.

One limitation on the use of pointer-type arguments should be noted: you cannot supply a given variable as a pointer argument more than once in any given function call. For example, suppose we have a function that takes two matrix-pointer arguments,

```
function scalar pointfunc (matrix *a, matrix *b)
```

Example 13.1: Performance comparison: values versus pointer

```
function scalar a(matrix X)
  return rows(X)
end function

function scalar b(const matrix *X)
  return rows(X)
end function

set echo off
set messages off
X = zeros(2000,2000)
scalar r

set stopwatch
loop 500
  r = a(X)
endloop
fa = $stopwatch

set stopwatch
loop 500
  r = b(&X)
endloop
fb = $stopwatch

printf "Elapsed time:\n\
\ta: without pointers (copy) = %g seconds,\n \
\tb: with pointers (no copy) = %g seconds,\n", fa, fb
```

And suppose we have two matrices, x and y, at the caller level. The call

```
pointfunc(&x, &y)
```

is OK, but the call

```
pointfunc(&x, &x) # will not work
```

will generate an error. That's because the situation inside the function would become too confusing, with what is really the same object existing under two names.

List arguments

The use of a named list as an argument to a function gives a means of supplying a function with a set of variables whose number is unknown when the function is written—for example, sets of regressors or instruments. Within the function, the list can be passed on to commands such as `ols`.

A list argument can also be "unpacked" using a `foreach` loop construct, but this requires some care. For example, suppose you have a list X and want to calculate the standard deviation of each variable in the list. You can do:

```
loop foreach i X
   scalar sd_$i = sd(X.$i)
endloop
```

Please note: a special piece of syntax is needed in this context. If we wanted to perform the above task on a list in a regular script (not inside a function), we could do

```
loop foreach i X
   scalar sd_$i = sd($i)
endloop
```

where `$i` gets the name of the variable at position i in the list, and `sd($i)` gets its standard deviation. But inside a function, working on a list supplied as an argument, if we want to reference an individual variable in the list we must use the syntax *listname.varname*. Hence in the example above we write `sd(X.$i)`.

This is necessary to avoid possible collisions between the name-space of the function and the name-space of the caller script. For example, suppose we have a function that takes a list argument, and that defines a local variable called y. Now suppose that this function is passed a list containing a variable named y. If the two name-spaces were not separated either we'd get an error, or the external variable y would be silently over-written by the local one. It is important, therefore, that list-argument variables should not be "visible" by name within functions. To "get hold of" such variables you need to use the form of identification just mentioned: the name of the list, followed by a dot, followed by the name of the variable.

Constancy of list arguments When a named list of variables is passed to a function, the function is actually provided with a copy of the list. The function may modify this copy (for instance, adding or removing members), but the original list at the level of the caller is not modified.

Optional list arguments If a list argument to a function is optional, this should be indicated by appending a default value of `null`, as in

```
function scalar myfunc (scalar y, list X[null])
```

In that case, if the caller gives `null` as the list argument (or simply omits the last argument) the named list X inside the function will be empty. This possibility can be detected using the `nelem()` function, which returns 0 for an empty list.

String arguments

String arguments can be used, for example, to provide flexibility in the naming of variables created within a function. In the following example the function mavg returns a list containing two moving averages constructed from an input series, with the names of the newly created variables governed by the string argument.

```
function list mavg (series y, string vname)
   list retlist = null
   string newname = sprintf("%s_2", vname)
   retlist += genseries(newname, (y+y(-1)) / 2)
   newname = sprintf("%s_4", vname)
   retlist += genseries(newname, (y+y(-1)+y(-2)+y(-3)) / 4)
   return retlist
end function

open data9-9
list malist = mavg(nocars, "nocars")
print malist --byobs
```

The last line of the script will print two variables named nocars_2 and nocars_4. For details on the handling of named strings, see chapter 14.

If a string argument is considered optional, it may be given a null default value, as in

```
function scalar foo (series y, string vname[null])
```

Retrieving the names of arguments

The variables given as arguments to a function are known inside the function by the names of the corresponding parameters. For example, within the function whose signature is

```
function void somefun (series y)
```

we have the series known as y. It may be useful, however, to be able to determine the names of the variables provided as arguments. This can be done using the function argname, which takes the name of a function parameter as its single argument and returns a string. Here is a simple illustration:

```
function void namefun (series y)
  printf "the series given as 'y' was named %s\n", argname(y)
end function

open data9-7
namefun(QNC)
```

This produces the output

```
the series given as 'y' was named QNC
```

Please note that this will not always work: the arguments given to functions may be anonymous variables, created on the fly, as in somefun(log(QNC)) or somefun(CPI/100). In that case the argname function returns an empty string. Function writers who wish to make use of this facility should check the return from argname using the strlen() function: if this returns 0, no name was found.

Return values

Functions can return nothing (just printing a result, perhaps), or they can return a single variable—a scalar, series, list, matrix, string, or bundle (see section 10.7). The return value, if any, is specified via a statement within the function body beginning with the keyword return, followed by either the name of a variable (which must be of the type announced on the first line of the function definition) or an expression which produces a value of the correct type.

Having a function return a list or bundle is a way of permitting the "return" of more than one variable. For example, you can define several series inside a function and package them as a list; in this case they are not destroyed when the function exits. Here is a simple example, which also illustrates the possibility of setting the descriptive labels for variables generated in a function.

```
function list make_cubes (list xlist)
   list cubes = null
   loop foreach i xlist --quiet
      series $i3 = (xlist.$i)^3
      setinfo $i3 -d "cube of $i"
      list cubes += $i3
   endloop
   return cubes
end function

open data4-1
list xlist = price sqft
list cubelist = make_cubes(xlist)
print xlist cubelist --byobs
labels
```

A `return` statement causes the function to return (exit) at the point where it appears within the body of the function. A function may also exit when (a) the end of the function code is reached (in the case of a function with no return value), (b) a gretl error occurs, or (c) a `funcerr` statement is reached.

The `funcerr` keyword—which may be followed by a string enclosed in double quotes, or the name of a string variable, or nothing—causes a function to exit with an error flagged. If a string is provided (either literally or via a variable), this is printed on exit, otherwise a generic error message is printed. This mechanism enables the author of a function to pre-empt an ordinary execution error and/or offer a more specific and helpful error message. For example,

```
if nelem(xlist) = 0
   funcerr "xlist must not be empty"
endif
```

A function may contain more than one `return` statement, as in

```
function scalar multi (bool s)
   if s
      return 1000
   else
      return 10
   endif
end function
```

However, it is recommended programming practice to have a single return point from a function unless this is very inconvenient. The simple example above would be better written as

```
function scalar multi (bool s)
   return s ? 1000 : 10
end function
```

Error checking

When gretl first reads and "compiles" a function definition there is minimal error-checking: the only checks are that the function name is acceptable, and, so far as the body is concerned, that you are not trying to define a function inside a function (see Section 13.1). Otherwise, if the function body contains invalid commands this will become apparent only when the function is called and its commands are executed.

Debugging

The usual mechanism whereby gretl echoes commands and reports on the creation of new variables is by default suppressed when a function is being executed. If you want more verbose output from a particular function you can use either or both of the following commands within the function:

```
set echo on
set messages on
```

Alternatively, you can achieve this effect for all functions via the command `set debug 1`. Usually when you set the value of a state variable using the `set` command, the effect applies only to the current level of function execution. For instance, if you do `set messages on` within function `f1`, which in turn calls function `f2`, then messages will be printed for `f1` but not `f2`. The debug variable, however, acts globally; all functions become verbose regardless of their level.

Further, you can do `set debug 2`: in addition to command echo and the printing of messages, this is equivalent to setting `max_verbose` (which produces verbose output from the BFGS maximizer) at all levels of function execution.

13.5 Function packages

Since gretl 1.6.0 there has been a mechanism to package functions and make them available to other users of gretl.

Creating a package via the command line

The mechanism described above, for creating function packages using the GUI, is likely to be convenient for small to medium-sized packages but may be too cumbersome for ambitious packages that include a large hierarchy of private functions. To facilitate the building of such packages gretl offers the `makepkg` command.

To use `makepkg` you create three files: a driver script that loads all the functions you want to package and invokes `makepkg`; a small, plain-text specification file that contains the required package details (author, version, etc.); and (in the simplest case) a plain text help file. You run the driver script and gretl writes the package (`.gfn`) file.

We first illustrate with a simple notional package. We have a gretl script file named `foo.inp` that contains a function, `foo`, that we want to package. Our driver script would then look like this

```
include foo.inp
makepkg foo.gfn
```

Note that the `makepkg` command takes one argument, the name of the package file to be created. The package *specification file* should have the same basename but the extension `.spec`. In this case gretl will therefore look for `foo.spec`. It should look something like this:

```
# foo.spec
author = A. U. Thor
version = 1.0
date = 2011-02-01
description = Does something with time series
public = foo
help = foohelp.txt
sample-script = example.inp
min-version = 1.9.3
data-requirement = needs-time-series-data
```

As you can see, the format of each line in this file is `key = value`, with two qualifications: blank lines are permitted (and ignored, as are comment lines that start with #).

All the fields included in the above example are required, with the exception of `data-requirement`, though the order in which they appear is immaterial. Here's a run-down of the basic fields:

- `author`: the name(s) of the author(s). Accented or other non-ASCII characters should be given as UTF-8.
- `version`: the version number of the package, which should be limited to two integers separated by a period.
- `date`: the release date of the current verson of the package, in ISO 8601 format: YYYY-MM-DD.
- `description`: a brief description of the functionality offered by the package. This will be displayed in the GUI function packages window so it should be just one short line.
- `public`: the listing of public functions.
- `help`: the name of a plain text (UTF-8) file containing help; all packages must provide help.
- `sample-script`: the name of a sample script that illustrates use of the package; all packages must supply a sample script.
- `min-version`: the minimum version of gretl required for the package to work correctly. If you're unsure about this, the conservative thing is to give the current gretl version.

The `public` field indicates which function or functions are to be made directly available to users (as opposed to private "helper" functions). In the example above there is just one public function. Note that any functions in memory when `makepkg` is invoked, other than those designated as public, are assumed to be private functions that should also be included in the package. That is, the list of private functions (if any) is implicit.

The `data-requirement` field should be specified if the package requires time-series or panel data, or alternatively if no dataset is required. If the `data-requirement` field is omitted, the assumption is that the package needs a dataset in place, but it doesn't matter what kind; if the packaged functions do not use any series or lists this requirement can be explicitly relaxed. Valid values for this field are:

`needs-time-series-data`	(any time-series data OK)
`needs-qm-data`	(must be quarterly or monthly)
`needs-panel-data`	(must be a panel)
`no-data-ok`	(no dataset is needed)

For a more complex example, let's look at the `gig` (GARCH-in-gretl) package. The driver script for building `gig` looks something like this:

```
set echo off
set messages off
include gig_mle.inp
include gig_setup.inp
include gig_estimate.inp
include gig_printout.inp
include gig_plot.inp
makepkg gig.gfn
```

In this case the functions to be packaged (of which there are many) are distributed across several script files, each of which is the target of an `include` command. The `set` commands at the top are included to cut down on the verbosity of the output.

The content of `gig.spec` is as follows:

```
author = Riccardo "Jack" Lucchetti and Stefano Balietti
version = 2.0
date = 2010-12-21
description = An assortment of univariate GARCH models
public = GUI_gig \
    gig_setup gig_set_dist gig_set_pq gig_set_vQR \
    gig_print gig_estimate \
    gig_plot gig_dplot \
    gig_bundle_print GUI_gig_plot

gui-main = GUI_gig
bundle-print = gig_bundle_print
bundle-plot = GUI_gig_plot
help = gig.pdf
sample-script = examples/example1.inp
min-version = 1.9.3
data-requirement = needs-time-series-data
```

Note that backslash continuation can be used for the elements of the public function listing.

In addition to the fields shown in the simple example above, gig.spec includes three optional fields: gui-main, bundle-print and bundle-plot. These keywords are used to designate certain functions as playing a special role in the gretl graphical interface. A function picked out in this way must be in the public list and must satisfy certain further requirements.

- gui-main: this specifies a function as the one which will be presented automatically to GUI users (instead of users' being faced with a choice of interfaces). This makes sense only for packages that have multiple public functions. In addition, the gui-main function must return a bundle (see section 10.7).
- bundle-print: this picks out a function that should be used to print the contents of a bundle returned by the gui-main function. It must take a pointer-to-bundle as its first argument. The second argument, if present, should be an int switch, with two or more valid values, that controls the printing in some way. Any further arguments must have default values specified so that they can be omitted.
- bundle-plot: selects a function for the role of producing a plot or graph based on the contents of a returned bundle. The requirements on this function are as for bundle-print.

The "GUI special" tags support a user-friendly mode of operation. On a successful call to gui-main, gretl opens a window displaying the contents of the returned bundle (formatted via bundle-print). Menus in this window give the user the option of saving the entire bundle (in which case it's represented as an icon in the "icon view" window) or of extracting specific elements from the bundle (series or matrices, for example).

If the package has a bundle-plot function, the bundle window also has a Graph menu. In gig, for example, the bundle-plot function has this signature:

```
function void GUI_gig_plot(bundle *model, int ptype[0:1:0] \
                           "Plot type" {"Time series", "Density"})
```

The ptype switch is used to choose between a time-series plot of the residual and its conditional variance, and a kernel density plot of the innovation against the theoretical distribution it is supposed to follow. The use of the value-labels Time series and Density means that the Graph menu will display these two choices.

One other feature of the gig spec file is noteworthy: the help field specifies gig.pdf, documentation in PDF format. Unlike plain-text help, this cannot be rolled into the gfn (XML) file produced by the makepkg command; rather, both gig.gfn and gig.pdf are packaged into a zip archive for distribution. This represents a form of package which is new in gretl 1.9.4. More details will be made available before long.

13.6 Memo: updating old-style functions

As mentioned at the start of this chapter, different rules were in force for defining functions prior to gretl 1.8.4. It is straightforward to convert an old function to the new style. The only thing that must be changed is the declaration of the function's return type. Previously this was placed inline in the `return` statement, whereas now it is placed right after the `function` keyword. For example:

```
# old style
function triple (series x)
  y = 3*x
  # note the "series" below: don't do that any more!
  return series y
end function

# new style
function series triple (series x)
  y = 3*x
  return y
end function
```

Note that if a function has no return value the keyword `void` must be used:

```
function void hello (string name)
  printf "Hello from %s\n", name
end function
```

Note also that the role of the `return` statement has changed (and its use has become more flexible):

- The `return` statement now causes the function to return directly, and you can have more than one such statement, wrapped in conditionals. Before there could only be one `return` statement, and its role was just to specify the type available for assignment by the caller.
- The final element in the `return` statement can now be an expression that evaluates to a value of the advertised return type; before, it had to be the name of a pre-defined variable.

Chapter 14

Named lists and strings

14.1 Named lists

Many gretl commands take one or more lists of series as arguments. To make this easier to handle in the context of command scripts, and in particular within user-defined functions, gretl offers the possibility of *named lists*.

Creating and modifying named lists

A named list is created using the keyword `list`, followed by the name of the list, an equals sign, and an expression that forms a list. The most basic sort of expression that works in this context is a space-separated list of variables, given either by name or by ID number. For example,

```
list xlist = 1 2 3 4
list reglist = income price
```

Note that the variables in question must be of the series type.

Two abbreviations are available in defining lists:

- You can use the wildcard character, "*", to create a list of variables by name. For example, `dum*` can be used to indicate all variables whose names begin with `dum`.
- You can use two dots to indicate a range of variables. For example `income..price` indicates the set of variables whose ID numbers are greater than or equal to that of `income` and less than or equal to that of `price`.

In addition there are two special forms:

- If you use the keyword `null` on the right-hand side, you get an empty list.
- If you use the keyword `dataset` on the right, you get a list containing all the series in the current dataset (except the pre-defined `const`).

The name of the list must start with a letter, and must be composed entirely of letters, numbers or the underscore character. The maximum length of the name is 31 characters; list names cannot contain spaces.

Once a named list has been created, it will be "remembered" for the duration of the gretl session (unless you delete it), and can be used in the context of any gretl command where a list of variables is expected. One simple example is the specification of a list of regressors:

```
list xlist = x1 x2 x3 x4
ols y 0 xlist
```

To get rid of a list, you use the following syntax:

```
list xlist delete
```

Be careful: `delete xlist` will delete the series contained in the list, so it implies data loss (which may not be what you want). On the other hand, `list xlist delete` will simply "undefine" the `xlist` identifier; the series themselves will not be affected.

Similarly, to print the names of the members of a list you have to invert the usual print command, as in

```
list xlist print
```

If you just say `print xlist` the list will be expanded and the values of all the member series will be printed.

Lists can be modified in various ways. To *redefine* an existing list altogether, use the same syntax as for creating a list. For example

```
list xlist = 1 2 3
xlist = 4 5 6
```

After the second assignment, `xlist` contains just variables 4, 5 and 6.

To *append* or *prepend* variables to an existing list, we can make use of the fact that a named list stands in for a "longhand" list. For example, we can do

```
list xlist = xlist 5 6 7
xlist = 9 10 xlist 11 12
```

Another option for appending a term (or a list) to an existing list is to use +=, as in

```
xlist += cpi
```

To drop a variable from a list, use -=:

```
xlist -= cpi
```

In most contexts where lists are used in gretl, it is expected that they do not contain any duplicated elements. If you form a new list by simple concatenation, as in `list L3 = L1 L2` (where `L1` and `L2` are existing lists), it's possible that the result may contain duplicates. To guard against this you can form a new list as the union of two existing ones:

```
list L3 = L1 || L2
```

The result is a list that contains all the members of `L1`, plus any members of `L2` that are not already in `L1`.

In the same vein, you can construct a new list as the intersection of two existing ones:

```
list L3 = L1 && L2
```

Here `L3` contains all the elements that are present in both `L1` and `L2`.

You can also subtract one list from another:

```
list L3 = L1 - L2
```

The result contains all the elements of `L1` that are not present in `L2`.

Lists and matrices

Another way of forming a list is by assignment from a matrix. The matrix in question must be interpretable as a vector containing ID numbers of data series. It may be either a row or a column vector, and each of its elements must have an integer part that is no greater than the number of variables in the data set. For example:

```
matrix m = {1,2,3,4}
list L = m
```

The above is OK provided the data set contains at least 4 variables.

Querying a list

You can determine the number of variables or elements in a list using the function `nelem()`.

```
list xlist = 1 2 3
nl = nelem(xlist)
```

The (scalar) variable `nl` will be assigned a value of 3 since `xlist` contains 3 members.

You can determine whether a given series is a member of a specified list using the function `inlist()`, as in

```
scalar k = inlist(L, y)
```

where `L` is a list and `y` a series. The series may be specified by name or ID number. The return value is the (1-based) position of the series in the list, or zero if the series is not present in the list.

Generating lists of transformed variables

Given a named list of series, you are able to generate lists of transformations of these series using the functions `log`, `lags`, `diff`, `ldiff`, `sdiff` or `dummify`. For example

```
list xlist = x1 x2 x3
list lxlist = log(xlist)
list difflist = diff(xlist)
```

When generating a list of *lags* in this way, you specify the maximum lag order inside the parentheses, before the list name and separated by a comma. For example

```
list xlist = x1 x2 x3
list laglist = lags(2, xlist)
```

or

```
scalar order = 4
list laglist = lags(order, xlist)
```

These commands will populate `laglist` with the specified number of lags of the variables in `xlist`. You can give the name of a single series in place of a list as the second argument to `lags`: this is equivalent to giving a list with just one member.

The `dummify` function creates a set of dummy variables coding for all but one of the distinct values taken on by the original variable, which should be discrete. (The smallest value is taken as the omitted catgory.) Like lags, this function returns a list even if the input is a single series.

Another useful operation you can perform with lists is creating *interaction* variables. Suppose you have a discrete variable x_i, taking values from 1 to n and a variable z_i, which could be continuous or discrete. In many cases, you want to "split" z_i into a set of n variables via the rule

$$z_i^{(j)} = \begin{cases} z_i & \text{when} \quad x_i = j \\ 0 & \text{otherwise;} \end{cases}$$

in practice, you create dummies for the x_i variable first and then you multiply them all by z_i; these are commonly called the *interactions* between x_i and z_i. In gretl you can do

```
list H = D ^ Z
```

where `D` is a list of discrete series (or a single discrete series), `Z` is a list (or a single series)[1]; all the interactions will be created and listed together under the name `H`.

An example is provided in script 14.1

[1] Warning: this construct does *not* work if neither `D` nor `Z` are of the the list type.

Example 14.1: Usage of interaction lists

Input:

```
open mroz87.gdt --quiet

# the coding below makes it so that
# KIDS = 0 -> no kids
# KIDS = 1 -> young kids only
# KIDS = 2 -> young or older kids

series KIDS = (KL6 > 0) + ((KL6 > 0) || (K618 > 0))

list D = CIT KIDS # interaction discrete variables
list X = WE WA    # variables to "split"
list INTER = D ^ X

smpl 1 6

print D X INTER -o
```

Output (selected portions):

```
            CIT         KIDS           WE           WA     WE_CIT_0

1             0            2           12           32           12
2             1            1           12           30            0
3             0            2           12           35           12
4             0            1           12           34           12
5             1            2           14           31            0
6             1            0           12           54            0

       WE_CIT_1     WA_CIT_0     WA_CIT_1    WE_KIDS_0    WE_KIDS_1

1             0           32            0            0            0
2            12            0           30            0           12
3             0           35            0            0            0
4             0           34            0            0           12
5            14            0           31            0            0
6            12            0           54           12            0

      WE_KIDS_2    WA_KIDS_0    WA_KIDS_1    WA_KIDS_2

1            12            0            0           32
2             0            0           30            0
3            12            0            0           35
4             0            0           34            0
5            14            0            0           31
6             0           54            0            0
```

Generating series from lists

There are various ways of retrieving or generating individual series from a named list. The most basic method is indexing into the list. For example,

```
series x3 = Xlist[3]
```

will retrieve the third element of the list `Xlist` under the name `x3` (or will generate an error if `Xlist` has less then three members).

In addition gretl offers several functions that apply to a list and return a series. In most cases, these functions also apply to single series and behave as natural extensions when applied to lists, but this is not always the case.

	YpcFR	YpcGE	YpcIT	NFR	NGE	NIT
1997	114.9	124.6	119.3	59830.635	82034.771	56890.372
1998	115.3	122.7	120.0	60046.709	82047.195	56906.744
1999	115.0	122.4	117.8	60348.255	82100.243	56916.317
2000	115.6	118.8	117.2	60750.876	82211.508	56942.108
2001	116.0	116.9	118.1	61181.560	82349.925	56977.217
2002	116.3	115.5	112.2	61615.562	82488.495	57157.406
2003	112.1	116.9	111.0	62041.798	82534.176	57604.658
2004	110.3	116.6	106.9	62444.707	82516.260	58175.310
2005	112.4	115.1	105.1	62818.185	82469.422	58607.043
2006	111.9	114.2	103.3	63195.457	82376.451	58941.499

Table 14.1: GDP per capita and population in 3 European countries (Source: Eurostat)

For recognizing and handling missing values, gretl offers several functions (see the *Gretl Command Reference* for details). In this context, it is worth remarking that the `ok()` function can be used with a list argument. For example,

```
list xlist = x1 x2 x3
series xok = ok(xlist)
```

After these commands, the series `xok` will have value 1 for observations where none of `x1`, `x2`, or `x3` has a missing value, and value 0 for any observations where this condition is not met.

The functions `max`, `min`, `mean`, `sd`, `sum` and `var` behave "horizontally" rather than "vertically" when their argument is a list. For instance, the following commands

```
list Xlist = x1 x2 x3
series m = mean(Xlist)
```

produce a series `m` whose i-th element is the average of $x_{1,i}, x_{2,i}$ and $x_{3,i}$; missing values, if any, are implicitly discarded.

In addition, gretl provides three functions for weighted operations: `wmean`, `wsd` and `wvar`. Consider as an illustration Table 14.1: the first three columns are GDP per capita for France, Germany and Italy; columns 4 to 6 contain the population for each country. If we want to compute an aggregate indicator of per capita GDP, all we have to do is

```
list Ypc = YpcFR YpcGE YpcIT
list N = NFR NGE NIT
y = wmean(Ypc, N)
```

so for example

$$y_{1996} = \frac{114.9 \times 59830.635 + 124.6 \times 82034.771 + 119.3 \times 56890.372}{59830.635 + 82034.771 + 56890.372} = 120.163$$

See the *Gretl Command Reference* for more details.

14.2 Named strings

For some purposes it may be useful to save a string (that is, a sequence of characters) as a named variable that can be reused.

Some examples of the definition of a string variable are shown below.

```
string s1 = "some stuff I want to save"
string s2 = getenv("HOME")
string s3 = s1 + 11
```

The first field after the type-name `string` is the name under which the string should be saved, then comes an equals sign, then comes a specification of the string to be saved. This can be the keyword `null`, to produce an empty string, or may take any of the following forms:

- a string literal (enclosed in double quotes); or
- the name of an existing string variable; or
- a function that returns a string (see below); or
- any of the above followed by + and an integer offset.

The role of the integer offset is to use a substring of the preceding element, starting at the given character offset. An empty string is returned if the offset is greater than the length of the string in question.

To add to the end of an existing string you can use the operator ~=, as in

```
string s1 = "some stuff I want to "
string s1 ~= "save"
```

or you can use the ~ operator to join two or more strings, as in

```
string s1 = "sweet"
string s2 = "Home, " ~ s1 ~ " home."
```

Note that when you define a string variable using a string literal, no characters are treated as "special" (other than the double quotes that delimit the string). Specifically, the backslash is not used as an escape character. So, for example,

```
string s = "\"
```

is a valid assignment, producing a string that contains a single backslash character. If you wish to use backslash-escapes to denote newlines, tabs, embedded double-quotes and so on, use `sprintf` instead.

The `sprintf` command can also be used to define a string variable. This command works in exactly the same way as the `printf` command except that the "format" string must be preceded by an identifier: either the name of an existing string variable or a new name to which the string should be assigned. For example,

```
scalar x = 8
sprintf foo "var%d", x
```

String variables and string substitution

String variables can be used in two ways in scripting: the name of the variable can be typed "as is", or it may be preceded by the "at" sign, @. In the first variant the named string is treated as a variable in its own right, while the second calls for "string substitution". The context determines which of these variants is appropriate.

In the following contexts the names of string variables should be given in plain form (without the "at" sign):

- When such a variable appears among the arguments to the commands `printf` or `sprintf`.
- When such a variable is given as the argument to a function.
- On the right-hand side of a `string` assignment.

Here is an illustration of the use of a named string argument with `printf`:

```
? string vstr = "variance"
Generated string vstr
? printf "vstr: %12s\n", vstr
vstr:      variance
```

String substitution can be used in contexts where a string variable is not acceptable as such. If gretl encounters the symbol @ followed directly by the name of a string variable, this notation is treated as a "macro": the value of the variable is sustituted literally into the command line before the regular parsing of the command is carried out.

One common use of string substitution is when you want to construct and use the name of a series programatically. For example, suppose you want to create 10 random normal series named norm1 to norm10. This can be accomplished as follows.

```
string sname = null
loop i=1..10
  sprintf sname "norm%d", i
  series @sname = normal()
endloop
```

Note that plain sname could not be used in the second line within the loop: the effect would be to attempt to overwrite the string variable named sname with a series of the same name. What we want is for the current *value* of sname to be dumped directly into the command that defines a series, and the "@" notation achieves that.

Another typical use of string substitution is when you want the options used with a particular command to vary depending on some condition. For example,

```
function void use_optstr (series y, list xlist, int verbose)
   string optstr = verbose ? "" : "--simple-print"
   ols y xlist @optstr
end function

open data4-1
list X = const sqft
use_optstr(price, X, 1)
use_optstr(price, X, 0)
```

When printing the value of a string variable using the print command, the plain variable name should generally be used, as in

```
string s = "Just testing"
print s
```

The following variant is equivalent, though clumsy.

```
string s = "Just testing"
print "@s"
```

But note that this next variant does something quite different.

```
string s = "Just testing"
print @s
```

After string substitution, the print command reads

```
print Just testing
```

which attempts to print the values of two variables, Just and testing.

`gretldir`	the gretl installation directory
`workdir`	user's current gretl working directory
`dotdir`	the directory gretl uses for temporary files
`gnuplot`	path to, or name of, the gnuplot executable
`tramo`	path to, or name of, the tramo executable
`x12a`	path to, or name of, the x-12-arima executable
`tramodir`	tramo data directory
`x12adir`	x-12-arima data directory

Table 14.2: Built-in string variables

Built-in strings

Apart from any strings that the user may define, some string variables are defined by gretl itself. These may be useful for people writing functions that include shell commands. The built-in strings are as shown in Table 14.2.

To access these as ordinary string variables, prepend a dollar sign (as in `$dotdir`); to use them in string-substitution mode, prepend the at-sign (`@dotdir`).

Reading strings from the environment

In addition, it is possible to read into gretl's named strings, values that are defined in the external environment. To do this you use the function `getenv`, which takes the name of an environment variable as its argument. For example:

```
? string user = getenv("USER")
Generated string user
? string home = getenv("HOME")
Generated string home
? printf "%s's home directory is %s\n", user, home
cottrell's home directory is /home/cottrell
```

To check whether you got a non-empty value from a given call to `getenv`, you can use the function `strlen`, which retrieves the length of the string, as in

```
? string temp = getenv("TEMP")
Generated string temp
? scalar x = strlen(temp)
Generated scalar x = 0
```

Capturing strings via the shell

If shell commands are enabled in gretl, you can capture the output from such commands using the syntax

`string` *stringname* = `$(`*shellcommand*`)`

That is, you enclose a shell command in parentheses, preceded by a dollar sign.

Reading from a file into a string

You can read the content of a file into a string variable using the syntax

`string` *stringname* = `readfile(`*filename*`)`

The *filename* field may be given as a string variable. For example

```
? sprintf fname "%s/QNC.rts", $x12adir
Generated string fname
? string foo = readfile(fname)
Generated string foo
```

More string functions

Gretl offers several functions for creating or manipulating strings. You can find these listed and explained in the *Function Reference* under the category Strings.

Chapter 15

Matrix manipulation

Together with the other two basic types of data (series and scalars), gretl offers a quite comprehensive array of matrix methods. This chapter illustrates the peculiarities of matrix syntax and discusses briefly some of the more complex matrix functions. For a full listing of matrix functions and a comprehensive account of their syntax, please refer to the *Gretl Command Reference*.

15.1 Creating matrices

Matrices can be created using any of these methods:

1. By direct specification of the scalar values that compose the matrix — in numerical form, by reference to pre-existing scalar variables, or using computed values.
2. By providing a list of data series.
3. By providing a *named list* of series.
4. Using a formula of the same general type that is used with the genr command, whereby a new matrix is defined in terms of existing matrices and/or scalars, or via some special functions.

To specify a matrix *directly in terms of scalars*, the syntax is, for example:

```
matrix A = { 1, 2, 3 ; 4, 5, 6 }
```

The matrix is defined by rows; the elements on each row are separated by commas and the rows are separated by semi-colons. The whole expression must be wrapped in braces. Spaces within the braces are not significant. The above expression defines a 2×3 matrix. Each element should be a numerical value, the name of a scalar variable, or an expression that evaluates to a scalar. Directly after the closing brace you can append a single quote (') to obtain the transpose.

To specify a matrix *in terms of data series* the syntax is, for example,

```
matrix A = { x1, x2, x3 }
```

where the names of the variables are separated by commas. Besides names of existing variables, you can use expressions that evaluate to a series. For example, given a series x you could do

```
matrix A = { x, x^2 }
```

Each variable occupies a column (and there can only be one variable per column). You cannot use the semicolon as a row separator in this case: if you want the series arranged in rows, append the transpose symbol. The range of data values included in the matrix depends on the current setting of the sample range.

Instead of giving an explicit list of variables, you may instead provide the *name of a saved list* (see Chapter 14), as in

```
list xlist = x1 x2 x3
matrix A = { xlist }
```

When you provide a named list, the data series are by default placed in columns, as is natural in an econometric context: if you want them in rows, append the transpose symbol.

As a special case of constructing a matrix from a list of variables, you can say

```
matrix A = { dataset }
```

This builds a matrix using all the series in the current dataset, apart from the constant (variable 0). When this dummy list is used, it must be the sole element in the matrix definition `{...}`. You can, however, create a matrix that includes the constant along with all other variables using horizontal concatenation (see below), as in

```
matrix A = {const}~{dataset}
```

By default, when you build a matrix from series that include missing values the data rows that contain NAs are skipped. But you can modify this behavior via the command `set skip_missing off`. In that case NAs are converted to NaN ("Not a Number"). In the IEEE floating-point standard, arithmetic operations involving NaN always produce NaN. Alternatively, you can take greater control over the observations (data rows) that are included in the matrix using the "set" variable `matrix_mask`, as in

```
set matrix_mask msk
```

where `msk` is the name of a series. Subsequent commands that form matrices from series or lists will include only observations for which `msk` has non-zero (and non-missing) values. You can remove this mask via the command `set matrix_mask null`.

☞ Names of matrices must satisfy the same requirements as names of gretl variables in general: the name can be no longer than 31 characters, must start with a letter, and must be composed of nothing but letters, numbers and the underscore character.

15.2 Empty matrices

The syntax

```
matrix A = {}
```

creates an empty matrix—a matrix with zero rows and zero columns.

The main purpose of the concept of an empty matrix is to enable the user to define a starting point for subsequent concatenation operations. For instance, if X is an already defined matrix of any size, the commands

```
matrix A = {}
matrix B = A ~ X
```

result in a matrix B identical to X.

Function	*Return value*	*Function*	*Return value*
`A', transp(A)`	A	`rows(A)`	0
`cols(A)`	0	`rank(A)`	0
`det(A)`	NA	`ldet(A)`	NA
`tr(A)`	NA	`onenorm(A)`	NA
`infnorm(A)`	NA	`rcond(A)`	NA

Table 15.1: Valid functions on an empty matrix, A

From an algebraic point of view, one can make sense of the idea of an empty matrix in terms of vector spaces: if a matrix is an ordered set of vectors, then A={} is the empty set. As a consequence, operations involving addition and multiplications don't have any clear meaning (arguably, they have none at all), but operations involving the cardinality of this set (that is, the dimension of the space spanned by A) are meaningful.

Legal operations on empty matrices are listed in Table 15.1. (All other matrix operations generate an error when an empty matrix is given as an argument.) In line with the above interpretation, some matrix functions return an empty matrix under certain conditions: the functions `diag, vec, vech, unvech` when the arguments is an empty matrix; the functions `I, ones, zeros, mnormal, muniform` when one or more of the arguments is 0; and the function `nullspace` when its argument has full column rank.

15.3 Selecting sub-matrices

You can select sub-matrices of a given matrix using the syntax

A[*rows,cols*]

where *rows* can take any of these forms:

1.	empty	selects all rows
2.	a single integer	selects the single specified row
3.	two integers separated by a colon	selects a range of rows
4.	the name of a matrix	selects the specified rows

With regard to option 2, the integer value can be given numerically, as the name of an existing scalar variable, or as an expression that evaluates to a scalar. With option 4, the index matrix given in the *rows* field must be either $p \times 1$ or $1 \times p$, and should contain integer values in the range 1 to n, where n is the number of rows in the matrix from which the selection is to be made.

The *cols* specification works in the same way, *mutatis mutandis.* Here are some examples.

```
matrix B = A[1,]
matrix B = A[2:3,3:5]
matrix B = A[2,2]
matrix idx = { 1, 2, 6 }
matrix B = A[idx,]
```

The first example selects row 1 from matrix A; the second selects a 2×3 submatrix; the third selects a scalar; and the fourth selects rows 1, 2, and 6 from matrix A.

If the matrix in question is $n \times 1$ or $1 \times m$, it is OK to give just one index specifier and omit the comma. For example, `A[2]` selects the second element of A if A is a vector. Otherwise the comma is mandatory.

In addition there is a pre-defined index specification, `diag`, which selects the principal diagonal of a square matrix, as in `B[diag]`, where B is square.

You can use selections of this sort on either the right-hand side of a matrix-generating formula or the left. Here is an example of use of a selection on the right, to extract a 2×2 submatrix B from a 3×3 matrix A:

```
matrix A = { 1, 2, 3; 4, 5, 6; 7, 8, 9 }
matrix B = A[1:2,2:3]
```

And here are examples of selection on the left. The second line below writes a 2×2 identity matrix into the bottom right corner of the 3×3 matrix A. The fourth line replaces the diagonal of A with 1s.

```
matrix A = { 1, 2, 3; 4, 5, 6; 7, 8, 9 }
matrix A[2:3,2:3] = I(2)
matrix d = { 1, 1, 1 }
matrix A[diag] = d
```

15.4 Matrix operators

The following binary operators are available for matrices:

+	addition
-	subtraction
*	ordinary matrix multiplication
'	pre-multiplication by transpose
\	matrix "left division" (see below)
/	matrix "right division" (see below)
~	column-wise concatenation
|	row-wise concatenation
**	Kronecker product
=	test for equality
!=	test for inequality

In addition, the following operators ("dot" operators) apply on an element-by-element basis:

```
.+   .-   .*   ./   .^   .=   .>   .<   .>=   .<=   .!=
```

Here are explanations of the less obvious cases.

For matrix addition and subtraction, in general the two matrices have to be of the same dimensions but an exception to this rule is granted if one of the operands is a 1×1 matrix or scalar. The scalar is implicitly promoted to the status of a matrix of the correct dimensions, all of whose elements are equal to the given scalar value. For example, if A is an $m \times n$ matrix and k a scalar, then the commands

```
matrix C = A + k
matrix D = A - k
```

both produce $m \times n$ matrices, with elements $c_{ij} = a_{ij} + k$ and $d_{ij} = a_{ij} - k$ respectively.

By "pre-multiplication by transpose" we mean, for example, that

```
matrix C = X'Y
```

produces the product of X-transpose and Y. In effect, the expression `X'Y` is shorthand for `X'*Y`, which is also valid. Consider, however, that in the special case $X = Y$, the two are not exactly equivalent: the former expression uses a specialized, optimized algorithm which has the double advantage of being more efficient computationally and of ensuring that the result will be free by construction of machine precision artifacts that may render it numerically non-symmetric. This, however, is unlikely to affect you unless your X matrix is rather large (at least several hundreds rows/columns).

In matrix "left division", the statement

```
matrix X = A \ B
```

is interpreted as a request to find the matrix X that solves $AX = B$. If B is a square matrix, this is in principle equivalent to $A^{-1}B$, which fails if A is singular; the numerical method employed here is the LU decomposition. If A is a $T \times k$ matrix with $T > k$, then X is the least-squares solution, $X = (A'A)^{-1}A'B$, which fails if $A'A$ is singular; the numerical method employed here is the QR decomposition. Otherwise, the operation necessarily fails.

For matrix "right division", as in `X = A / B`, X is the matrix that solves $XB = A$, in principle equivalent to AB^{-1}.

In "dot" operations a binary operation is applied element by element; the result of this operation is obvious if the matrices are of the same size. However, there are several other cases where such operators may be applied. For example, if we write

```
matrix C = A .- B
```

then the result C depends on the dimensions of A and B. Let A be an $m \times n$ matrix and let B be $p \times q$; the result is as follows:

Case	*Result*
Dimensions match ($m = p$ and $n = q$)	$c_{ij} = a_{ij} - b_{ij}$
A is a column vector; rows match ($m = p$; $n = 1$)	$c_{ij} = a_i - b_{ij}$
B is a column vector; rows match ($m = p$; $q = 1$)	$c_{ij} = a_{ij} - b_i$
A is a row vector; columns match ($m = 1$; $n = q$)	$c_{ij} = a_j - b_{ij}$
B is a row vector; columns match ($m = p$; $q = 1$)	$c_{ij} = a_{ij} - b_j$
A is a column vector; B is a row vector ($n = 1$; $p = 1$)	$c_{ij} = a_i - b_j$
A is a row vector; B is a column vector ($m = 1$; $q = 1$)	$c_{ij} = a_j - b_i$
A is a scalar ($m = 1$ and $n = 1$)	$c_{ij} = a - b_{ij}$
B is a scalar ($p = 1$ and $q = 1$)	$c_{ij} = a_{ij} - b$

If none of the above conditions are satisfied the result is undefined and an error is flagged.

Note that this convention makes it unnecessary, in most cases, to use diagonal matrices to perform transformations by means of ordinary matrix multiplication: if $Y = XV$, where V is diagonal, it is computationally much more convenient to obtain Y via the instruction

```
matrix Y = X .* v
```

where v is a row vector containing the diagonal of V.

In *column-wise concatenation* of an $m \times n$ matrix A and an $m \times p$ matrix B, the result is an $m \times (n + p)$ matrix. That is,

```
matrix C = A ~ B
```

produces $C = \begin{bmatrix} A & B \end{bmatrix}$.

Row-wise concatenation of an $m \times n$ matrix A and an $p \times n$ matrix B produces an $(m + p) \times n$ matrix. That is,

```
matrix C = A | B
```

produces $C = \begin{bmatrix} A \\ B \end{bmatrix}$.

15.5 Matrix-scalar operators

For matrix A and scalar k, the operators shown in Table 15.2 are available. (Addition and subtraction were discussed in section 15.4 but we include them in the table for completeness.) In addition, for square A and integer $k \geq 0$, `B = A^k` produces a matrix B which is A raised to the power k.

15.6 Matrix functions

Most of the gretl functions available for scalars and series also apply to matrices in an element-by-element fashion, and as such their behavior should be pretty obvious. This is the case for functions such as `log`, `exp`, `sin`, etc. These functions have the effects documented in relation to the `genr` command. For example, if a matrix A is already defined, then

```
matrix B = sqrt(A)
```

Expression	*Effect*
`matrix B = A * k`	$b_{ij} = ka_{ij}$
`matrix B = A / k`	$b_{ij} = a_{ij}/k$
`matrix B = k / A`	$b_{ij} = k/a_{ij}$
`matrix B = A + k`	$b_{ij} = a_{ij} + k$
`matrix B = A - k`	$b_{ij} = a_{ij} - k$
`matrix B = k - A`	$b_{ij} = k - a_{ij}$
`matrix B = A % k`	$b_{ij} = a_{ij}$ modulo k

Table 15.2: Matrix-scalar operators

generates a matrix such that $b_{ij} = \sqrt{a_{ij}}$. All such functions require a single matrix as argument, or an expression which evaluates to a single matrix.[1]

In this section, we review some aspects of `genr` functions that apply specifically to matrices. A full account of each function is available in the *Gretl Command Reference.*

Matrix reshaping

In addition to the methods discussed in sections 15.1 and 15.3, a matrix can also be created by re-arranging the elements of a pre-existing matrix. This is accomplished via the `mshape` function. It takes three arguments: the input matrix, A, and the rows and columns of the target matrix, r and c respectively. Elements are read from A and written to the target in column-major order. If A contains fewer elements than $n = r \times c$, they are repeated cyclically; if A has more elements, only the first n are used.

For example:

```
matrix a = mnormal(2,3)
a
matrix b = mshape(a,3,1)
b
matrix b = mshape(a,5,2)
b
```

produces

```
?  a
a

      1.2323       0.99714      -0.39078
     0.54363       0.43928      -0.48467

?  matrix b = mshape(a,3,1)
Generated matrix b
?  b
b

      1.2323
     0.54363
     0.99714

?  matrix b = mshape(a,5,2)
Replaced matrix b
?  b
b
```

[1]Note that to find the "matrix square root" you need the `cholesky` function (see below); moreover, the `exp` function computes the exponential element by element, and therefore does *not* return the matrix exponential unless the matrix is diagonal—to get the matrix exponential, use `mexp`.

Creation and I/O						
	colnames	diag	diagcat	I	lower	mnormal
	mread	muniform	ones	rownames	seq	unvech
	upper	vec	vech	zeros		
Shape/size/arrangement						
	cols	dsort	mreverse	mshape	msortby	rows
	selifc	selifr	sort	trimr		
Matrix algebra						
	cdiv	cholesky	cmult	det	eigengen	eigensym
	eigsolve	fft	ffti	ginv	hdprod	infnorm
	inv	invpd	ldet	mexp	nullspace	onenorm
	polroots	psdroot	qform	qrdecomp	rank	rcond
	svd	toepsolv	tr	transp	varsimul	
Statistics/transformations						
	aggregate	cdemean	corr	corrgm	cov	fcstats
	ghk	halton	imaxc	imaxr	iminc	iminr
	irf	iwishart	kdensity	kpsscrit	maxc	maxr
	mcorr	mcov	mcovg	meanc	meanr	minc
	minr	mlag	mols	mpols	mrls	mxtab
	pergm	princomp	prodc	prodr	quadtable	quantile
	ranking	resample	sdc	sumc	sumr	uniq
	values					
Data utilities						
	isconst	isdummy	mwrite	ok	pexpand	pshrink
	replace					
Filters						
	filter	kfilter	ksimul	ksmooth	lrvar	
Numerical methods						
	BFGSmax	fdjac	NRmax	simann		
Strings						
	colname					
Transformations						
	chowlin	cum	lincomb			

Table 15.3: Matrix functions by category

```
   1.2323     -0.48467
  0.54363       1.2323
  0.99714      0.54363
  0.43928      0.99714
 -0.39078      0.43928
```

Complex multiplication and division

Gretl has no native provision for complex numbers. However, basic operations can be performed on vectors of complex numbers by using the convention that a vector of n complex numbers is represented as a $n \times 2$ matrix, where the first column contains the real part and the second the imaginary part.

Addition and subtraction are trivial; the functions `cmult` and `cdiv` compute the complex product and division, respectively, of two input matrices, A and B, representing complex numbers. These matrices must have the same number of rows, n, and either one or two columns. The first column contains the real part and the second (if present) the imaginary part. The return value is an $n \times 2$ matrix, or, if the result has no imaginary part, an n-vector.

For example, suppose you have $z_1 = [1 + 2i, 3 + 4i]'$ and $z_2 = [1, i]'$:

```
? z1 = {1,2;3,4}
 z1 = {1,2;3,4}
Generated matrix z1
? z2 = I(2)
 z2 = I(2)
Generated matrix z2
? conj_z1 = z1 .* {1,-1}
 conj_z1 = z1 .* {1,-1}
Generated matrix conj_z1
? eval cmult(z1,z2)
 eval cmult(z1,z2)
   1    2
  -4    3

? eval cmult(z1,conj_z1)
 eval cmult(z1,conj_z1)
   5
  25
```

Multiple returns and the `null` keyword

Some functions take one or more matrices as arguments and compute one or more matrices; these are:

`eigensym`	Eigen-analysis of symmetric matrix
`eigengen`	Eigen-analysis of general matrix
`mols`	Matrix OLS
`qrdecomp`	QR decomposition
`svd`	Singular value decomposition (SVD)

The general rule is: the "main" result of the function is always returned as the result proper. Auxiliary returns, if needed, are retrieved using pre-existing matrices, which are passed to the function as pointers (see 13.4). If such values are not needed, the pointer may be substituted with the keyword `null`.

The syntax for `qrdecomp`, `eigensym` and `eigengen` is of the form

```
matrix B = func(A, &C)
```

The first argument, A, represents the input data, that is, the matrix whose decomposition or analysis is required. The second argument must be either the name of an existing matrix preceded by & (to indicate the "address" of the matrix in question), in which case an auxiliary result is written to that matrix, or the keyword null, in which case the auxiliary result is not produced, or is discarded.

In case a non-null second argument is given, the specified matrix will be over-written with the auxiliary result. (It is not required that the existing matrix be of the right dimensions to receive the result.)

The function eigensym computes the eigenvalues, and optionally the right eigenvectors, of a symmetric $n \times n$ matrix. The eigenvalues are returned directly in a column vector of length n; if the eigenvectors are required, they are returned in an $n \times n$ matrix. For example:

```
matrix V
matrix E = eigensym(M, &V)
matrix E = eigensym(M, null)
```

In the first case E holds the eigenvalues of M and V holds the eigenvectors. In the second, E holds the eigenvalues but the eigenvectors are not computed.

The function eigengen computes the eigenvalues, and optionally the eigenvectors, of a general $n \times n$ matrix. The eigenvalues are returned directly in an $n \times 2$ matrix, the first column holding the real components and the second column the imaginary components.

If the eigenvectors are required (that is, if the second argument to eigengen is not null), they are returned in an $n \times n$ matrix. The column arrangement of this matrix is somewhat non-trivial: the eigenvectors are stored in the same order as the eigenvalues, but the real eigenvectors occupy one column, whereas complex eigenvectors take two (the real part comes first); the total number of columns is still n, because the conjugate eigenvector is skipped. Example 15.1 provides a (hopefully) clarifying example (see also subsection 15.6).

The qrdecomp function computes the QR decomposition of an $m \times n$ matrix A: $A = QR$, where Q is an $m \times n$ orthogonal matrix and R is an $n \times n$ upper triangular matrix. The matrix Q is returned directly, while R can be retrieved via the second argument. Here are two examples:

```
matrix R
matrix Q = qrdecomp(M, &R)
matrix Q = qrdecomp(M, null)
```

In the first example, the triangular R is saved as R; in the second, R is discarded. The first line above shows an example of a "simple declaration" of a matrix: R is declared to be a matrix variable but is not given any explicit value. In this case the variable is initialized as a 1×1 matrix whose single element equals zero.

The syntax for svd is

```
matrix B = func(A, &C, &D)
```

The function svd computes all or part of the singular value decomposition of the real $m \times n$ matrix A. Let $k = \min(m, n)$. The decomposition is

$$A = U\Sigma V'$$

where U is an $m \times k$ orthogonal matrix, Σ is an $k \times k$ diagonal matrix, and V is an $k \times n$ orthogonal matrix.[2] The diagonal elements of Σ are the singular values of A; they are real and non-negative, and are returned in descending order. The first k columns of U and V are the left and right singular vectors of A.

The svd function returns the singular values, in a vector of length k. The left and/or right singular vectors may be obtained by supplying non-null values for the second and/or third arguments respectively. For example:

[2]This is not the only definition of the SVD: some writers define U as $m \times m$, Σ as $m \times n$ (with k non-zero diagonal elements) and V as $n \times n$.

Example 15.1: Complex eigenvalues and eigenvectors

```
set seed 34756

matrix v
A = mnormal(3,3)

/* do the eigen-analysis */
l = eigengen(A,&v)
/* eigenvalue 1 is real, 2 and 3 are complex conjugates */
print l
print v

/*
  column 1 contains the first eigenvector (real)
*/

B = A*v[,1]
c = l[1,1] * v[,1]
/* B should equal c */
print B
print c


/*
  columns 2:3 contain the real and imaginary parts
  of eigenvector 2
*/

B = A*v[,2:3]
c = cmult(ones(3,1)*(l[2,]),v[,2:3])
/* B should equal c */
print B
print c
```

```
matrix s = svd(A, &U, &V)
matrix s = svd(A, null, null)
matrix s = svd(A, null, &V)
```

In the first case both sets of singular vectors are obtained, in the second case only the singular values are obtained; and in the third, the right singular vectors are obtained but U is not computed. *Please note*: when the third argument is non-null, it is actually V' that is provided. To reconstitute the original matrix from its SVD, one can do:

```
matrix s = svd(A, &U, &V)
matrix B = (U.*s)*V
```

Finally, the syntax for `mols` is

```
matrix B = mols(Y, X, &U)
```

This function returns the OLS estimates obtained by regressing the $T \times n$ matrix Y on the $T \times k$ matrix X, that is, a $k \times n$ matrix holding $(X'X)^{-1}X'Y$. The Cholesky decomposition is used. The matrix U, if not `null`, is used to store the residuals.

Reading and writing matrices from/to text files

The two functions `mread` and `mwrite` can be used for basic matrix input/output. This can be useful to enable gretl to exchange data with other programs.

The `mread` function accepts one string parameter: the name of the (plain text) file from which the matrix is to be read. The file in question may start with any number of comment lines, defined as lines that start with the hash mark, "#"; such lines are ignored. Beyond that, the content must conform to the following rules:

1. The first non-comment line must contain two integers, separated by a space or a tab, indicating the number of rows and columns, respectively.
2. The columns must be separated by spaces or tab characters.
3. The decimal separator must be the dot "`.`" character.

Should an error occur (such as the file being badly formatted or inaccessible), an empty matrix (see section 15.2) is returned.

The complementary function `mwrite` produces text files formatted as described above. The column separator is the tab character, so import into spreadsheets should be straightforward. Usage is illustrated in example 15.2. Matrices stored via the `mwrite` command can be easily read by other programs; the following table summarizes the appropriate commands for reading a matrix `A` from a file called `a.mat` in some widely-used programs.[3] Note that the Python example requires that the `numpy` module is loaded.

Program	Sample code
GAUSS	`tmp[] = load a.mat;`
	`A = reshape(tmp[3:rows(tmp)],tmp[1],tmp[2]);`
Octave	`fd = fopen("a.mat");`
	`[r,c] = fscanf(fd, "%d %d", "C");`
	`A = reshape(fscanf(fd, "%g", r*c),c,r)';`
	`fclose(fd);`
Ox	`decl A = loadmat("a.mat");`
R	`A <- as.matrix(read.table("a.mat", skip=1))`
Python	`A = numpy.loadtxt('a.mat', skiprows=1)`
Julia	`A = readdlm("a.mat", skipstart=1)`

Example 15.2: Matrix input/output via text files

```
nulldata 64
scalar n = 3
string f1 = "a.csv"
string f2 = "b.csv"

matrix a = mnormal(n,n)
matrix b = inv(a)

err = mwrite(a, f1)

if err != 0
  fprintf "Failed to write %s\n", f1
else
  err = mwrite(b, f2)
endif

if err != 0
  fprintf "Failed to write %s\n", f2
else
  c = mread(f1)
  d = mread(f2)
  a = c*d
  printf "The following matrix should be an identity matrix\n"
  print a
endif
```

Optionally, the `mwrite` and `mread` functions can use gzip compression: this is invoked if the name of the matrix file has the suffix "`.gz`." In this case the elements of the matrix are written in a single column. Note, however, that compression should not be applied when writing matrices for reading by third-party software unless you are sure that the software can handle compressed data.

15.7 Matrix accessors

In addition to the matrix functions discussed above, various "accessor" strings allow you to create copies of internal matrices associated with models previously estimated. These are set out in Table 15.4.

Many of the accessors in Table 15.4 behave somewhat differently depending on the sort of model that is referenced, as follows:

- Single-equation models: `$sigma` gets a scalar (the standard error of the regression); `$coeff` and `$stderr` get column vectors; `$uhat` and `$yhat` get series.
- System estimators: `$sigma` gets the cross-equation residual covariance matrix; `$uhat` and `$yhat` get matrices with one column per equation. The format of `$coeff` and `$stderr` depends on the nature of the system: for VARs and VECMs (where the matrix of regressors is the same for all equations) these return matrices with one column per equation, but for other system estimators they return a big column vector.
- VARs and VECMs: `$vcv` is not available, but $X'X^{-1}$ (where X is the common matrix of regressors) is available as `$xtxinv`.

[3]Matlab users may find the Octave example helpful, since the two programs are mostly compatible with one another.

`$coeff`	matrix of estimated coefficients
`$compan`	companion matrix (after VAR or VECM estimation)
`$jalpha`	matrix α (loadings) from Johansen's procedure
`$jbeta`	matrix β (cointegration vectors) from Johansen's procedure
`$jvbeta`	covariance matrix for the unrestricted elements of β from Johansen's procedure
`$rho`	autoregressive coefficients for error process
`$sigma`	residual covariance matrix
`$stderr`	matrix of estimated standard errors
`$uhat`	matrix of residuals
`$vcv`	covariance matrix of parameter estimates
`$vma`	VMA matrices in stacked form (see section 26.2)
`$yhat`	matrix of fitted values

Table 15.4: Matrix accessors for model data

If the accessors are given without any prefix, they retrieve results from the last model estimated, if any. Alternatively, they may be prefixed with the name of a saved model plus a period (`.`), in which case they retrieve results from the specified model. Here are some examples:

```
matrix u = $uhat
matrix b = m1.$coeff
matrix v2 = m1.$vcv[1:2,1:2]
```

The first command grabs the residuals from the last model; the second grabs the coefficient vector from model `m1`; and the third (which uses the mechanism of sub-matrix selection described above) grabs a portion of the covariance matrix from model `m1`.

If the model in question a VAR or VECM (only) `$compan` and `$vma` return the companion matrix and the VMA matrices in stacked form, respectively (see section 26.2 for details). After a vector error correction model is estimated via Johansen's procedure, the matrices `$jalpha` and `$jbeta` are also available. These have a number of columns equal to the chosen cointegration rank; therefore, the product

```
matrix Pi = $jalpha * $jbeta'
```

returns the reduced-rank estimate of $A(1)$. Since β is automatically identified via the Phillips normalization (see section 27.5), its unrestricted elements do have a proper covariance matrix, which can be retrieved through the `$jvbeta` accessor.

15.8 Namespace issues

Matrices share a common namespace with data series and scalar variables. In other words, no two objects of any of these types can have the same name. It is an error to attempt to change the type of an existing variable, for example:

```
scalar x = 3
matrix x = ones(2,2) # wrong!
```

It is possible, however, to delete or rename an existing variable then reuse the name for a variable of a different type:

```
scalar x = 3
delete x
matrix x = ones(2,2) # OK
```

15.9 Creating a data series from a matrix

Section 15.1 above describes how to create a matrix from a data series or set of series. You may sometimes wish to go in the opposite direction, that is, to copy values from a matrix into a regular data series. The syntax for this operation is

`series` *`sname`* `=` *`mspec`*

where *`sname`* is the name of the series to create and *`mspec`* is the name of the matrix to copy from, possibly followed by a matrix selection expression. Here are two examples.

```
series s = x
series u1 = U[,1]
```

It is assumed that `x` and `U` are pre-existing matrices. In the second example the series `u1` is formed from the first column of the matrix `U`.

For this operation to work, the matrix (or matrix selection) must be a vector with length equal to either the full length of the current dataset, n, or the length of the current sample range, n'. If $n' < n$ then only n' elements are drawn from the matrix; if the matrix or selection comprises n elements, the n' values starting at element t_1 are used, where t_1 represents the starting observation of the sample range. Any values in the series that are not assigned from the matrix are set to the missing code.

15.10 Matrices and lists

To facilitate the manipulation of named lists of variables (see Chapter 14), it is possible to convert between matrices and lists. In section 15.1 above we mentioned the facility for creating a matrix from a list of variables, as in

```
matrix M = { listname }
```

That formulation, with the name of the list enclosed in braces, builds a matrix whose columns hold the variables referenced in the list. What we are now describing is a different matter: if we say

```
matrix M = listname
```

(without the braces), we get a row vector whose elements are the ID numbers of the variables in the list. This special case of matrix generation cannot be embedded in a compound expression. The syntax must be as shown above, namely simple assignment of a list to a matrix.

To go in the other direction, you can include a matrix on the right-hand side of an expression that defines a list, as in

```
list Xl = M
```

where `M` is a matrix. The matrix must be suitable for conversion; that is, it must be a row or column vector containing non-negative integer values, none of which exceeds the highest ID number of a series in the current dataset.

Example 15.3 illustrates the use of this sort of conversion to "normalize" a list, moving the constant (variable 0) to first position.

15.11 Deleting a matrix

To delete a matrix, just write

```
delete M
```

where `M` is the name of the matrix to be deleted.

Example 15.3: Manipulating a list

```
function void normalize_list (matrix *x)
  # If the matrix (representing a list) contains var 0,
  # but not in first position, move it to first position

  if (x[1] != 0)
     scalar k = cols(x)
     loop for (i=2; i<=k; i++) --quiet
        if (x[i] = 0)
            x[i] = x[1]
            x[1] = 0
            break
        endif
     endloop
  endif
end function

open data9-7
list Xl = 2 3 0 4
matrix x = Xl
normalize_list(&x)
list Xl = x
```

15.12 Printing a matrix

To print a matrix, the easiest way is to give the name of the matrix in question on a line by itself, which is equivalent to using the `print` command:

```
matrix M = mnormal(100,2)
M
print M
```

You can get finer control on the formatting of output by using the `printf` command, as illustrated in the interactive session below:

```
? matrix Id = I(2)
 matrix Id = I(2)
Generated matrix Id
? print Id
 print Id
Id (2 x 2)

  1   0
  0   1

? printf "%10.3f", Id
     1.000     0.000
     0.000     1.000
```

For presentation purposes you may wish to give titles to the columns of a matrix. For this you can use the `colnames` function: the first argument is a matrix and the second is either a named list of variables, whose names will be used as headings, or a string that contains as many space-separated substrings as the matrix has columns. For example,

```
? matrix M = mnormal(3,3)
? colnames(M, "foo bar baz")
? print M
```

```
M (3 x 3)

         foo          bar          baz
      1.7102     -0.76072     0.089406
    -0.99780      -1.9003     -0.25123
    -0.91762     -0.39237      -1.6114
```

15.13 Example: OLS using matrices

Example 15.4 shows how matrix methods can be used to replicate gretl's built-in OLS functionality.

Example 15.4: OLS via matrix methods

```
open data4-1
matrix X = { const, sqft }
matrix y = { price }
matrix b = invpd(X'X) * X'y
print "estimated coefficient vector"
b
matrix u = y - X*b
scalar SSR = u'u
scalar s2 = SSR / (rows(X) - rows(b))
matrix V = s2 * inv(X'X)
V
matrix se = sqrt(diag(V))
print "estimated standard errors"
se
# compare with built-in function
ols price const sqft --vcv
```

Chapter 16

Cheat sheet

This chapter explains how to perform some common—and some not so common—tasks in gretl's scripting language, hansl. Some but not all of the techniques listed here are also available through the graphical interface. Although the graphical interface may be more intuitive and less intimidating at first, we encourage users to take advantage of the power of gretl's scripting language as soon as they feel comfortable with the program.

16.1 Dataset handling

"Weird" periodicities

Problem: You have data sampled each 3 minutes from 9am onwards; you'll probably want to specify the hour as 20 periods.

Solution:

```
setobs 20 9:1 --special
```

Comment: Now functions like `sdiff()` ("seasonal" difference) or estimation methods like seasonal ARIMA will work as expected.

Generating a panel dataset of given dimensions

Problem: You want to generate via `nulldata` a panel dataset and specify in advance the number of units and the time length of your series via two scalar variables.

Solution:

```
scalar n_units = 100
scalar T = 12
scalar NT = T * n_units

nulldata NT --preserve
setobs T 1:1 --stacked-time-series
```

Comment: The essential ingredient that we use here is the `--preserve` option: it protects existing scalars (and matrices, for that matter) from being trashed by `nulldata`, thus making it possible to use the scalar T in the `setobs` command.

Help, my data are backwards!

Problem: Gretl expects time series data to be in chronological order (most recent observation last), but you have imported third-party data that are in reverse order (most recent first).

Solution:

```
setobs 1 1 --cross-section
series sortkey = -obs
dataset sortby sortkey
setobs 1 1950 --time-series
```

Comment: The first line is required only if the data currently have a time series interpretation: it removes that interpretation, because (for fairly obvious reasons) the `dataset sortby` operation is not allowed for time series data. The following two lines reverse the data, using the negative of the built-in index variable obs. The last line is just illustrative: it establishes the data as annual time series, starting in 1950.

If you have a dataset that is mostly the right way round, but a particular variable is wrong, you can reverse that variable as follows:

```
x = sortby(-obs, x)
```

Dropping missing observations selectively

Problem: You have a dataset with many variables and want to restrict the sample to those observations for which there are no missing observations for the variables x1, x2 and x3.

Solution:

```
list X = x1 x2 x3
smpl --no-missing X
```

Comment: You can now save the file via a store command to preserve a subsampled version of the dataset. Alternative solutions based on the ok function, such as

```
list X = x1 x2 x3
series sel = ok(X)
smpl sel --restrict
```

are perhaps less obvious, but more flexible. Pick your poison.

"By" operations

Problem: You have a discrete variable d and you want to run some commands (for example, estimate a model) by splitting the sample according to the values of d.

Solution:

```
matrix vd = values(d)
m = rows(vd)
loop i=1..m
  scalar sel = vd[i]
  smpl d==sel --restrict --replace
  ols y const x
endloop
smpl --full
```

Comment: The main ingredient here is a loop. You can have gretl perform as many instructions as you want for each value of d, as long as they are allowed inside a loop. Note, however, that if all you want is descriptive statistics, the summary command does have a --by option.

Adding a time series to a panel

Problem: You have a panel dataset (comprising observations of n indidivuals in each of T periods) and you want to add a variable which is available in straight time-series form. For example, you want to add annual CPI data to a panel in order to deflate nominal income figures.

In gretl a panel is represented in stacked time-series format, so in effect the task is to create a new variable which holds n stacked copies of the original time series. Let's say the panel comprises 500 individuals observed in the years 1990, 1995 and 2000 ($n = 500$, $T = 3$), and we have these CPI data in the ASCII file cpi.txt:

```
date cpi
1990 130.658
1995 152.383
2000 172.192
```

What we need is for the CPI variable in the panel to repeat these three values 500 times.

Solution: Simple! With the panel dataset open in gretl,

```
append cpi.txt
```

Comment: If the length of the time series is the same as the length of the time dimension in the panel (3 in this example), gretl will perform the stacking automatically. Rather than using the append command you could use the "Append data" item under the File menu in the GUI program. For this to work, your main dataset must be recognized as a panel. This can be arranged via the setobs command or the "Dataset structure" item under the Data menu.

16.2 Creating/modifying variables

Generating a dummy variable for a specific observation

Problem: Generate $d_t = 0$ for all observation but one, for which $d_t = 1$.

Solution:

```
series d = (t=="1984:2")
```

Comment: The internal variable t is used to refer to observations in string form, so if you have a cross-section sample you may just use d = (t=="123"). If the dataset has observation labels you can use the corresponding label. For example, if you open the dataset mrw.gdt, supplied with gretl among the examples, a dummy variable for Italy could be generated via

```
series DIta = (t=="Italy")
```

Note that this method does not require scripting at all. In fact, you might as well use the GUI Menu "Add/Define new variable" for the same purpose, with the same syntax.

Generating a discrete variable out of a set of dummies

Problem: The dummify function (also available as a command) generates a set of mutually exclusive dummies from a discrete variable. The reverse functionality, however, seems to be absent.

Solution:

```
series x = lincomb(D, seq(1, nelem(D)))
```

Comment: Suppose you have a list D of mutually exclusive dummies, that is a full set of 0/1 variables coding for the value of some characteristic, such that the sum of the values of the elements of D is 1 at each observation. This is, by the way, exactly what the dummify command produces. The reverse job of dummify can be performed neatly by using the lincomb function.

The code above multiplies the first dummy variable in the list D by 1, the second one by 2, and so on. Hence, the return value is a series whose value is i if and only if the i-th member of D has value 1.

If you want your coding to start from 0 instead of 1, you'll have to modify the code snippet above into

```
series x = lincomb(D, seq(0, nelem(D)-1))
```

Recoding a variable

Problem: You want to perform a 1-to-1 recode on a variable. For example, consider tennis points: you may have a variable x holding values 1 to 3 and you want to recode it to 15, 30, 40.

Solution 1:

```
series x = replace(x, 1, 15)
series x = replace(x, 2, 30)
series x = replace(x, 3, 40)
```

Solution 2:

```
matrix tennis = {15, 30, 40}
series x = replace(x, seq(1,3), tennis)
```

Comment: There are many equivalent ways to achieve the same effect, but for simple cases such as this, the `replace` function is simple and transparent. If you don't mind using matrices, scripts using `replace` can also be remarkably compact. Note that `replace` also performs n-to-1 ("surjective") replacements, such as

```
series x = replace{z, {2, 3, 5, 11, 22, 33}, 1)
```

which would turn all entries equal to 2, 3, 5, 11, 22 or 33 to 1 and leave the other ones unchanged.

Generating a "subset of values" dummy

Problem: You have a dataset which contains a fine-grained coding for some qualitative variable and you want to "collapse" this to a relatively small set of dummy variables. Examples: you have place of work by US state and you want a small set of regional dummies; or you have detailed occupational codes from a census dataset and you want a manageable number of occupational category dummies.

Let's call the source series `src` and one of the target dummies `D1`. And let's say that the values of `src` to be grouped under `D1` are 2, 13, 14 and 25.

"Longhand" solution:

```
series D1 = src==2 || src==13 || src==14 || src==25
```

Comment: The above works fine if the number of distinct values in the source to be condensed into each dummy variable is fairly small, but it becomes cumbersome if a single dummy must comprise dozens of source values.

Clever solution:

```
matrix sel = {2,13,14,25}
series D1 = maxr({src} .= vec(sel)') .> 0
```

Comment: The subset of values to be grouped together can be written out as a matrix relatively compactly (first line). The magic that turns this into the desired series (second line) relies on the versatility of the "dot" (element-wise) matrix operators. The expression "`{src}`" gets a column-vector version of the input series—call this x—and "`vec(sel)'`" gets the input matrix as a row vector, in case it's a column vector or a matrix with both dimensions greater than 1—call this s. If x is $n \times 1$ and s is $1 \times m$, the "`.=`" operator produces an $n \times m$ result, each element (i, j) of which equals 1 if $x_i = s_j$, otherwise 0. The `maxr()` function along with the "`.>`" operator (see chapter 15 for both) then produces the result we want.

Of course, whichever procedure you use, you have to repeat for each of the dummy series you want to create (but keep reading—the "proper" solution is probably what you want if you plan to create several dummies).

Further comment: Note that the clever solution depends on converting what is "naturally" a vector result into a series. This will fail if there are missing values in `src`, since (by default) missing values will be skipped when converting `src` to x, and so the number of rows in the result will fall short of the number of observations in the dataset. One fix is then to subsample the dataset to exclude missing values before employing this method; another is to adjust the `skip_missing` setting via the `set` command (see the *Gretl Command Reference*).

Proper solution:

The best solution, in terms of both computational efficiency and code clarity, would be using a "conversion table" and the `replace` function, to produce a series on which the `dummify` command can be used. For example, suppose we want to convert from a series called `fips` holding FIPS codes[1] for the 50 US states plus the District of Columbia to a series holding codes for the four standard US regions. We could create a 2×51 matrix—call it `srmap`—with the 51 FIPS codes on the first row and the corresponding region codes on the second, and then do

```
series region = replace(fips, srmap[1,], srmap[2,])
```

Generating an ARMA(1,1)

Problem: Generate $y_t = 0.9y_{t-1} + \varepsilon_t - 0.5\varepsilon_{t-1}$, with $\varepsilon_t \sim NIID(0,1)$.

Recommended solution:

```
alpha = 0.9
theta = -0.5
series y = filter(normal(), {1, theta}, alpha)
```

"Bread and butter" solution:

```
alpha = 0.9
theta = -0.5
series e = normal()
series y = 0
series y = alpha * y(-1) + e + theta * e(-1)
```

Comment: The `filter` function is specifically designed for this purpose so in most cases you'll want to take advantage of its speed and flexibility. That said, in some cases you may want to generate the series in a manner which is more transparent (maybe for teaching purposes).

In the second solution, the statement `series y = 0` is necessary because the next statement evaluates `y` recursively, so `y[1]` must be set. Note that you must use the keyword `series` here instead of writing `genr y = 0` or simply `y = 0`, to ensure that `y` is a series and not a scalar.

Recoding a variable by classes

Problem: You want to recode a variable by classes. For example, you have the age of a sample of individuals (x_i) and you need to compute age classes (y_i) as

$$\begin{aligned} y_i &= 1 \quad \text{for} \quad x_i < 18 \\ y_i &= 2 \quad \text{for} \quad 18 \le x_i < 65 \\ y_i &= 3 \quad \text{for} \quad x_i \ge 65 \end{aligned}$$

Solution:

```
series y = 1 + (x >= 18) + (x >= 65)
```

Comment: True and false expressions are evaluated as 1 and 0 respectively, so they can be manipulated algebraically as any other number. The same result could also be achieved by using the conditional assignment operator (see below), but in most cases it would probably lead to more convoluted constructs.

[1]FIPS is the Federal Information Processing Standard: it assigns numeric codes from 1 to 56 to the US states and outlying areas.

Conditional assignment

Problem: Generate y_t via the following rule:

$$y_t = \begin{cases} x_t & \text{for} \quad d_t > a \\ z_t & \text{for} \quad d_t \leq a \end{cases}$$

Solution:

```
series y = (d > a) ? x : z
```

Comment: There are several alternatives to the one presented above. One is a brute force solution using loops. Another one, more efficient but still suboptimal, would be

```
series y = (d>a)*x + (d<=a)*z
```

However, the ternary conditional assignment operator is not only the most numerically efficient way to accomplish what we want, it is also remarkably transparent to read when one gets used to it. Some readers may find it helpful to note that the conditional assignment operator works exactly the same way as the `=IF()` function in spreadsheets.

Generating a time index for panel datasets

Problem: gretl has a `$unit` accessor, but not the equivalent for time. What should I use?

Solution:

```
series x = time
```

Comment: The special construct `genr time` and its variants are aware of whether a dataset is a panel.

Generating the "hat" values after an OLS regression

Problem: I've just run an OLS regression, and now I need the so-called the leverage values (also known as the "hat" values). I know you can access residuals and fitted values through "dollar" accessors, but nothing like that seems to be available for "hat" values.

Solution: "Hat" values are can be thought of as the diagonal of the projection matrix P_X, or more explicitly as

$$h_i = \mathbf{x}_i'(X'X)^{-1}\mathbf{x}_i$$

where X is the matrix of regressors and $\mathbf{x}_i'$ is its i-th row.

The reader is invited to study the code below, which offers four different solutions to the problem:

```
open data4-1.gdt --quiet
list X = const sqft bedrms baths
ols price X

# method 1
leverage --save --quiet
series h1 = lever

# these are necessary for what comes next
matrix mX  = {X}
matrix iXX = invpd(mX'mX)

# method 2
series h2 = diag(qform(mX, iXX))
# method 3
```

```
series h3 = sumr(mX .* (mX*iXX))
# method 4
series h4 = NA
loop i=1..$nobs --quiet
    matrix x = mX[i,]'
    h4[i] = x'iXX*x
endloop

# verify
print h1 h2 h3 h4 --byobs
```

Comment: Solution 1 is the preferable one: it relies on the built-in `leverage` command, which computes the requested series quite efficiently, taking care of missing values, possible restrictions to the sample, etcetera.

However, three more are shown for didactical purposes, mainly to show the user how to manipulate matrices. Solution 2 first constructs the P_X matrix explicitly, via the `qform` function, and then takes its diagonal; this is definitely *not* recommended (despite its compactness), since you generate a much bigger matrix than you actually need and waste a lot of memory and CPU cycles in the process. It doesn't matter very much in the present case, since the sample size is very small, but with a big dataset this could be a very bad idea.

Solution 3 is more clever, and relies on the fact that, if you define $Z = X \cdot (X'X)^{-1}$, then h_i could also be written as

$$h_i = \mathbf{x}_i'\mathbf{z}_i = \sum_{i=1}^{k} x_{ik} z_{i_k}$$

which is in turn equivalent to the sum of the elements of the i-th row of $X \odot Z$, where $\odot$ is the element-by-element product. In this case, your clever usage of matrix algebra would produce a solution computationally much superior to solution 2.

Solution 4 is the most old-fashioned one, and employs an indexed loop. While this wastes practically no memory and employs no more CPU cycles in algebraic operations than strictly necessary, it imposes a much greater burden on the hansl interpreter, since handling a loop is conceptually more complex than a single operation. In practice, you'll find that for any realistically-sized problem, solution 4 is much slower that solution 3.

16.3 Neat tricks

Interaction dummies

Problem: You want to estimate the model $y_i = \mathbf{x}_i\beta_1 + \mathbf{z}_i\beta_2 + d_i\beta_3 + (d_i \cdot \mathbf{z}_i)\beta_4 + \varepsilon_t$, where d_i is a dummy variable while $\mathbf{x}_i$ and $\mathbf{z}_i$ are vectors of explanatory variables.

Solution: As of version 1.9.12, gretl provides the ^ operator to make this operation easy. See section 14.1 for details (especially example script 14.1). But back in my day, we used loops to do that! Here's how:

```
list X = x1 x2 x3
list Z = z1 z2
list dZ = null
loop foreach i Z
  series d$i = d * $i
  list dZ = dZ d$i
endloop

ols y X Z d dZ
```

Comment: It's amazing what string substitution can do for you, isn't it?

Realized volatility

Problem: Given data by the minute, you want to compute the "realized volatility" for the hour as $RV_t = \frac{1}{60}\sum_{\tau=1}^{60} y_{t:\tau}^2$. Imagine your sample starts at time 1:1.

Solution:

```
smpl --full
genr time
series minute = int(time/60) + 1
series second = time % 60
setobs minute second --panel
series rv = psd(y)^2
setobs 1 1
smpl second==1 --restrict
store foo rv
```

Comment: Here we trick gretl into thinking that our dataset is a panel dataset, where the minutes are the "units" and the seconds are the "time"; this way, we can take advantage of the special function `psd()`, panel standard deviation. Then we simply drop all observations but one per minute and save the resulting data (`store foo rv` translates as "store in the gretl datafile `foo.gdt` the series `rv`").

Looping over two paired lists

Problem: Suppose you have two lists with the same number of elements, and you want to apply some command to corresponding elements over a loop.

Solution:

```
list L1 = a b c
list L2 = x y z

k1 = 1
loop foreach i L1 --quiet
    k2 = 1
    loop foreach j L2 --quiet
        if k1 == k2
            ols $i 0 $j
        endif
        k2++
    endloop
    k1++
endloop
```

Comment: The simplest way to achieve the result is to loop over all possible combinations and filter out the unneeded ones via an `if` condition, as above. That said, in some cases variable names can help. For example, if

```
list Lx = x1 x2 x3
list Ly = y1 y2 y3
```

then we could just loop over the integers—quite intuitive and certainly more elegant:

```
loop i=1..3
  ols y$i const x$i
endloop
```

Dropping collinear variables

Problem: I'm writing my own estimator and I'd like to have a function that automatically drops collinear variables from a list.

Solution: The neatest way to achieve this is to use *Gram-Schmidt orthogonalization*, which is performed by the matrix function `qrdecomp()`. The function below defines a user-level function that (i) converts the list to a matrix (ii) spots collinear columns and (iii) removes them if any.

```
function list autodrop(list X, scalar criterion[1.0e-12])
    list drop = null
    matrix R
    qrdecomp({X}, &R)

    scalar ndrop = 0
    loop i=1..nelem(X) --quiet
        if criterion > abs(R[i,i])
            drop += X[i]
            ndrop++
        endif
    endloop

    if ndrop > 0
        printf "%d collinear variables dropped (%s)\n", \
          ndrop, varname(drop)
        return X - drop
    else
        return X
    endif
end function
```

Comment: The function also accepts a second optional argument which sets a criterion for deciding how strict the decision rule must be. Usage example:

```
nulldata 20

x1 = normal()
x2 = normal()
x3 = x1 + x2
x4 = x1 + 1

list A = const x1 x2 x3 x4
list C = autodrop(A)
```

Running the code above will create a list `C` containing just `const`, `x1` and `x2`.

Comparing two lists

Problem: How can I tell if two lists contain the same variables (not necessarily in the same order)?

Solution: Under many respects, lists are like sets, so it makes sense to use the so-called "symmetric difference" operator; this, in turn can be defined as

$$A \triangle B = (A \backslash B) \cup (B \backslash A).$$

In practice: we first check if there are series in A but not in B, then we perform the reverse check. If the union of the two results is an empty set, then the lists contain the same variables. The hansl syntax for this would be something like

```
scalar NotTheSame = nelem((A-B) || (B-A)) > 0
```

Part II

Econometric methods

Chapter 17

Robust covariance matrix estimation

17.1 Introduction

Consider (once again) the linear regression model

$$y = X\beta + u \tag{17.1}$$

where y and u are T-vectors, X is a $T \times k$ matrix of regressors, and β is a k-vector of parameters. As is well known, the estimator of β given by Ordinary Least Squares (OLS) is

$$\hat{\beta} = (X'X)^{-1}X'y \tag{17.2}$$

If the condition $E(u|X) = 0$ is satisfied, this is an unbiased estimator; under somewhat weaker conditions the estimator is biased but consistent. It is straightforward to show that when the OLS estimator is unbiased (that is, when $E(\hat{\beta} - \beta) = 0$), its variance is

$$\mathrm{Var}(\hat{\beta}) = E\left((\hat{\beta} - \beta)(\hat{\beta} - \beta)'\right) = (X'X)^{-1}X'\Omega X(X'X)^{-1} \tag{17.3}$$

where $\Omega = E(uu')$ is the covariance matrix of the error terms.

Under the assumption that the error terms are independently and identically distributed (iid) we can write $\Omega = \sigma^2 I$, where σ^2 is the (common) variance of the errors (and the covariances are zero). In that case (17.3) simplifies to the "classical" formula,

$$\mathrm{Var}(\hat{\beta}) = \sigma^2(X'X)^{-1} \tag{17.4}$$

If the iid assumption is not satisfied, two things follow. First, it is possible in principle to construct a more efficient estimator than OLS—for instance some sort of Feasible Generalized Least Squares (FGLS). Second, the simple "classical" formula for the variance of the least squares estimator is no longer correct, and hence the conventional OLS standard errors—which are just the square roots of the diagonal elements of the matrix defined by (17.4)—do not provide valid means of statistical inference.

In the recent history of econometrics there are broadly two approaches to the problem of non-iid errors. The "traditional" approach is to use an FGLS estimator. For example, if the departure from the iid condition takes the form of time-series dependence, and if one believes that this could be modeled as a case of first-order autocorrelation, one might employ an AR(1) estimation method such as Cochrane-Orcutt, Hildreth-Lu, or Prais-Winsten. If the problem is that the error variance is non-constant across observations, one might estimate the variance as a function of the independent variables and then perform weighted least squares, using as weights the reciprocals of the estimated variances.

While these methods are still in use, an alternative approach has found increasing favor: that is, use OLS but compute standard errors (or more generally, covariance matrices) that are robust with respect to deviations from the iid assumption. This is typically combined with an emphasis on using large datasets—large enough that the researcher can place some reliance on the (asymptotic) consistency property of OLS. This approach has been enabled by the availability of cheap computing power. The computation of robust standard errors and the handling of very large datasets were daunting tasks at one time, but now they are unproblematic. The other point favoring the newer methodology is that while FGLS offers an efficiency advantage in principle, it often involves making additional statistical assumptions which may or may not be justified, which may not be easy to test rigorously, and which may threaten the consistency of the estimator—for example, the "common factor restriction" that is implied by traditional FGLS "corrections" for autocorrelated errors.

James Stock and Mark Watson's *Introduction to Econometrics* illustrates this approach at the level of undergraduate instruction: many of the datasets they use comprise thousands or tens of thousands of observations; FGLS is downplayed; and robust standard errors are reported as a matter of course. In fact, the discussion of the classical standard errors (labeled "homoskedasticity-only") is confined to an Appendix.

Against this background it may be useful to set out and discuss all the various options offered by gretl in respect of robust covariance matrix estimation. The first point to notice is that gretl produces "classical" standard errors by default (in all cases apart from GMM estimation). In script mode you can get robust standard errors by appending the `--robust` flag to estimation commands. In the GUI program the model specification dialog usually contains a "Robust standard errors" check box, along with a "configure" button that is activated when the box is checked. The configure button takes you to a configuration dialog (which can also be reached from the main menu bar: Tools → Preferences → General → HCCME). There you can select from a set of possible robust estimation variants, and can also choose to make robust estimation the default.

The specifics of the available options depend on the nature of the data under consideration—cross-sectional, time series or panel—and also to some extent the choice of estimator. (Although we introduced robust standard errors in the context of OLS above, they may be used in conjunction with other estimators too.) The following three sections of this chapter deal with matters that are specific to the three sorts of data just mentioned. Note that additional details regarding covariance matrix estimation in the context of GMM are given in chapter 22.

We close this introduction with a brief statement of what "robust standard errors" can and cannot achieve. They can provide for asymptotically valid statistical inference in models that are basically correctly specified, but in which the errors are not iid. The "asymptotic" part means that they may be of little use in small samples. The "correct specification" part means that they are not a magic bullet: if the error term is correlated with the regressors, so that the parameter estimates themselves are biased and inconsistent, robust standard errors will not save the day.

17.2 Cross-sectional data and the HCCME

With cross-sectional data, the most likely departure from iid errors is heteroskedasticity (non-constant variance).[1] In some cases one may be able to arrive at a judgment regarding the likely form of the heteroskedasticity, and hence to apply a specific correction. The more common case, however, is where the heteroskedasticity is of unknown form. We seek an estimator of the covariance matrix of the parameter estimates that retains its validity, at least asymptotically, in face of unspecified heteroskedasticity. It is not obvious, a priori, that this should be possible, but White (1980) showed that

$$\widehat{\mathrm{Var}}_{\mathrm{h}}(\hat{\beta}) = (X'X)^{-1}X'\hat{\Omega}X(X'X)^{-1} \tag{17.5}$$

does the trick. (As usual in statistics, we need to say "under certain conditions", but the conditions are not very restrictive.) $\hat{\Omega}$ is in this context a diagonal matrix, whose non-zero elements may be estimated using squared OLS residuals. White referred to (17.5) as a heteroskedasticity-consistent covariance matrix estimator (HCCME).

Davidson and MacKinnon (2004, chapter 5) offer a useful discussion of several variants on White's HCCME theme. They refer to the original variant of (17.5)—in which the diagonal elements of $\hat{\Omega}$ are estimated directly by the squared OLS residuals, $\hat{u}_t^2$—as HC_0. (The associated standard errors are often called "White's standard errors".) The various refinements of White's proposal share a common point of departure, namely the idea that the squared OLS residuals are likely to be "too small" on average. This point is quite intuitive. The OLS parameter estimates, $\hat{\beta}$, satisfy by design the criterion that the sum of squared residuals,

$$\sum \hat{u}_t^2 = \sum \left(y_t - X_t\hat{\beta}\right)^2$$

[1] In some specialized contexts spatial autocorrelation may be an issue. Gretl does not have any built-in methods to handle this and we will not discuss it here.

is minimized for given X and y. Suppose that $\hat{\beta} \neq \beta$. This is almost certain to be the case: even if OLS is not biased, it would be a miracle if the $\hat{\beta}$ calculated from any finite sample were exactly equal to β. But in that case the sum of squares of the true, unobserved *errors*, $\sum u_t^2 = \sum(y_t - X_t\beta)^2$ is bound to be greater than $\sum \hat{u}_t^2$. The elaborated variants on HC_0 take this point on board as follows:

- HC_1: Applies a degrees-of-freedom correction, multiplying the HC_0 matrix by $T/(T-k)$.
- HC_2: Instead of using $\hat{u}_t^2$ for the diagonal elements of $\hat{\Omega}$, uses $\hat{u}_t^2/(1-h_t)$, where $h_t = X_t(X'X)^{-1}X_t'$, the tth diagonal element of the projection matrix, P, which has the property that $P \cdot y = \hat{y}$. The relevance of h_t is that if the variance of all the u_t is σ^2, the expectation of $\hat{u}_t^2$ is $\sigma^2(1-h_t)$, or in other words, the ratio $\hat{u}_t^2/(1-h_t)$ has expectation σ^2. As Davidson and MacKinnon show, $0 \leq h_t < 1$ for all t, so this adjustment cannot reduce the the diagonal elements of $\hat{\Omega}$ and in general revises them upward.
- HC_3: Uses $\hat{u}_t^2/(1-h_t)^2$. The additional factor of $(1-h_t)$ in the denominator, relative to HC_2, may be justified on the grounds that observations with large variances tend to exert a lot of influence on the OLS estimates, so that the corresponding residuals tend to be under-estimated. See Davidson and MacKinnon for a fuller explanation.

The relative merits of these variants have been explored by means of both simulations and theoretical analysis. Unfortunately there is not a clear consensus on which is "best". Davidson and MacKinnon argue that the original HC_0 is likely to perform worse than the others; nonetheless, "White's standard errors" are reported more often than the more sophisticated variants and therefore, for reasons of comparability, HC_0 is the default HCCME in gretl.

If you wish to use HC_1, HC_2 or HC_3 you can arrange for this in either of two ways. In script mode, you can do, for example,

```
set hc_version 2
```

In the GUI program you can go to the HCCME configuration dialog, as noted above, and choose any of these variants to be the default.

17.3 Time series data and HAC covariance matrices

Heteroskedasticity may be an issue with time series data too, but it is unlikely to be the only, or even the primary, concern.

One form of heteroskedasticity is common in macroeconomic time series, but is fairly easily dealt with. That is, in the case of strongly trending series such as Gross Domestic Product, aggregate consumption, aggregate investment, and so on, higher levels of the variable in question are likely to be associated with higher variability in absolute terms. The obvious "fix", employed in many macroeconometric studies, is to use the logs of such series rather than the raw levels. Provided the *proportional* variability of such series remains roughly constant over time, the log transformation is effective.

Other forms of heteroskedasticity may resist the log transformation, but may demand a special treatment distinct from the calculation of robust standard errors. We have in mind here *autoregressive conditional* heteroskedasticity, for example in the behavior of asset prices, where large disturbances to the market may usher in periods of increased volatility. Such phenomena call for specific estimation strategies, such as GARCH (see chapter 25).

Despite the points made above, some residual degree of heteroskedasticity may be present in time series data: the key point is that in most cases it is likely to be combined with serial correlation (autocorrelation), hence demanding a special treatment. In White's approach, $\hat{\Omega}$, the estimated covariance matrix of the u_t, remains conveniently diagonal: the variances, $E(u_t^2)$, may differ by t but the covariances, $E(u_t u_s)$, are all zero. Autocorrelation in time series data means that at least some of the the off-diagonal elements of $\hat{\Omega}$ should be non-zero. This introduces a substantial complication and requires another piece of terminology; estimates of the covariance matrix that are asymptotically valid in face of both heteroskedasticity and autocorrelation of the error process are termed HAC (heteroskedasticity and autocorrelation consistent).

The issue of HAC estimation is treated in more technical terms in chapter 22. Here we try to convey some of the intuition at a more basic level. We begin with a general comment: residual autocorrelation is not so much a property of the data, as a symptom of an inadequate model. Data may be persistent though time, and if we fit a model that does not take this aspect into account properly, we end up with a model with autocorrelated disturbances. Conversely, it is often possible to mitigate or even eliminate the problem of autocorrelation by including relevant lagged variables in a time series model, or in other words, by specifying the dynamics of the model more fully. HAC estimation should *not* be seen as the first resort in dealing with an autocorrelated error process.

That said, the "obvious" extension of White's HCCME to the case of autocorrelated errors would seem to be this: estimate the off-diagonal elements of $\hat{\Omega}$ (that is, the autocovariances, $E(u_t u_s)$) using, once again, the appropriate OLS residuals: $\hat{\omega}_{ts} = \hat{u}_t \hat{u}_s$. This is basically right, but demands an important amendment. We seek a *consistent* estimator, one that converges towards the true Ω as the sample size tends towards infinity. This can't work if we allow unbounded serial dependence. Bigger samples will enable us to estimate more of the true ω_{ts} elements (that is, for t and s more widely separated in time) but will *not* contribute ever-increasing information regarding the maximally separated ω_{ts} pairs, since the maximal separation itself grows with the sample size. To ensure consistency, we have to confine our attention to processes exhibiting temporally limited dependence, or in other words cut off the computation of the $\hat{\omega}_{ts}$ values at some maximum value of $p = t - s$ (where p is treated as an increasing function of the sample size, T, although it cannot increase in proportion to T).

The simplest variant of this idea is to truncate the computation at some finite lag order p, where p grows as, say, $T^{1/4}$. The trouble with this is that the resulting $\hat{\Omega}$ may not be a positive definite matrix. In practical terms, we may end up with negative estimated variances. One solution to this problem is offered by The Newey–West estimator (Newey and West, 1987), which assigns declining weights to the sample autocovariances as the temporal separation increases.

To understand this point it is helpful to look more closely at the covariance matrix given in (17.5), namely,

$$(X'X)^{-1}(X'\hat{\Omega}X)(X'X)^{-1}$$

This is known as a "sandwich" estimator. The bread, which appears on both sides, is $(X'X)^{-1}$. This is a $k \times k$ matrix, and is also the key ingredient in the computation of the classical covariance matrix. The filling in the sandwich is

$$\underset{(k\times k)}{\hat{\Sigma}} = \underset{(k\times T)}{X'} \; \underset{(T\times T)}{\hat{\Omega}} \; \underset{(T\times k)}{X}$$

Since $\Omega = E(uu')$, the matrix being estimated here can also be written as

$$\Sigma = E(X'u\,u'X)$$

which expresses Σ as the long-run covariance of the random k-vector $X'u$.

From a computational point of view, it is not necessary or desirable to store the (potentially very large) $T \times T$ matrix $\hat{\Omega}$ as such. Rather, one computes the sandwich filling by summation as

$$\hat{\Sigma} = \hat{\Gamma}(0) + \sum_{j=1}^{p} w_j \left(\hat{\Gamma}(j) + \hat{\Gamma}'(j)\right)$$

where the $k \times k$ sample autocovariance matrix $\hat{\Gamma}(j)$, for $j \geq 0$, is given by

$$\hat{\Gamma}(j) = \frac{1}{T} \sum_{t=j+1}^{T} \hat{u}_t \hat{u}_{t-j} X_t' X_{t-j}$$

and w_j is the weight given to the autocovariance at lag $j > 0$.

This leaves two questions. How exactly do we determine the maximum lag length or "bandwidth", p, of the HAC estimator? And how exactly are the weights w_j to be determined? We will return to the (difficult) question of the bandwidth shortly. As regards the weights, gretl offers three variants. The default is the Bartlett kernel, as used by Newey and West. This sets

$$w_j = \begin{cases} 1 - \frac{j}{p+1} & j \le p \\ 0 & j > p \end{cases}$$

so the weights decline linearly as j increases. The other two options are the Parzen kernel and the Quadratic Spectral (QS) kernel. For the Parzen kernel,

$$w_j = \begin{cases} 1 - 6a_j^2 + 6a_j^3 & 0 \le a_j \le 0.5 \\ 2(1 - a_j)^3 & 0.5 < a_j \le 1 \\ 0 & a_j > 1 \end{cases}$$

where $a_j = j/(p + 1)$, and for the QS kernel,

$$w_j = \frac{25}{12\pi^2 d_j^2} \left(\frac{\sin m_j}{m_j} - \cos m_j \right)$$

where $d_j = j/p$ and $m_j = 6\pi d_i/5$.

Figure 17.1 shows the weights generated by these kernels, for $p = 4$ and $j = 1$ to 9.

Figure 17.1: Three HAC kernels

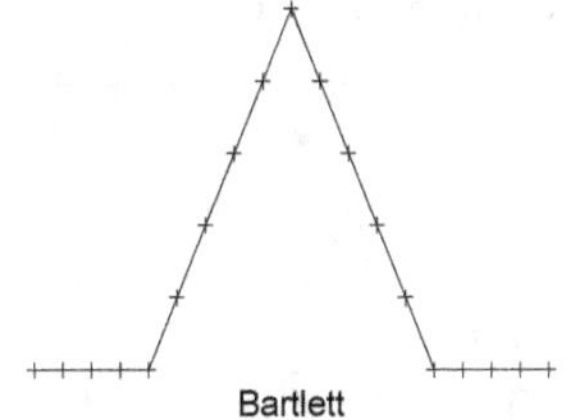

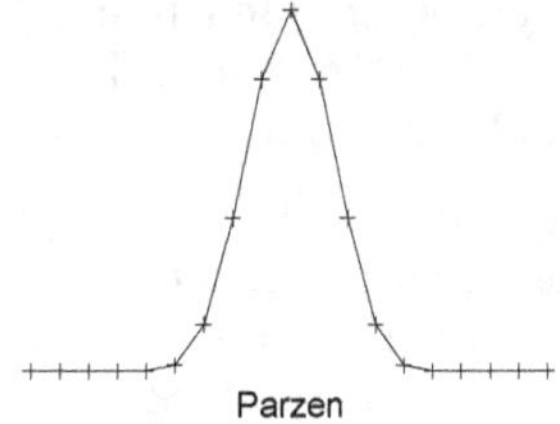

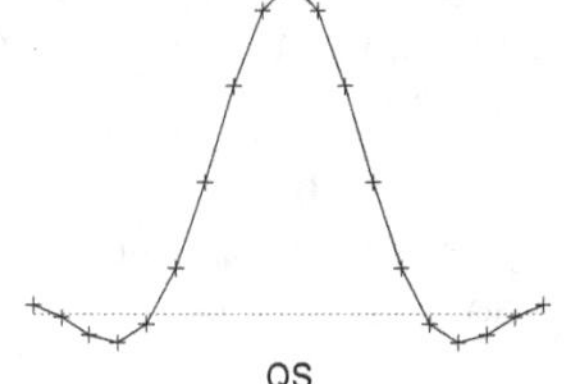

In gretl you select the kernel using the `set` command with the `hac_kernel` parameter:

```
set hac_kernel parzen
set hac_kernel qs
set hac_kernel bartlett
```

Selecting the HAC bandwidth

The asymptotic theory developed by Newey, West and others tells us in general terms how the HAC bandwidth, p, should grow with the sample size, T—that is, p should grow in proportion to some fractional power of T. Unfortunately this is of little help to the applied econometrician, working with a given dataset of fixed size. Various rules of thumb have been suggested, and gretl implements two such. The default is $p = 0.75T^{1/3}$, as recommended by Stock and Watson (2003). An alternative is $p = 4(T/100)^{2/9}$, as in Wooldridge (2002b). In each case one takes the integer part of the result. These variants are labeled nw1 and nw2 respectively, in the context of the `set` command with the `hac_lag` parameter. That is, you can switch to the version given by Wooldridge with

```
set hac_lag nw2
```

As shown in Table 17.1 the choice between nw1 and nw2 does not make a great deal of difference.

You also have the option of specifying a fixed numerical value for p, as in

```
set hac_lag 6
```

In addition you can set a distinct bandwidth for use with the Quadratic Spectral kernel (since this need not be an integer). For example,

```
set qs_bandwidth 3.5
```

T	p (nw1)	p (nw2)
50	2	3
100	3	4
150	3	4
200	4	4
300	5	5
400	5	5

Table 17.1: HAC bandwidth: two rules of thumb

Prewhitening and data-based bandwidth selection

An alternative approach is to deal with residual autocorrelation by attacking the problem from two sides. The intuition behind the technique known as *VAR prewhitening* (Andrews and Monahan, 1992) can be illustrated by a simple example. Let x_t be a sequence of first-order autocorrelated random variables

$$x_t = \rho x_{t-1} + u_t$$

The long-run variance of x_t can be shown to be

$$V_{LR}(x_t) = \frac{V_{LR}(u_t)}{(1-\rho)^2}$$

In most cases, u_t is likely to be less autocorrelated than x_t, so a smaller bandwidth should suffice. Estimation of $V_{LR}(x_t)$ can therefore proceed in three steps: (1) estimate ρ; (2) obtain a HAC estimate of $\hat{u}_t = x_t - \hat{\rho} x_{t-1}$; and (3) divide the result by $(1-\rho)^2$.

The application of the above concept to our problem implies estimating a finite-order Vector Autoregression (VAR) on the vector variables $\xi_t = X_t \hat{u}_t$. In general, the VAR can be of any order, but in most cases 1 is sufficient; the aim is not to build a watertight model for ξ_t, but just to "mop up" a substantial part of the autocorrelation. Hence, the following VAR is estimated

$$\xi_t = A\xi_{t-1} + \varepsilon_t$$

Then an estimate of the matrix $X'\Omega X$ can be recovered via

$$(I - \hat{A})^{-1}\hat{\Sigma}_{\varepsilon}(I - \hat{A}')^{-1}$$

where $\hat{\Sigma}_{\varepsilon}$ is any HAC estimator, applied to the VAR residuals.

You can ask for prewhitening in gretl using

```
set hac_prewhiten on
```

There is at present no mechanism for specifying an order other than 1 for the initial VAR.

A further refinement is available in this context, namely data-based bandwidth selection. It makes intuitive sense that the HAC bandwidth should not simply be based on the size of the sample, but should somehow take into account the time-series properties of the data (and also the kernel chosen). A nonparametric method for doing this was proposed by Newey and West (1994); a good concise account of the method is given in Hall (2005). This option can be invoked in gretl via

```
set hac_lag nw3
```

This option is the default when prewhitening is selected, but you can override it by giving a specific numerical value for `hac_lag`.

Even the Newey-West data-based method does not fully pin down the bandwidth for any particular sample. The first step involves calculating a series of residual covariances. The length of this series is given as a function of the sample size, but only up to a scalar multiple—for example, it is given as $O(T^{2/9})$ for the Bartlett kernel. Gretl uses an implied multiple of 1.

VARs: a special case

A well-specified vector autoregression (VAR) will generally include enough lags of the dependent variables to obviate the problem of residual autocorrelation, in which case HAC estimation is redundant—although there may still be a need to correct for heteroskedasticity. For that reason plain HCCME, and not HAC, is the default when the `--robust` flag is given in the context of the `var` command. However, if for some reason you need HAC you can force the issue by giving the option `--robust-hac`.

17.4 Special issues with panel data

Since panel data have both a time-series and a cross-sectional dimension one might expect that, in general, robust estimation of the covariance matrix would require handling both heteroskedasticity and autocorrelation (the HAC approach). In addition, some special features of panel data require attention.

- The variance of the error term may differ across the cross-sectional units.
- The covariance of the errors across the units may be non-zero in each time period.
- If the "between" variation is not removed, the errors may exhibit autocorrelation, not in the usual time-series sense but in the sense that the mean error for unit i may differ from that of unit j. (This is particularly relevant when estimation is by pooled OLS.)

Gretl currently offers two robust covariance matrix estimators specifically for panel data. These are available for models estimated via fixed effects, pooled OLS, and pooled two-stage least squares. The default robust estimator is that suggested by Arellano (2003), which is HAC provided the panel is of the "large n, small T" variety (that is, many units are observed in relatively few periods). The Arellano estimator is

$$\hat{\Sigma}_{\mathrm{A}} = (X'X)^{-1}\left(\sum_{i=1}^{n} X_i'\hat{u}_i\hat{u}_i'X_i\right)(X'X)^{-1}$$

where X is the matrix of regressors (with the group means subtracted, in the case of fixed effects) $\hat{u}_i$ denotes the vector of residuals for unit i, and n is the number of cross-sectional units.[2] Cameron and Trivedi (2005) make a strong case for using this estimator; they note that the ordinary White HCCME can produce misleadingly small standard errors in the panel context because it fails to take autocorrelation into account. In addition Stock and Watson (2008) show that the White HCCME is inconsistent in the fixed-effects panel context for fixed $T > 2$.

In cases where autocorrelation is not an issue the estimator proposed by Beck and Katz (1995) and discussed by Greene (2003, chapter 13) may be appropriate. This estimator, which takes into account contemporaneous correlation across the units and heteroskedasticity by unit, is

$$\hat{\Sigma}_{\mathrm{BK}} = (X'X)^{-1}\left(\sum_{i=1}^{n}\sum_{j=1}^{n} \hat{\sigma}_{ij}X_i'X_j\right)(X'X)^{-1}$$

The covariances $\hat{\sigma}_{ij}$ are estimated via

$$\hat{\sigma}_{ij} = \frac{\hat{u}_i'\hat{u}_j}{T}$$

where T is the length of the time series for each unit. Beck and Katz call the associated standard errors "Panel-Corrected Standard Errors" (PCSE). This estimator can be invoked in gretl via the command

```
set pcse on
```

The Arellano default can be re-established via

[2]This variance estimator is also known as the "clustered (over entities)" estimator.

```
set pcse off
```

(Note that regardless of the pcse setting, the robust estimator is not used unless the --robust flag is given, or the "Robust" box is checked in the GUI program.)

17.5 The cluster-robust estimator

One further variance estimator is available in gretl, namely the "cluster-robust" estimator. This may be appropriate (for cross-sectional data, mostly) when the observations naturally fall into groups or clusters, and one suspects that the error term may exhibit dependency within the clusters and/or have a variance that differs across clusters. Such clusters may be binary (e.g. employed versus unemployed workers), categorical with several values (e.g. products grouped by manufacturer) or ordinal (e.g. individuals with low, middle or high education levels).

For linear regression models estimated via least squares the cluster estimator is defined as

$$\hat{\Sigma}_C = (X'X)^{-1}\left(\sum_{j=1}^{m} X_j'\hat{u}_j\hat{u}_j'X_j\right)(X'X)^{-1}$$

where m denotes the number of clusters, and X_j and $\hat{u}_j$ denote, respectively, the matrix of regressors and the vector of residuals that fall within cluster j. As noted above, the Arellano variance estimator for panel data models is a special case of this, where the clustering is by panel unit.

For models estimated by the method of Maximum Likelihood (in which case the standard variance estimator is the inverse of the negative Hessian, H), the cluster estimator is

$$\hat{\Sigma}_C = H^{-1}\left(\sum_{j=1}^{m} G_j'G_j\right)H^{-1}$$

where G_j is the sum of the "score" (that is, the derivative of the loglikelihood with respect to the parameter estimates) across the observations falling within cluster j.

It is common to apply a degrees of freedom adjustment to these estimators (otherwise the variance may appear misleadingly small in comparison with other estimators, if the number of clusters is small). In the least squares case the factor is $(m/(m-1)) \times (n-1)/(n-k)$, where n is the total number of observations and k is the number of parameters estimated; in the case of ML estimation the factor is just $m/(m-1)$.

Availability and syntax

The cluster-robust estimator is currently available for models estimated via OLS and TSLS, and also for most ML estimators other than those specialized for time-series data: binary logit and probit, ordered logit and probit, multinomial logit, Tobit, interval regression, biprobit, count models and duration models. In all cases the syntax is the same: you give the option flag --cluster= followed by the name of the series to be used to define the clusters, as in

```
ols y 0 x1 x2 --cluster=cvar
```

The specified clustering variable must (a) be defined (not missing) at all observations used in estimating the model and (b) take on at least two distinct values over the estimation range. The clusters are defined as sets of observations having a common value for the clustering variable. It is generally expected that the number of clusters is substantially less than the total number of observations.

Chapter 18

Panel data

18.1 Estimation of panel models

Pooled Ordinary Least Squares

The simplest estimator for panel data is pooled OLS. In most cases this is unlikely to be adequate, but it provides a baseline for comparison with more complex estimators.

If you estimate a model on panel data using OLS an additional test item becomes available. In the GUI model window this is the item "panel diagnostics" under the `Tests` menu; the script counterpart is the `hausman` command.

To take advantage of this test, you should specify a model without any dummy variables representing cross-sectional units. The test compares pooled OLS against the principal alternatives, the fixed effects and random effects models. These alternatives are explained in the following section.

The fixed and random effects models

In the graphical interface these options are found under the menu item "Model/Panel/Fixed and random effects". In the command-line interface one uses the `panel` command, with or without the `--random-effects` option.

This section explains the nature of these models and comments on their estimation via gretl.

The pooled OLS specification may be written as

$$y_{it} = X_{it}\beta + u_{it} \tag{18.1}$$

where y_{it} is the observation on the dependent variable for cross-sectional unit i in period t, X_{it} is a $1 \times k$ vector of independent variables observed for unit i in period t, β is a $k \times 1$ vector of parameters, and u_{it} is an error or disturbance term specific to unit i in period t.

The fixed and random effects models have in common that they decompose the unitary pooled error term, u_{it}. For the *fixed effects* model we write $u_{it} = \alpha_i + \varepsilon_{it}$, yielding

$$y_{it} = X_{it}\beta + \alpha_i + \varepsilon_{it} \tag{18.2}$$

That is, we decompose u_{it} into a unit-specific and time-invariant component, α_i, and an observation-specific error, ε_{it}.[1] The α_is are then treated as fixed parameters (in effect, unit-specific y-intercepts), which are to be estimated. This can be done by including a dummy variable for each cross-sectional unit (and suppressing the global constant). This is sometimes called the Least Squares Dummy Variables (LSDV) method. Alternatively, one can subtract the group mean from each of variables and estimate a model without a constant. In the latter case the dependent variable may be written as

$$\tilde{y}_{it} = y_{it} - \bar{y}_i$$

The "group mean", $\bar{y}_i$, is defined as

$$\bar{y}_i = \frac{1}{T_i}\sum_{t=1}^{T_i} y_{it}$$

[1]It is possible to break a third component out of u_{it}, namely w_t, a shock that is time-specific but common to all the units in a given period. In the interest of simplicity we do not pursue that option here.

where T_i is the number of observations for unit i. An exactly analogous formulation applies to the independent variables. Given parameter estimates, $\hat{\beta}$, obtained using such de-meaned data we can recover estimates of the α_is using

$$\hat{\alpha}_i = \frac{1}{T_i} \sum_{t=1}^{T_i} \left(y_{it} - X_{it}\hat{\beta} \right)$$

These two methods (LSDV, and using de-meaned data) are numerically equivalent. gretl takes the approach of de-meaning the data. If you have a small number of cross-sectional units, a large number of time-series observations per unit, and a large number of regressors, it is more economical in terms of computer memory to use LSDV. If need be you can easily implement this manually. For example,

```
genr unitdum
ols y x du_*
```

(See Chapter 9 for details on `unitdum`).

The $\hat{\alpha}_i$ estimates are not printed as part of the standard model output in gretl (there may be a large number of these, and typically they are not of much inherent interest). However you can retrieve them after estimation of the fixed effects model if you wish. In the graphical interface, go to the "Save" menu in the model window and select "per-unit constants". In command-line mode, you can do `series` *newname* = `$ahat`, where *newname* is the name you want to give the series.

For the *random effects* model we write $u_{it} = v_i + \varepsilon_{it}$, so the model becomes

$$y_{it} = X_{it}\beta + v_i + \varepsilon_{it} \tag{18.3}$$

In contrast to the fixed effects model, the v_is are not treated as fixed parameters, but as random drawings from a given probability distribution.

The celebrated Gauss–Markov theorem, according to which OLS is the best linear unbiased estimator (BLUE), depends on the assumption that the error term is independently and identically distributed (IID). In the panel context, the IID assumption means that $E(u_{it}^2)$, in relation to equation 18.1, equals a constant, σ_u^2, for all i and t, while the covariance $E(u_{is}u_{it})$ equals zero for all $s \neq t$ and the covariance $E(u_{jt}u_{it})$ equals zero for all $j \neq i$.

If these assumptions are not met—and they are unlikely to be met in the context of panel data—OLS is not the most efficient estimator. Greater efficiency may be gained using generalized least squares (GLS), taking into account the covariance structure of the error term.

Consider observations on a given unit i at two different times s and t. From the hypotheses above it can be worked out that $\text{Var}(u_{is}) = \text{Var}(u_{it}) = \sigma_v^2 + \sigma_\varepsilon^2$, while the covariance between u_{is} and u_{it} is given by $E(u_{is}u_{it}) = \sigma_v^2$.

In matrix notation, we may group all the T_i observations for unit i into the vector $\mathbf{y}_i$ and write it as

$$\mathbf{y}_i = \mathbf{X}_i\beta + \mathbf{u}_i \tag{18.4}$$

The vector $\mathbf{u}_i$, which includes all the disturbances for individual i, has a variance-covariance matrix given by

$$\text{Var}(\mathbf{u}_i) = \Sigma_i = \sigma_\varepsilon^2 I + \sigma_v^2 J \tag{18.5}$$

where J is a square matrix with all elements equal to 1. It can be shown that the matrix

$$K_i = I - \frac{\theta_i}{T_i} J,$$

where $\theta_i = 1 - \sqrt{\frac{\sigma_\varepsilon^2}{\sigma_\varepsilon^2 + T_i \sigma_v^2}}$, has the property

$$K_i \Sigma K_i' = \sigma_\varepsilon^2 I$$

It follows that the transformed system

$$K_i \mathbf{y}_i = K_i \mathbf{X}_i \beta + K_i \mathbf{u}_i \qquad (18.6)$$

satisfies the Gauss-Markov conditions, and OLS estimation of (18.6) provides efficient inference. But since

$$K_i \mathbf{y}_i = \mathbf{y}_i - \theta_i \bar{\mathbf{y}}_i$$

GLS estimation is equivalent to OLS using "quasi-demeaned" variables; that is, variables from which we subtract a fraction θ of their average.[2] Notice that for $\sigma_\varepsilon^2 \to 0$, $\theta \to 1$, while for $\sigma_v^2 \to 0$, $\theta \to 0$. This means that if all the variance is attributable to the individual effects, then the fixed effects estimator is optimal; if, on the other hand, individual effects are negligible, then pooled OLS turns out, unsurprisingly, to be the optimal estimator.

To implement the GLS approach we need to calculate θ, which in turn requires estimates of the variances σ_ε^2 and σ_v^2. (These are often referred to as the "within" and "between" variances respectively, since the former refers to variation within each cross-sectional unit and the latter to variation between the units). Several means of estimating these magnitudes have been suggested in the literature (see Baltagi, 1995); by default gretl uses the method of Swamy and Arora (1972): σ_ε^2 is estimated by the residual variance from the fixed effects model, and the sum $\sigma_\varepsilon^2 + T_i \sigma_v^2$ is estimated as T_i times the residual variance from the "between" estimator,

$$\bar{y}_i = \bar{X}_i \beta + e_i$$

The latter regression is implemented by constructing a data set consisting of the group means of all the relevant variables. Alternatively, if the `--nerlove` option is given, gretl uses the method suggested by Nerlove (1971). In this case σ_v^2 is estimated as the sample variance of the fixed effects,

$$\hat{\sigma}_v^2 = \frac{1}{n-1} \sum_{i=1}^{n} (\alpha_i - \bar{\alpha})^2$$

where n is the number of individuals and $\bar{\alpha}$ is the mean of the fixed effects.

Choice of estimator

Which panel method should one use, fixed effects or random effects?

One way of answering this question is in relation to the nature of the data set. If the panel comprises observations on a fixed and relatively small set of units of interest (say, the member states of the European Union), there is a presumption in favor of fixed effects. If it comprises observations on a large number of randomly selected individuals (as in many epidemiological and other longitudinal studies), there is a presumption in favor of random effects.

Besides this general heuristic, however, various statistical issues must be taken into account.

1. Some panel data sets contain variables whose values are specific to the cross-sectional unit but which do not vary over time. If you want to include such variables in the model, the fixed effects option is simply not available. When the fixed effects approach is implemented using dummy variables, the problem is that the time-invariant variables are perfectly collinear with the per-unit dummies. When using the approach of subtracting the group means, the issue is that after de-meaning these variables are nothing but zeros.

2. A somewhat analogous prohibition applies to the random effects estimator. This estimator is in effect a matrix-weighted average of pooled OLS and the "between" estimator. Suppose we have observations on n units or individuals and there are k independent variables of interest. If $k > n$, the "between" estimator is undefined—since we have only n effective observations—and hence so is the random effects estimator.

[2] In a balanced panel, the value of θ is common to all individuals, otherwise it differs depending on the value of T_i.

If one does not fall foul of one or other of the prohibitions mentioned above, the choice between fixed effects and random effects may be expressed in terms of the two econometric *desiderata*, efficiency and consistency.

From a purely statistical viewpoint, we could say that there is a tradeoff between robustness and efficiency. In the fixed effects approach, we do not make any hypotheses on the "group effects" (that is, the time-invariant differences in mean between the groups) beyond the fact that they exist—and that can be tested; see below. As a consequence, once these effects are swept out by taking deviations from the group means, the remaining parameters can be estimated.

On the other hand, the random effects approach attempts to model the group effects as drawings from a probability distribution instead of removing them. This requires that individual effects are representable as a legitimate part of the disturbance term, that is, zero-mean random variables, uncorrelated with the regressors.

As a consequence, the fixed-effects estimator "always works", but at the cost of not being able to estimate the effect of time-invariant regressors. The richer hypothesis set of the random-effects estimator ensures that parameters for time-invariant regressors can be estimated, and that estimation of the parameters for time-varying regressors is carried out more efficiently. These advantages, though, are tied to the validity of the additional hypotheses. If, for example, there is reason to think that individual effects may be correlated with some of the explanatory variables, then the random-effects estimator would be inconsistent, while fixed-effects estimates would still be valid. It is precisely on this principle that the Hausman test is built (see below): if the fixed- and random-effects estimates agree, to within the usual statistical margin of error, there is no reason to think the additional hypotheses invalid, and as a consequence, no reason *not* to use the more efficient RE estimator.

Testing panel models

If you estimate a fixed effects or random effects model in the graphical interface, you may notice that the number of items available under the "Tests" menu in the model window is relatively limited. Panel models carry certain complications that make it difficult to implement all of the tests one expects to see for models estimated on straight time-series or cross-sectional data.

Nonetheless, various panel-specific tests are printed along with the parameter estimates as a matter of course, as follows.

When you estimate a model using *fixed effects*, you automatically get an F-test for the null hypothesis that the cross-sectional units all have a common intercept. That is to say that all the α_is are equal, in which case the pooled model (18.1), with a column of 1s included in the X matrix, is adequate.

When you estimate using *random effects*, the Breusch–Pagan and Hausman tests are presented automatically.

The Breusch–Pagan test is the counterpart to the F-test mentioned above. The null hypothesis is that the variance of v_i in equation (18.3) equals zero; if this hypothesis is not rejected, then again we conclude that the simple pooled model is adequate.

The Hausman test probes the consistency of the GLS estimates. The null hypothesis is that these estimates are consistent—that is, that the requirement of orthogonality of the v_i and the X_i is satisfied. The test is based on a measure, H, of the "distance" between the fixed-effects and random-effects estimates, constructed such that under the null it follows the χ^2 distribution with degrees of freedom equal to the number of time-varying regressors in the matrix X. If the value of H is "large" this suggests that the random effects estimator is not consistent and the fixed-effects model is preferable.

There are two ways of calculating H, the matrix-difference method and the regression method. The procedure for the matrix-difference method is this:

- Collect the fixed-effects estimates in a vector $\tilde{\beta}$ and the corresponding random-effects estimates in $\hat{\beta}$, then form the difference vector $(\tilde{\beta} - \hat{\beta})$.
- Form the covariance matrix of the difference vector as $\text{Var}(\tilde{\beta} - \hat{\beta}) = \text{Var}(\tilde{\beta}) - \text{Var}(\hat{\beta}) = \Psi$,

where $\text{Var}(\tilde{\beta})$ and $\text{Var}(\hat{\beta})$ are estimated by the sample variance matrices of the fixed- and random-effects models respectively.[3]

- Compute $H = \left(\tilde{\beta} - \hat{\beta}\right)' \Psi^{-1} \left(\tilde{\beta} - \hat{\beta}\right)$.

Given the relative efficiencies of $\tilde{\beta}$ and $\hat{\beta}$, the matrix Ψ "should be" positive definite, in which case H is positive, but in finite samples this is not guaranteed and of course a negative χ^2 value is not admissible. The regression method avoids this potential problem. The procedure is:

- Treat the random-effects model as the restricted model, and record its sum of squared residuals as SSR_r.
- Estimate via OLS an unrestricted model in which the dependent variable is quasi-demeaned y and the regressors include both quasi-demeaned X (as in the RE model) and the demeaned variants of all the time-varying variables (i.e. the fixed-effects regressors); record the sum of squared residuals from this model as SSR_u.
- Compute $H = n\left(\text{SSR}_r - \text{SSR}_u\right)/\text{SSR}_u$, where n is the total number of observations used. On this variant H cannot be negative, since adding additional regressors to the RE model cannot raise the SSR.

By default gretl computes the Hausman test via the regression method, but it uses the matrix-difference method if you pass the option `--matrix-diff` to the `panel` command.

Robust standard errors

For most estimators, gretl offers the option of computing an estimate of the covariance matrix that is robust with respect to heteroskedasticity and/or autocorrelation (and hence also robust standard errors). In the case of panel data, robust covariance matrix estimators are available for the pooled and fixed effects model but not currently for random effects. Please see section 17.4 for details.

The constant in the fixed effects model

Users are sometimes puzzled by the constant or intercept reported by gretl on estimation of the fixed effects model: how can a constant remain when the group means have been subtracted from the data? The method of calculation of this term is a matter of convention, but the gretl authors decided to follow the convention employed by Stata; this involves adding the global mean back into the variables from which the group means have been removed.[4]

The method that gretl uses internally is exemplified in Example 18.1. The coefficients in the final OLS estimation, including the intercept, agree with those in the initial fixed effects model, though the standard errors differ due to a degrees of freedom correction in the fixed-effects covariance matrix. (Note that the `pmean` function returns the group mean of a series.)

R-squared in the fixed effects model

There is no uniquely "correct" way of calculating R^2 in the context of the fixed-effects model. It may be argued that a measure of the squared correlation between the dependent variable and the prediction yielded by the model is a desirable descriptive statistic to have, but which model and which (variant of the) dependent variable are we talking about?

Fixed-effects models can be thought of in two equally defensible ways. From one perspective they provide a nice, clean way of sweeping out individual effects by using the fact that in the linear model a sufficient statistic is easy to compute. Alternatively, they provide a clever way to estimate the "important" parameters of a model in which you want to include (for whatever reason) a full set of individual dummies. If you take the second of these perspectives, your

[3] Hausman (1978) showed that the covariance of the difference takes this simple form when $\hat{\beta}$ is an efficient estimator and $\tilde{\beta}$ is inefficient.

[4] See Gould (2013) for an extended explanation.

Example 18.1: Calculating the intercept in the fixed effects model

```
open abdata.gdt
panel n const w k ys --fixed-effects

depvar = n - pmean(n) + mean(n)
list indepvars = const
loop foreach i w k ys --quiet
  x_$i = $i - pmean($i) + mean($i)
  indepvars += x_$i
endloop
ols depvar indepvars
```

dependent variable is unmodified y and your model includes the unit dummies; the appropriate R^2 measure is then the squared correlation between y and the $\hat{y}$ computed using both the measured individual effects and the effects of the explicitly named regressors. This is reported by gretl as the "LSDV R-squared". If you take the first point of view, on the other hand, your dependent variable is really $y_{it} - \bar{y}_i$ and your model just includes the β terms, the coefficients of deviations of the x variables from their per-unit means. In this case, the relevant measure of R^2 is the so-called "within" R^2; this variant is printed by gretl for fixed-effects model in place of the adjusted R^2 (it being unclear in this case what exactly the "adjustment" should amount to anyway).

Residuals in the fixed and random effect models

After estimation of most kinds of models in gretl, you can retrieve a series containing the residuals using the `$uhat` accessor. This is true of the fixed and random effects models, but the exact meaning of gretl's `$uhat` in these cases requires a little explanation.

Consider first the fixed effects model:

$$y_{it} = X_{it}\beta + \alpha_i + \varepsilon_{it}$$

In this model gretl takes the "fitted value" (`$yhat`) to be $\hat{\alpha}_i + X_{it}\hat{\beta}$, and the residual (`$uhat`) to be y_{it} minus this fitted value. This makes sense because the fixed effects (the α_i terms) are taken as parameters to be estimated. However, it can be argued that the fixed effects are not really "explanatory" and if one defines the residual as the observed y_{it} value minus its "explained" component one might prefer to see just $y_{it} - X_{it}\hat{\beta}$. You can get this after fixed-effects estimation as follows:

```
series ue_fe = $uhat + $ahat - $coeff[1]
```

where `$ahat` gives the unit-specific intercept (as it would be calculated if one included all N unit dummies and omitted a common y-intercept), and `$coeff[1]` gives the "global" y-intercept.[5]

Now consider the random-effects model:

$$y_{it} = X_{it}\beta + v_i + \varepsilon_{it}$$

In this case gretl considers the error term to be $v_i + \varepsilon_{it}$ (since v_i is conceived as a random drawing) and the `$uhat` series is an estimate of this, namely

$$y_{it} - X_{it}\hat{\beta}$$

What if you want an estimate of just v_i (or just ε_{it}) in this case? This poses a signal-extraction problem: given the composite residual, how to recover an estimate of its components? The

[5] For anyone used to Stata, gretl's fixed-effects `$uhat` corresponds to what you get from Stata's "`predict, e`" after `xtreg`, while the second variant corresponds to Stata's "`predict, ue`".

solution is to ascribe to the individual effect, $\hat{v}_i$, a suitable fraction of the mean residual per individual, $\bar{\hat{u}}_i = \sum_{t=1}^{T_i} \hat{u}_{it}$. The "suitable fraction" is the proportion of the variance of the variance of $\bar{u}_i$ that is due to v_i, namely

$$\frac{\sigma_v^2}{\sigma_v^2 + \sigma_\varepsilon^2/T_i} = 1 - (1 - \theta_i)^2$$

After random effects estimation in gretl you can construct a series containing the $\hat{v}_i$s as follows:

```
# case 1: balanced panel
scalar theta = $["theta"]
series vhat = (1 - (1 - theta)^2) * pmean($uhat)

# case 2: unbalanced, Ti varies by individual
scalar s2v = $["s2v"]
scalar s2e = $["s2e"]
series frac = s2v / (s2v + s2e/pnobs($uhat))
series vhat = frac * pmean($uhat)
```

Having found vhat, an estimate of ε_{it} can then be obtained by subtraction from $uhat.

18.2 Autoregressive panel models

Special problems arise when a lag of the dependent variable is included among the regressors in a panel model. Consider a dynamic variant of the pooled model (eq. 18.1):

$$y_{it} = X_{it}\beta + \rho y_{it-1} + u_{it} \tag{18.7}$$

First, if the error u_{it} includes a group effect, v_i, then y_{it-1} is bound to be correlated with the error, since the value of v_i affects y_i at all t. That means that OLS applied to (18.7) will be inconsistent as well as inefficient. The fixed-effects model sweeps out the group effects and so overcomes this particular problem, but a subtler issue remains, which applies to both fixed and random effects estimation. Consider the de-meaned representation of fixed effects, as applied to the dynamic model,

$$\tilde{y}_{it} = \tilde{X}_{it}\beta + \rho\tilde{y}_{i,t-1} + \varepsilon_{it}$$

where $\tilde{y}_{it} = y_{it} - \bar{y}_i$ and $\varepsilon_{it} = u_{it} - \bar{u}_i$ (or $u_{it} - \alpha_i$, using the notation of equation 18.2). The trouble is that $\tilde{y}_{i,t-1}$ will be correlated with ε_{it} via the group mean, $\bar{y}_i$. The disturbance ε_{it} influences y_{it} directly, which influences $\bar{y}_i$, which, by construction, affects the value of $\tilde{y}_{it}$ for all t. The same issue arises in relation to the quasi-demeaning used for random effects. Estimators which ignore this correlation will be consistent only as $T \to \infty$ (in which case the marginal effect of ε_{it} on the group mean of y tends to vanish).

One strategy for handling this problem, and producing consistent estimates of β and ρ, was proposed by Anderson and Hsiao (1981). Instead of de-meaning the data, they suggest taking the first difference of (18.7), an alternative tactic for sweeping out the group effects:

$$\Delta y_{it} = \Delta X_{it}\beta + \rho\Delta y_{i,t-1} + \eta_{it} \tag{18.8}$$

where $\eta_{it} = \Delta u_{it} = \Delta(v_i + \varepsilon_{it}) = \varepsilon_{it} - \varepsilon_{i,t-1}$. We're not in the clear yet, given the structure of the error η_{it}: the disturbance $\varepsilon_{i,t-1}$ is an influence on both η_{it} and $\Delta y_{i,t-1} = y_{it} - y_{i,t-1}$. The next step is then to find an instrument for the "contaminated" $\Delta y_{i,t-1}$. Anderson and Hsiao suggest using either $y_{i,t-2}$ or $\Delta y_{i,t-2}$, both of which will be uncorrelated with η_{it} provided that the underlying errors, ε_{it}, are not themselves serially correlated.

The Anderson–Hsiao estimator is not provided as a built-in function in gretl, since gretl's sensible handling of lags and differences for panel data makes it a simple application of regression with instrumental variables—see Example 18.2, which is based on a study of country growth rates by Nerlove (1999).[6]

[6]Also see Clint Cummins' benchmarks page, http://www.stanford.edu/~clint/bench/.

Example 18.2: The Anderson-Hsiao estimator for a dynamic panel model

```
# Penn World Table data as used by Nerlove
open penngrow.gdt
# Fixed effects (for comparison)
panel Y 0 Y(-1) X
# Random effects (for comparison)
panel Y 0 Y(-1) X --random-effects
# take differences of all variables
diff Y X
# Anderson-Hsiao, using Y(-2) as instrument
tsls d_Y d_Y(-1) d_X ; 0 d_X Y(-2)
# Anderson-Hsiao, using d_Y(-2) as instrument
tsls d_Y d_Y(-1) d_X ; 0 d_X d_Y(-2)
```

Although the Anderson–Hsiao estimator is consistent, it is not most efficient: it does not make the fullest use of the available instruments for $\Delta y_{i,t-1}$, nor does it take into account the differenced structure of the error η_{it}. It is improved upon by the methods of Arellano and Bond (1991) and Blundell and Bond (1998). These methods are taken up in the next chapter.

Chapter 19

Dynamic panel models

Since gretl version 1.9.2, the primary command for estimating dynamic panel models has been `dpanel`. The closely related `arbond` command predated `dpanel`, and is still present, but whereas `arbond` only supports the so-called "difference" estimator (Arellano and Bond, 1991), `dpanel` is addition offers the "system" estimator (Blundell and Bond, 1998), which has become the method of choice in the applied literature.

19.1 Introduction

Notation

A dynamic linear panel data model can be represented as follows (in notation based on Arellano (2003)):

$$y_{it} = \alpha y_{i,t-1} + \beta' x_{it} + \eta_i + v_{it} \tag{19.1}$$

The main idea on which the difference estimator is based is to get rid of the individual effect via differencing. First-differencing eq. (19.1) yields

$$\Delta y_{it} = \alpha \Delta y_{i,t-1} + \beta' \Delta x_{it} + \Delta v_{it} = \gamma' W_{it} + \Delta v_{it}, \tag{19.2}$$

in obvious notation. The error term of (19.2) is, by construction, autocorrelated and also correlated with the lagged dependent variable, so an estimator that takes both issues into account is needed. The endogeneity issue is solved by noting that all values of $y_{i,t-k}$, with $k > 1$ can be used as instruments for $\Delta y_{i,t-1}$: unobserved values of $y_{i,t-k}$ (because they could be missing, or pre-sample) can safely be substituted with 0. In the language of GMM, this amounts to using the relation

$$E(\Delta v_{it} \cdot y_{i,t-k}) = 0, \quad k > 1 \tag{19.3}$$

as an orthogonality condition.

Autocorrelation is dealt with by noting that if v_{it} is white noise, the covariance matrix of the vector whose typical element is Δv_{it} is proportional to a matrix H that has 2 on the main diagonal, -1 on the first subdiagonals and 0 elsewhere. One-step GMM estimation of equation (19.2) amounts to computing

$$\hat{\gamma} = \left[\left(\sum_{i=1}^{N} \boldsymbol{W}_i' \boldsymbol{Z}_i \right) \boldsymbol{A}_N \left(\sum_{i=1}^{N} \boldsymbol{Z}_i' \boldsymbol{W}_i \right) \right]^{-1} \left(\sum_{i=1}^{N} \boldsymbol{W}_i' \boldsymbol{Z}_i \right) \boldsymbol{A}_N \left(\sum_{i=1}^{N} \boldsymbol{Z}_i' \Delta \boldsymbol{y}_i \right) \tag{19.4}$$

where

$$\Delta \boldsymbol{y}_i = \begin{bmatrix} \Delta y_{i,3} & \cdots & \Delta y_{i,T} \end{bmatrix}'$$

$$\boldsymbol{W}_i = \begin{bmatrix} \Delta y_{i,2} & \cdots & \Delta y_{i,T-1} \\ \Delta x_{i,3} & \cdots & \Delta x_{i,T} \end{bmatrix}'$$

$$\boldsymbol{Z}_i = \begin{bmatrix} y_{i1} & 0 & 0 & \cdots & 0 & \Delta x_{i3} \\ 0 & y_{i1} & y_{i2} & \cdots & 0 & \Delta x_{i4} \\ & & \vdots & & & \\ 0 & 0 & 0 & \cdots & y_{i,T-2} & \Delta x_{iT} \end{bmatrix}'$$

and

$$A_N = \left(\sum_{i=1}^{N} Z_i' H Z_i \right)^{-1}$$

Once the 1-step estimator is computed, the sample covariance matrix of the estimated residuals can be used instead of H to obtain 2-step estimates, which are not only consistent but asymptotically efficient. (In principle the process may be iterated, but nobody seems to be interested.) Standard GMM theory applies, except for one thing: Windmeijer (2005) has computed finite-sample corrections to the asymptotic covariance matrix of the parameters, which are nowadays almost universally used.

The difference estimator is consistent, but has been shown to have poor properties in finite samples when α is near one. People these days prefer the so-called "system" estimator, which complements the differenced data (with lagged levels used as instruments) with data in levels (using lagged differences as instruments). The system estimator relies on an extra orthogonality condition which has to do with the earliest value of the dependent variable $y_{i,1}$. The interested reader is referred to Blundell and Bond (1998, pp. 124–125) for details, but here it suffices to say that this condition is satisfied in mean-stationary models and brings an improvement in efficiency that may be substantial in many cases.

The set of orthogonality conditions exploited in the system approach is not very much larger than with the difference estimator, the reason being that most of the possible orthogonality conditions associated with the equations in levels are redundant, given those already used for the equations in differences.

The key equations of the system estimator can be written as

$$\tilde{\gamma} = \left[\left(\sum_{i=1}^{N} \tilde{W}' \tilde{Z} \right) A_N \left(\sum_{i=1}^{N} \tilde{Z}' \tilde{W} \right) \right]^{-1} \left(\sum_{i=1}^{N} \tilde{W}' \tilde{Z} \right) A_N \left(\sum_{i=1}^{N} \tilde{Z}' \Delta \tilde{y}_i \right) \tag{19.5}$$

where

$$\Delta \tilde{y}_i = \begin{bmatrix} \Delta y_{i3} & \cdots & \Delta y_{iT} & y_{i3} & \cdots & y_{iT} \end{bmatrix}'$$

$$\tilde{W}_i = \begin{bmatrix} \Delta y_{i2} & \cdots & \Delta y_{i,T-1} & y_{i2} & \cdots & y_{i,T-1} \\ \Delta x_{i3} & \cdots & \Delta x_{iT} & x_{i3} & \cdots & x_{iT} \end{bmatrix}'$$

$$\tilde{Z}_i = \begin{bmatrix} y_{i1} & 0 & 0 & \cdots & 0 & 0 & \cdots & 0 & \Delta x_{i,3} \\ 0 & y_{i1} & y_{i2} & \cdots & 0 & 0 & \cdots & 0 & \Delta x_{i,4} \\ & & \vdots & & & & & & \\ 0 & 0 & 0 & \cdots & y_{i,T-2} & 0 & \cdots & 0 & \Delta x_{iT} \\ & & \vdots & & & & & & \\ 0 & 0 & 0 & \cdots & 0 & \Delta y_{i2} & \cdots & 0 & x_{i3} \\ & & \vdots & & & & & & \\ 0 & 0 & 0 & \cdots & 0 & 0 & \cdots & \Delta y_{i,T-1} & x_{iT} \end{bmatrix}'$$

and

$$A_N = \left(\sum_{i=1}^{N} \tilde{Z}' H^* \tilde{Z} \right)^{-1}$$

In this case choosing a precise form for the matrix H^* for the first step is no trivial matter. Its north-west block should be as similar as possible to the covariance matrix of the vector Δv_{it}, so the same choice as the "difference" estimator is appropriate. Ideally, the south-east block should be proportional to the covariance matrix of the vector $\iota \eta_i + \boldsymbol{v}$, that is $\sigma_v^2 I + \sigma_\eta^2 \iota \iota'$; but since σ_η^2 is unknown and any positive definite matrix renders the estimator consistent, people just use I. The off-diagonal blocks should, in principle, contain the covariances between Δv_{is} and v_{it}, which would be an identity matrix if v_{it} is white noise. However, since the south-east block is typically given a conventional value anyway, the benefit in making this choice is not obvious. Some packages use I; others use a zero matrix. Asymptotically, it should not matter, but on real datasets the difference between the resulting estimates can be noticeable.

Rank deficiency

Both the difference estimator (19.4) and the system estimator (19.5) depend, for their existence, on the invertibility of $\boldsymbol{A}_N$. This matrix may turn out to be singular for several reasons. However, this does not mean that the estimator is not computable: in some cases, adjustments are possible such that the estimator does exist, but the user should be aware that in these cases not all software packages use the same strategy and replication of results may prove difficult or even impossible.

A first reason why $\boldsymbol{A}_N$ may be singular could be the unavailability of instruments, chiefly because of missing observations. This case is easy to handle. If a particular row of $\tilde{\boldsymbol{Z}}_i$ is zero for all units, the corresponding orthogonality condition (or the corresponding instrument if you prefer) is automatically dropped; of course, the overidentification rank is adjusted for testing purposes.

Even if no instruments are zero, however, $\boldsymbol{A}_N$ could be rank deficient. A trivial case occurs if there are collinear instruments, but a less trivial case may arise when T (the total number of time periods available) is not much smaller than N (the number of units), as, for example, in some macro datasets where the units are countries. The total number of potentially usable orthogonality conditions is $O(T^2)$, which may well exceed N in some cases. Of course $\boldsymbol{A}_N$ is the sum of N matrices which have, at most, rank $2T-3$ and therefore it could well happen that the sum is singular.

In all these cases, what we consider the "proper" way to go is to substitute the pseudo-inverse of $\boldsymbol{A}_N$ (Moore–Penrose) for its regular inverse. Again, our choice is shared by some software packages, but not all, so replication may be hard.

Treatment of missing values

Textbooks seldom bother with missing values, but in some cases their treatment may be far from obvious. This is especially true if missing values are interspersed between valid observations. For example, consider the plain difference estimator with one lag, so

$$y_t = \alpha y_{t-1} + \eta + \epsilon_t$$

where the i index is omitted for clarity. Suppose you have an individual with $t = 1 \dots 5$, for which y_3 is missing. It may seem that the data for this individual are unusable, because differencing y_t would produce something like

t	1	2	3	4	5
y_t	$*$	$*$	$\circ$	$*$	$*$
Δy_t	$\circ$	$*$	$\circ$	$\circ$	$*$

where $*$ = nonmissing and $\circ$ = missing. Estimation seems to be unfeasible, since there are no periods in which Δy_t and Δy_{t-1} are both observable.

However, we can use a k-difference operator and get

$$\Delta_k y_t = \alpha \Delta_k y_{t-1} + \Delta_k \epsilon_t$$

where $\Delta_k = 1 - L^k$ and past levels of y_t are perfectly valid instruments. In this example, we can choose $k = 3$ and use y_1 as an instrument, so this unit is in fact perfectly usable.

Not all software packages seem to be aware of this possibility, so replicating published results may prove tricky if your dataset contains individuals with "gaps" between valid observations.

19.2 Usage

One of the concepts underlying the syntax of dpanel is that you get default values for several choices you may want to make, so that in a "standard" situation the command itself is very short to write (and read). The simplest case of the model (19.1) is a plain AR(1) process:

$$y_{i,t} = \alpha y_{i,t-1} + \eta_i + v_{it}. \tag{19.6}$$

If you give the command

```
dpanel 1 ; y
```

gretl assumes that you want to estimate (19.6) via the difference estimator (19.4), using as many orthogonality conditions as possible. The scalar 1 between dpanel and the semicolon indicates that only one lag of y is included as an explanatory variable; using 2 would give an AR(2) model. The syntax that gretl uses for the non-seasonal AR and MA lags in an ARMA model is also supported in this context.[1] For example, if you want the first and third lags of y (but not the second) included as explanatory variables you can say

```
dpanel {1 3} ; y
```

or you can use a pre-defined matrix for this purpose:

```
matrix ylags = {1, 3}
dpanel ylags ; y
```

To use a single lag of y other than the first you need to employ this mechanism:

```
dpanel {3} ; y # only lag 3 is included
dpanel 3 ; y   # compare: lags 1, 2 and 3 are used
```

To use the system estimator instead, you add the --system option, as in

```
dpanel 1 ; y --system
```

The level orthogonality conditions and the corresponding instrument are appended automatically (see eq. 19.5).

Regressors

If we want to introduce additional regressors, we list them after the dependent variable in the same way as other gretl commands, such as ols.

For the difference orthogonality relations, dpanel takes care of transforming the regressors in parallel with the dependent variable. Note that this differs from gretl's arbond command, where only the dependent variable is differenced automatically; it brings us more in line with other software.

One case of potential ambiguity is when an intercept is specified but the difference-only estimator is selected, as in

```
dpanel 1 ; y const
```

In this case the default dpanel behavior, which agrees with Stata's xtabond2, is to drop the constant (since differencing reduces it to nothing but zeros). However, for compatibility with the DPD package for Ox, you can give the option --dpdstyle, in which case the constant is retained (equivalent to including a linear trend in equation 19.1). A similar point applies to the period-specific dummy variables which can be added in dpanel via the --time-dummies option: in the differences-only case these dummies are entered in differenced form by default, but when the --dpdstyle switch is applied they are entered in levels.

The standard gretl syntax applies if you want to use lagged explanatory variables, so for example the command

[1] This represents an enhancement over the arbond command.

```
dpanel 1 ; y const x(0 to -1) --system
```

would result in estimation of the model

$$y_{it} = \alpha y_{i,t-1} + \beta_0 + \beta_1 x_{it} + \beta_2 x_{i,t-1} + \eta_i + v_{it}.$$

Instruments

The default rules for instruments are:

- lags of the dependent variable are instrumented using all available orthogonality conditions; and
- additional regressors are considered exogenous, so they are used as their own instruments.

If a different policy is wanted, the instruments should be specified in an additional list, separated from the regressors list by a semicolon. The syntax closely mirrors that for the `tsls` command, but in this context it is necessary to distinguish between "regular" instruments and what are often called "GMM-style" instruments (that is, instruments that are handled in the same block-diagonal manner as lags of the dependent variable, as described above).

"Regular" instruments are transformed in the same way as regressors, and the contemporaneous value of the transformed variable is used to form an orthogonality condition. Since regressors are treated as exogenous by default, it follows that these two commands estimate the same model:

```
dpanel 1 ; y z
dpanel 1 ; y z ; z
```

The instrument specification in the second case simply confirms what is implicit in the first: that `z` is exogenous. Note, though, that if you have some additional variable `z2` which you want to add as a regular instrument, it then becomes necessary to include `z` in the instrument list if it is to be treated as exogenous:

```
dpanel 1 ; y z ; z2   # z is now implicitly endogenous
dpanel 1 ; y z ; z z2 # z is treated as exogenous
```

The specification of "GMM-style" instruments is handled by the special constructs `GMM()` and `GMMlevel()`. The first of these relates to instruments for the equations in differences, and the second to the equations in levels. The syntax for `GMM()` is

`GMM(`*name*`,` *minlag*`,` *maxlag*`)`

where *name* is replaced by the name of a series (or the name of a list of series), and *minlag* and *maxlag* are replaced by the minimum and maximum lags to be used as instruments. The same goes for `GMMlevel()`.

One common use of `GMM()` is to limit the number of lagged levels of the dependent variable used as instruments for the equations in differences. It's well known that although exploiting all possible orthogonality conditions yields maximal asymptotic efficiency, in finite samples it may be preferable to use a smaller subset (but see also Okui (2009)). For example, the specification

```
dpanel 1 ; y ; GMM(y, 2, 4)
```

ensures that no lags of y_t earlier than $t - 4$ will be used as instruments.

A second use of `GMM()` is to exploit more fully the potential block-diagonal orthogonality conditions offered by an exogenous regressor, or a related variable that does not appear as a regressor. For example, in

```
dpanel 1 ; y x ; GMM(z, 2, 6)
```

the variable x is considered an endogenous regressor, and up to 5 lags of z are used as instruments.

Note that in the following script fragment

```
dpanel 1 ; y z
dpanel 1 ; y z ; GMM(z,0,0)
```

the two estimation commands should not be expected to give the same result, as the sets of orthogonality relationships are subtly different. In the latter case, you have $T-2$ separate orthogonality relationships pertaining to z_{it}, none of which has any implication for the other ones; in the former case, you only have one. In terms of the $\boldsymbol{Z}_i$ matrix, the first form adds a single row to the bottom of the instruments matrix, while the second form adds a diagonal block with $T-2$ columns; that is,

$$\begin{bmatrix} z_{i3} & z_{i4} & \cdots & z_{it} \end{bmatrix}$$

versus

$$\begin{bmatrix} z_{i3} & 0 & \cdots & 0 \\ 0 & z_{i4} & \cdots & 0 \\ & \ddots & \ddots & \\ 0 & 0 & \cdots & z_{it} \end{bmatrix}$$

19.3 Replication of DPD results

In this section we show how to replicate the results of some of the pioneering work with dynamic panel-data estimators by Arellano, Bond and Blundell. As the DPD manual (Doornik, Arellano and Bond, 2006) explains, it is difficult to replicate the original published results exactly, for two main reasons: not all of the data used in those studies are publicly available; and some of the choices made in the original software implementation of the estimators have been superseded. Here, therefore, our focus is on replicating the results obtained using the current DPD package and reported in the DPD manual.

The examples are based on the program files abest1.ox, abest3.ox and bbest1.ox. These are included in the DPD package, along with the Arellano-Bond database files abdata.bn7 and abdata.in7.[2] The Arellano-Bond data are also provided with gretl, in the file abdata.gdt. In the following we do not show the output from DPD or gretl; it is somewhat voluminous, and is easily generated by the user. As of this writing the results from Ox/DPD and gretl are identical in all relevant respects for all of the examples shown.[3]

A complete Ox/DPD program to generate the results of interest takes this general form:

```
#include <oxstd.h>
#import <packages/dpd/dpd>

main()
{
    decl dpd = new DPD();

    dpd.Load("abdata.in7");
    dpd.SetYear("YEAR");

    // model-specific code here

    delete dpd;
}
```

[2] See http://www.doornik.com/download.html.

[3] To be specific, this is using Ox Console version 5.10, version 1.24 of the DPD package, and gretl built from CVS as of 2010-10-23, all on Linux.

In the examples below we take this template for granted and show just the model-specific code.

Example 1

The following Ox/DPD code—drawn from abest1.ox—replicates column (b) of Table 4 in Arellano and Bond (1991), an instance of the differences-only or GMM-DIF estimator. The dependent variable is the log of employment, n; the regressors include two lags of the dependent variable, current and lagged values of the log real-product wage, w, the current value of the log of gross capital, k, and current and lagged values of the log of industry output, ys. In addition the specification includes a constant and five year dummies; unlike the stochastic regressors, these deterministic terms are not differenced. In this specification the regressors w, k and ys are treated as exogenous and serve as their own instruments. In DPD syntax this requires entering these variables twice, on the X_VAR and I_VAR lines. The GMM-type (block-diagonal) instruments in this example are the second and subsequent lags of the level of n. Both 1-step and 2-step estimates are computed.

```
dpd.SetOptions(FALSE); // don't use robust standard errors
dpd.Select(Y_VAR, {"n", 0, 2});
dpd.Select(X_VAR, {"w", 0, 1, "k", 0, 0, "ys", 0, 1});
dpd.Select(I_VAR, {"w", 0, 1, "k", 0, 0, "ys", 0, 1});

dpd.Gmm("n", 2, 99);
dpd.SetDummies(D_CONSTANT + D_TIME);

print("\n\n***** Arellano & Bond (1991), Table 4 (b)");
dpd.SetMethod(M_1STEP);
dpd.Estimate();
dpd.SetMethod(M_2STEP);
dpd.Estimate();
```

Here is gretl code to do the same job:

```
open abdata.gdt
list X = w w(-1) k ys ys(-1)
dpanel 2 ; n X const --time-dummies --asy --dpdstyle
dpanel 2 ; n X const --time-dummies --asy --two-step --dpdstyle
```

Note that in gretl the switch to suppress robust standard errors is --asymptotic, here abbreviated to --asy.[4] The --dpdstyle flag specifies that the constant and dummies should not be differenced, in the context of a GMM-DIF model. With gretl's dpanel command it is not necessary to specify the exogenous regressors as their own instruments since this is the default; similarly, the use of the second and all longer lags of the dependent variable as GMM-type instruments is the default and need not be stated explicitly.

Example 2

The DPD file abest3.ox contains a variant of the above that differs with regard to the choice of instruments: the variables w and k are now treated as predetermined, and are instrumented GMM-style using the second and third lags of their levels. This approximates column (c) of Table 4 in Arellano and Bond (1991). We have modified the code in abest3.ox slightly to allow the use of robust (Windmeijer-corrected) standard errors, which are the default in both DPD and gretl with 2-step estimation:

```
dpd.Select(Y_VAR, {"n", 0, 2});
dpd.Select(X_VAR, {"w", 0, 1, "k", 0, 0, "ys", 0, 1});
dpd.Select(I_VAR, {"ys", 0, 1});
dpd.SetDummies(D_CONSTANT + D_TIME);
```

[4]Option flags in gretl can always be truncated, down to the minimal unique abbreviation.

```
dpd.Gmm("n", 2, 99);
dpd.Gmm("w", 2, 3);
dpd.Gmm("k", 2, 3);

print("\n***** Arellano & Bond (1991), Table 4 (c)\n");
print("       (but using different instruments!!)\n");
dpd.SetMethod(M_2STEP);
dpd.Estimate();
```

The gretl code is as follows:

```
open abdata.gdt
list X = w w(-1) k ys ys(-1)
list Ivars = ys ys(-1)
dpanel 2 ; n X const ; GMM(w,2,3) GMM(k,2,3) Ivars --time --two-step --dpd
```

Note that since we are now calling for an instrument set other then the default (following the second semicolon), it is necessary to include the Ivars specification for the variable ys. However, it is not necessary to specify GMM(n,2,99) since this remains the default treatment of the dependent variable.

Example 3

Our third example replicates the DPD output from bbest1.ox: this uses the same dataset as the previous examples but the model specifications are based on Blundell and Bond (1998), and involve comparison of the GMM-DIF and GMM-SYS ("system") estimators. The basic specification is slightly simplified in that the variable ys is not used and only one lag of the dependent variable appears as a regressor. The Ox/DPD code is:

```
dpd.Select(Y_VAR, {"n", 0, 1});
dpd.Select(X_VAR, {"w", 0, 1, "k", 0, 1});
dpd.SetDummies(D_CONSTANT + D_TIME);

print("\n\n***** Blundell & Bond (1998), Table 4: 1976-86 GMM-DIF");
dpd.Gmm("n", 2, 99);
dpd.Gmm("w", 2, 99);
dpd.Gmm("k", 2, 99);
dpd.SetMethod(M_2STEP);
dpd.Estimate();

print("\n\n***** Blundell & Bond (1998), Table 4: 1976-86 GMM-SYS");
dpd.GmmLevel("n", 1, 1);
dpd.GmmLevel("w", 1, 1);
dpd.GmmLevel("k", 1, 1);
dpd.SetMethod(M_2STEP);
dpd.Estimate();
```

Here is the corresponding gretl code:

```
open abdata.gdt
list X = w w(-1) k k(-1)
list Z = w k

# Blundell & Bond (1998), Table 4: 1976-86 GMM-DIF
dpanel 1 ; n X const ; GMM(Z,2,99) --time --two-step --dpd

# Blundell & Bond (1998), Table 4: 1976-86 GMM-SYS
dpanel 1 ; n X const ; GMM(Z,2,99) GMMlevel(Z,1,1) \
 --time --two-step --dpd --system
```

Note the use of the `--system` option flag to specify GMM-SYS, including the default treatment of the dependent variable, which corresponds to `GMMlevel(n,1,1)`. In this case we also want to use lagged differences of the regressors `w` and `k` as instruments for the levels equations so we need explicit `GMMlevel` entries for those variables. If you want something other than the default treatment for the dependent variable as an instrument for the levels equations, you should give an explicit `GMMlevel` specification for that variable—and in that case the `--system` flag is redundant (but harmless).

For the sake of completeness, note that if you specify at least one `GMMlevel` term, `dpanel` will then include equations in levels, but it will not automatically add a default `GMMlevel` specification for the dependent variable unless the `--system` option is given.

19.4 Cross-country growth example

The previous examples all used the Arellano–Bond dataset; for this example we use the dataset `CEL.gdt`, which is also included in the gretl distribution. As with the Arellano–Bond data, there are numerous missing values. Details of the provenance of the data can be found by opening the dataset information window in the gretl GUI (Data menu, Dataset info item). This is a subset of the Barro–Lee 138-country panel dataset, an approximation to which is used in Caselli, Esquivel and Lefort (1996) and Bond, Hoeffler and Temple (2001).[5] Both of these papers explore the dynamic panel-data approach in relation to the issues of growth and convergence of per capita income across countries.

The dependent variable is growth in real GDP per capita over successive five-year periods; the regressors are the log of the initial (five years prior) value of GDP per capita, the log-ratio of investment to GDP, s, in the prior five years, and the log of annual average population growth, n, over the prior five years plus 0.05 as stand-in for the rate of technical progress, g, plus the rate of depreciation, δ (with the last two terms assumed to be constant across both countries and periods). The original model is

$$\Delta_5 y_{it} = \beta y_{i,t-5} + \alpha s_{it} + \gamma(n_{it} + g + \delta) + \nu_t + \eta_i + \epsilon_{it} \tag{19.7}$$

which allows for a time-specific disturbance ν_t. The Solow model with Cobb-Douglas production function implies that $\gamma = -\alpha$, but this assumption is not imposed in estimation. The time-specific disturbance is eliminated by subtracting the period mean from each of the series.

Equation (19.7) can be transformed to an AR(1) dynamic panel-data model by adding $y_{i,t-5}$ to both sides, which gives

$$y_{it} = (1 + \beta)y_{i,t-5} + \alpha s_{it} + \gamma(n_{it} + g + \delta) + \eta_i + \epsilon_{it} \tag{19.8}$$

where all variables are now assumed to be time-demeaned.

In (rough) replication of Bond *et al.* (2001) we now proceed to estimate the following two models: (a) equation (19.8) via GMM-DIF, using as instruments the second and all longer lags of y_{it}, s_{it} and $n_{it} + g + \delta$; and (b) equation (19.8) via GMM-SYS, using $\Delta y_{i,t-1}$, $\Delta s_{i,t-1}$ and $\Delta(n_{i,t-1} + g + \delta)$ as additional instruments in the levels equations. We report robust standard errors throughout. (As a purely notational matter, we now use "$t - 1$" to refer to values five years prior to t, as in Bond *et al.* (2001)).

The gretl script to do this job is shown below. Note that the final transformed versions of the variables (logs, with time-means subtracted) are named `ly` (y_{it}), `linv` (s_{it}) and `lngd` ($n_{it} + g + \delta$).

```
open CEL.gdt

ngd = n + 0.05
ly = log(y)
linv = log(s)
```

[5] We say an "approximation" because we have not been able to replicate exactly the OLS results reported in the papers cited, though it seems from the description of the data in Caselli *et al.* (1996) that we ought to be able to do so. We note that Bond *et al.* (2001) used data provided by Professor Caselli yet did not manage to reproduce the latter's results.

```
lngd = log(ngd)

# take out time means
loop i=1..8 --quiet
  smpl (time == i) --restrict --replace
  ly -= mean(ly)
  linv -= mean(linv)
  lngd -= mean(lngd)
endloop

smpl --full
list X = linv lngd
# 1-step GMM-DIF
dpanel 1 ; ly X ; GMM(X,2,99)
# 2-step GMM-DIF
dpanel 1 ; ly X ; GMM(X,2,99) --two-step
# GMM-SYS
dpanel 1 ; ly X ; GMM(X,2,99) GMMlevel(X,1,1) --two-step --sys
```

For comparison we estimated the same two models using Ox/DPD and the Stata command `xtabond2`. (In each case we constructed a comma-separated values dataset containing the data as transformed in the gretl script shown above, using a missing-value code appropriate to the target program.) For reference, the commands used with Stata are reproduced below:

```
insheet using CEL.csv
tsset unit time
xtabond2 ly L.ly linv lngd, gmm(L.ly, lag(1 99)) gmm(linv, lag(2 99))
  gmm(lngd, lag(2 99)) rob nolev
xtabond2 ly L.ly linv lngd, gmm(L.ly, lag(1 99)) gmm(linv, lag(2 99))
  gmm(lngd, lag(2 99)) rob nolev twostep
xtabond2 ly L.ly linv lngd, gmm(L.ly, lag(1 99)) gmm(linv, lag(2 99))
  gmm(lngd, lag(2 99)) rob nocons twostep
```

For the GMM-DIF model all three programs find 382 usable observations and 30 instruments, and yield identical parameter estimates and robust standard errors (up to the number of digits printed, or more); see Table 19.1.[6]

	1-step		2-step	
	coeff	std. error	coeff	std. error
`ly(-1)`	0.577564	0.1292	0.610056	0.1562
`linv`	0.0565469	0.07082	0.100952	0.07772
`lngd`	−0.143950	0.2753	−0.310041	0.2980

Table 19.1: GMM-DIF: Barro-Lee data

Results for GMM-SYS estimation are shown in Table 19.2. In this case we show two sets of gretl results: those labeled "gretl(1)" were obtained using gretl's `--dpdstyle` option, while those labeled "gretl(2)" did not use that option—the intent being to reproduce the H matrices used by Ox/DPD and `xtabond2` respectively.

In this case all three programs use 479 observations; gretl and `xtabond2` use 41 instruments and produce the same estimates (when using the same H matrix) while Ox/DPD nominally uses 66.[7] It is noteworthy that with GMM-SYS plus "messy" missing observations, the results depend on the precise array of instruments used, which in turn depends on the details of the implementation of the estimator.

[6]The coefficient shown for `ly(-1)` in the Tables is that reported directly by the software; for comparability with the original model (eq. 19.7) it is necesary to subtract 1, which produces the expected negative value indicating conditional convergence in per capita income.

[7]This is a case of the issue described in section 19.1: the full A_N matrix turns out to be singular and special measures must be taken to produce estimates.

	gretl(1)	Ox/DPD	gretl(2)	xtabond2
`ly(-1)`	0.9237 (0.0385)	0.9167 (0.0373)	0.9073 (0.0370)	0.9073 (0.0370)
`linv`	0.1592 (0.0449)	0.1636 (0.0441)	0.1856 (0.0411)	0.1856 (0.0411)
`lngd`	−0.2370 (0.1485)	−0.2178 (0.1433)	−0.2355 (0.1501)	−0.2355 (0.1501)

Table 19.2: 2-step GMM-SYS: Barro–Lee data (standard errors in parentheses)

19.5 Auxiliary test statistics

We have concentrated above on the parameter estimates and standard errors. It may be worth adding a few words on the additional test statistics that typically accompany both GMM-DIF and GMM-SYS estimation. These include the Sargan test for overidentification, one or more Wald tests for the joint significance of the regressors (and time dummies, if applicable) and tests for first- and second-order autocorrelation of the residuals from the equations in differences.

As in Ox/DPD, the Sargan test statistic reported by gretl is

$$S = \left(\sum_{i=1}^{N} \hat{\boldsymbol{v}}_i^{*\prime} \boldsymbol{Z}_i \right) \boldsymbol{A}_N \left(\sum_{i=1}^{N} \boldsymbol{Z}_i' \hat{\boldsymbol{v}}_i^{*} \right)$$

where the $\hat{\boldsymbol{v}}_i^*$ are the transformed (e.g. differenced) residuals for unit i. Under the null hypothesis that the instruments are valid, S is asymptotically distributed as chi-square with degrees of freedom equal to the number of overidentifying restrictions.

In general we see a good level of agreement between gretl, DPD and `xtabond2` with regard to these statistics, with a few relatively minor exceptions. Specifically, `xtabond2` computes both a "Sargan test" and a "Hansen test" for overidentification, but what it calls the Hansen test is, apparently, what DPD calls the Sargan test. (We have had difficulty determining from the `xtabond2` documentation (Roodman, 2006) exactly how its Sargan test is computed.) In addition there are cases where the degrees of freedom for the Sargan test differ between DPD and gretl; this occurs when the $\boldsymbol{A}_N$ matrix is singular (section 19.1). In concept the df equals the number of instruments minus the number of parameters estimated; for the first of these terms gretl uses the rank of $\boldsymbol{A}_N$, while DPD appears to use the full dimension of this matrix.

19.6 Memo: `dpanel` options

flag	*effect*
`--asymptotic`	Suppresses the use of robust standard errors
`--two-step`	Calls for 2-step estimation (the default being 1-step)
`--system`	Calls for GMM-SYS, with default treatment of the dependent variable, as in `GMMlevel(y,1,1)`
`--time-dummies`	Includes period-specific dummy variables
`--dpdstyle`	Compute the H matrix as in DPD; also suppresses differencing of automatic time dummies and omission of intercept in the GMM-DIF case
`--verbose`	When `--two-step` is selected, prints the 1-step estimates first
`--vcv`	Calls for printing of the covariance matrix
`--quiet`	Suppresses the printing of results

The time dummies option supports the qualifier `noprint`, as in

```
--time-dummies=noprint
```

This means that although the dummies are included in the specification their coefficients, standard errors and so on are not printed.

Chapter 20

Nonlinear least squares

20.1 Introduction and examples

Gretl supports nonlinear least squares (NLS) using a variant of the Levenberg-Marquardt algorithm. The user must supply a specification of the regression function; prior to giving this specification the parameters to be estimated must be "declared" and given initial values. Optionally, the user may supply analytical derivatives of the regression function with respect to each of the parameters. If derivatives are not given, the user must instead give a list of the parameters to be estimated (separated by spaces or commas), preceded by the keyword `params`. The tolerance (criterion for terminating the iterative estimation procedure) can be adjusted using the `set` command.

The syntax for specifying the function to be estimated is the same as for the `genr` command. Here are two examples, with accompanying derivatives.

```
# Consumption function from Greene
nls C = alpha + beta * Y^gamma
    deriv alpha = 1
    deriv beta = Y^gamma
    deriv gamma = beta * Y^gamma * log(Y)
end nls

# Nonlinear function from Russell Davidson
nls y = alpha + beta * x1 + (1/beta) * x2
    deriv alpha = 1
    deriv beta = x1 - x2/(beta*beta)
end nls --vcv
```

Note the command words `nls` (which introduces the regression function), `deriv` (which introduces the specification of a derivative), and `end nls`, which terminates the specification and calls for estimation. If the `--vcv` flag is appended to the last line the covariance matrix of the parameter estimates is printed.

20.2 Initializing the parameters

The parameters of the regression function must be given initial values prior to the `nls` command. This can be done using the `genr` command (or, in the GUI program, via the menu item "Variable, Define new variable").

In some cases, where the nonlinear function is a generalization of (or a restricted form of) a linear model, it may be convenient to run an `ols` and initialize the parameters from the OLS coefficient estimates. In relation to the first example above, one might do:

```
ols C 0 Y
genr alpha = $coeff(0)
genr beta = $coeff(Y)
genr gamma = 1
```

And in relation to the second example one might do:

```
ols y 0 x1 x2
genr alpha = $coeff(0)
genr beta = $coeff(x1)
```

20.3 NLS dialog window

It is probably most convenient to compose the commands for NLS estimation in the form of a gretl script but you can also do so interactively, by selecting the item "Nonlinear Least Squares" under the "Model, Nonlinear models" menu. This opens a dialog box where you can type the function specification (possibly prefaced by `genr` lines to set the initial parameter values) and the derivatives, if available. An example of this is shown in Figure 20.1. Note that in this context you do not have to supply the `nls` and `end nls` tags.

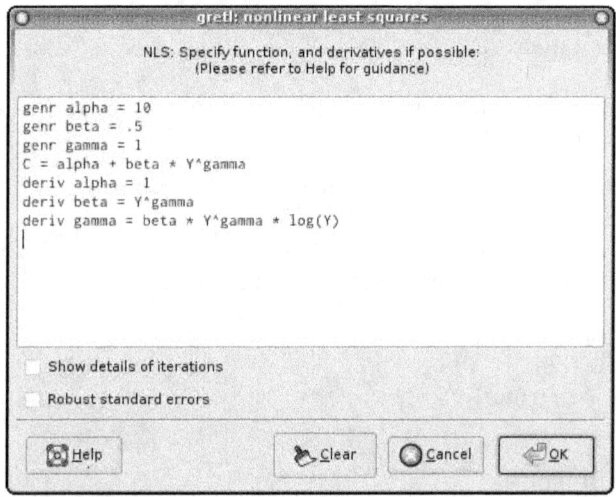

Figure 20.1: NLS dialog box

20.4 Analytical and numerical derivatives

If you are able to figure out the derivatives of the regression function with respect to the parameters, it is advisable to supply those derivatives as shown in the examples above. If that is not possible, gretl will compute approximate numerical derivatives. However, the properties of the NLS algorithm may not be so good in this case (see section 20.7).

This is done by using the `params` statement, which should be followed by a list of identifiers containing the parameters to be estimated. In this case, the examples above would read as follows:

```
# Greene
nls C = alpha + beta * Y^gamma
    params alpha beta gamma
end nls
```

```
# Davidson
nls y = alpha + beta * x1 + (1/beta) * x2
    params alpha beta
end nls
```

If analytical derivatives are supplied, they are checked for consistency with the given nonlinear function. If the derivatives are clearly incorrect estimation is aborted with an error message. If the derivatives are "suspicious" a warning message is issued but estimation proceeds. This warning may sometimes be triggered by incorrect derivatives, but it may also be triggered by a high degree of collinearity among the derivatives.

Note that you cannot mix analytical and numerical derivatives: you should supply expressions for all of the derivatives or none.

20.5 Controlling termination

The NLS estimation procedure is an iterative process. Iteration is terminated when the criterion for convergence is met or when the maximum number of iterations is reached, whichever comes first.

Let k denote the number of parameters being estimated. The maximum number of iterations is $100 \times (k+1)$ when analytical derivatives are given, and $200 \times (k+1)$ when numerical derivatives are used.

Let ϵ denote a small number. The iteration is deemed to have converged if at least one of the following conditions is satisfied:

- Both the actual and predicted relative reductions in the error sum of squares are at most ϵ.
- The relative error between two consecutive iterates is at most ϵ.

This default value of ϵ is the machine precision to the power 3/4,[1] but it can be adjusted using the `set` command with the parameter `nls_toler`. For example

```
set nls_toler .0001
```

will relax the value of ϵ to 0.0001.

20.6 Details on the code

The underlying engine for NLS estimation is based on the `minpack` suite of functions, available from netlib.org. Specifically, the following `minpack` functions are called:

`lmder`	Levenberg-Marquardt algorithm with analytical derivatives
`chkder`	Check the supplied analytical derivatives
`lmdif`	Levenberg-Marquardt algorithm with numerical derivatives
`fdjac2`	Compute final approximate Jacobian when using numerical derivatives
`dpmpar`	Determine the machine precision

On successful completion of the Levenberg-Marquardt iteration, a Gauss-Newton regression is used to calculate the covariance matrix for the parameter estimates. If the `--robust` flag is given a robust variant is computed. The documentation for the `set` command explains the specific options available in this regard.

Since NLS results are asymptotic, there is room for debate over whether or not a correction for degrees of freedom should be applied when calculating the standard error of the regression (and the standard errors of the parameter estimates). For comparability with OLS, and in light of the reasoning given in Davidson and MacKinnon (1993), the estimates shown in gretl *do* use a degrees of freedom correction.

20.7 Numerical accuracy

Table 20.1 shows the results of running the gretl NLS procedure on the 27 Statistical Reference Datasets made available by the U.S. National Institute of Standards and Technology (NIST) for testing nonlinear regression software.[2] For each dataset, two sets of starting values for the parameters are given in the test files, so the full test comprises 54 runs. Two full tests were performed, one using all analytical derivatives and one using all numerical approximations. In each case the default tolerance was used.[3]

[1] On a 32-bit Intel Pentium machine a likely value for this parameter is 1.82×10^{-12}.

[2] For a discussion of gretl's accuracy in the estimation of linear models, see Appendix D.

[3] The data shown in the table were gathered from a pre-release build of gretl version 1.0.9, compiled with gcc 3.3, linked against glibc 2.3.2, and run under Linux on an i686 PC (IBM ThinkPad A21m).

Out of the 54 runs, gretl failed to produce a solution in 4 cases when using analytical derivatives, and in 5 cases when using numeric approximation. Of the four failures in analytical derivatives mode, two were due to non-convergence of the Levenberg–Marquardt algorithm after the maximum number of iterations (on MGH09 and Bennett5, both described by NIST as of "Higher difficulty") and two were due to generation of range errors (out-of-bounds floating point values) when computing the Jacobian (on BoxBOD and MGH17, described as of "Higher difficulty" and "Average difficulty" respectively). The additional failure in numerical approximation mode was on MGH10 ("Higher difficulty", maximum number of iterations reached).

The table gives information on several aspects of the tests: the number of outright failures, the average number of iterations taken to produce a solution and two sorts of measure of the accuracy of the estimates for both the parameters and the standard errors of the parameters.

For each of the 54 runs in each mode, if the run produced a solution the parameter estimates obtained by gretl were compared with the NIST certified values. We define the "minimum correct figures" for a given run as the number of significant figures to which the *least accurate* gretl estimate agreed with the certified value, for that run. The table shows both the average and the worst case value of this variable across all the runs that produced a solution. The same information is shown for the estimated standard errors.[4]

The second measure of accuracy shown is the percentage of cases, taking into account all parameters from all successful runs, in which the gretl estimate agreed with the certified value to at least the 6 significant figures which are printed by default in the gretl regression output.

Table 20.1: Nonlinear regression: the NIST tests

	Analytical derivatives	*Numerical derivatives*
Failures in 54 tests	4	5
Average iterations	32	127
Mean of min. correct figures, parameters	8.120	6.980
Worst of min. correct figures, parameters	4	3
Mean of min. correct figures, standard errors	8.000	5.673
Worst of min. correct figures, standard errors	5	2
Percent correct to at least 6 figures, parameters	96.5	91.9
Percent correct to at least 6 figures, standard errors	97.7	77.3

Using analytical derivatives, the worst case values for both parameters and standard errors were improved to 6 correct figures on the test machine when the tolerance was tightened to 1.0e−14. Using numerical derivatives, the same tightening of the tolerance raised the worst values to 5 correct figures for the parameters and 3 figures for standard errors, at a cost of one additional failure of convergence.

Note the overall superiority of analytical derivatives: on average solutions to the test problems were obtained with substantially fewer iterations and the results were more accurate (most notably for the estimated standard errors). Note also that the six-digit results printed by gretl are not 100 percent reliable for difficult nonlinear problems (in particular when using numerical

[4]For the standard errors, I excluded one outlier from the statistics shown in the table, namely Lanczos1. This is an odd case, using generated data with an almost-exact fit: the standard errors are 9 or 10 orders of magnitude smaller than the coefficients. In this instance gretl could reproduce the certified standard errors to only 3 figures (analytical derivatives) and 2 figures (numerical derivatives).

derivatives). Having registered this caveat, the percentage of cases where the results were good to six digits or better seems high enough to justify their printing in this form.

Chapter 21

Maximum likelihood estimation

21.1 Generic ML estimation with gretl

Maximum likelihood estimation is a cornerstone of modern inferential procedures. Gretl provides a way to implement this method for a wide range of estimation problems, by use of the `mle` command. We give here a few examples.

To give a foundation for the examples that follow, we start from a brief reminder on the basics of ML estimation. Given a sample of size T, it is possible to define the density function[1] for the whole sample, namely the joint distribution of all the observations $f(\mathbf{Y};\theta)$, where $\mathbf{Y} = \{y_1, \ldots, y_T\}$. Its shape is determined by a k-vector of unknown parameters θ, which we assume is contained in a set Θ, and which can be used to evaluate the probability of observing a sample with any given characteristics.

After observing the data, the values $\mathbf{Y}$ are given, and this function can be evaluated for any legitimate value of θ. In this case, we prefer to call it the *likelihood* function; the need for another name stems from the fact that this function works as a density when we use the y_ts as arguments and θ as parameters, whereas in this context θ is taken as the function's argument, and the data $\mathbf{Y}$ only have the role of determining its shape.

In standard cases, this function has a unique maximum. The location of the maximum is unaffected if we consider the logarithm of the likelihood (or log-likelihood for short): this function will be denoted as

$$\ell(\theta) = \log f(\mathbf{Y};\theta)$$

The log-likelihood functions that gretl can handle are those where $\ell(\theta)$ can be written as

$$\ell(\theta) = \sum_{t=1}^{T} \ell_t(\theta)$$

which is true in most cases of interest. The functions $\ell_t(\theta)$ are called the log-likelihood contributions.

Moreover, the location of the maximum is obviously determined by the data $\mathbf{Y}$. This means that the value

$$\hat{\theta}(\mathbf{Y}) = \underset{\theta\in\Theta}{\operatorname{Argmax}}\, \ell(\theta) \qquad (21.1)$$

is some function of the observed data (a statistic), which has the property, under mild conditions, of being a consistent, asymptotically normal and asymptotically efficient estimator of θ.

Sometimes it is possible to write down explicitly the function $\hat{\theta}(\mathbf{Y})$; in general, it need not be so. In these circumstances, the maximum can be found by means of numerical techniques. These often rely on the fact that the log-likelihood is a smooth function of θ, and therefore on the maximum its partial derivatives should all be 0. The *gradient vector*, or *score vector*, is a function that enjoys many interesting statistical properties in its own right; it will be denoted here as $\mathbf{g}(\theta)$. It is a k-vector with typical element

$$g_i(\theta) = \frac{\partial \ell(\theta)}{\partial \theta_i} = \sum_{t=1}^{T} \frac{\partial \ell_t(\theta)}{\partial \theta_i}$$

[1] We are supposing here that our data are a realization of continuous random variables. For discrete random variables, everything continues to apply by referring to the probability function instead of the density. In both cases, the distribution may be conditional on some exogenous variables.

Gradient-based methods can be shortly illustrated as follows:

1. pick a point $\theta_0 \in \Theta$;
2. evaluate $\mathbf{g}(\theta_0)$;
3. if $\mathbf{g}(\theta_0)$ is "small", stop. Otherwise, compute a direction vector $d(\mathbf{g}(\theta_0))$;
4. evaluate $\theta_1 = \theta_0 + d(\mathbf{g}(\theta_0))$;
5. substitute θ_0 with θ_1;
6. restart from 2.

Many algorithms of this kind exist; they basically differ from one another in the way they compute the direction vector $d(\mathbf{g}(\theta_0))$, to ensure that $\ell(\theta_1) > \ell(\theta_0)$ (so that we eventually end up on the maximum).

The default method gretl uses to maximize the log-likelihood is a gradient-based algorithm known as the **BFGS** (Broyden, Fletcher, Goldfarb and Shanno) method. This technique is used in most econometric and statistical packages, as it is well-established and remarkably powerful. Clearly, in order to make this technique operational, it must be possible to compute the vector $\mathbf{g}(\theta)$ for any value of θ. In some cases this vector can be written explicitly as a function of $\mathbf{Y}$. If this is not possible or too difficult the gradient may be evaluated numerically. The alternative **Newton-Raphson** algorithm is also available, which is more effective under some circumstances but is also more fragile; see section 21.8 and chapter 31 for details.

The choice of the starting value, θ_0, is crucial in some contexts and inconsequential in others. In general, however, it is advisable to start the algorithm from "sensible" values whenever possible. If a consistent estimator is available, this is usually a safe and efficient choice: this ensures that in large samples the starting point will be likely close to $\hat{\theta}$ and convergence can be achieved in few iterations.

The maximum number of iterations allowed for the BFGS procedure, and the relative tolerance for assessing convergence, can be adjusted using the `set` command: the relevant variables are `bfgs_maxiter` (default value 500) and `bfgs_toler` (default value, the machine precision to the power 3/4).

Covariance matrix and standard errors

By default the covariance matrix of the parameter estimates is based on the Outer Product of the Gradient. That is,

$$\widehat{\mathrm{Var}}_{\mathrm{OPG}}(\hat{\theta}) = \left(G'(\hat{\theta})G(\hat{\theta})\right)^{-1} \tag{21.2}$$

where $G(\hat{\theta})$ is the $T \times k$ matrix of contributions to the gradient. Two other options are available. If the `--hessian` flag is given, the covariance matrix is computed from a numerical approximation to the Hessian at convergence. If the `--robust` option is selected, the quasi-ML "sandwich" estimator is used:

$$\widehat{\mathrm{Var}}_{\mathrm{QML}}(\hat{\theta}) = H(\hat{\theta})^{-1}G'(\hat{\theta})G(\hat{\theta})H(\hat{\theta})^{-1}$$

where H denotes the numerical approximation to the Hessian.

21.2 Gamma estimation

Suppose we have a sample of T independent and identically distributed observations from a Gamma distribution. The density function for each observation x_t is

$$f(x_t) = \frac{\alpha^p}{\Gamma(p)} x_t^{p-1} \exp(-\alpha x_t) \tag{21.3}$$

The log-likelihood for the entire sample can be written as the logarithm of the joint density of all the observations. Since these are independent and identical, the joint density is the product of the individual densities, and hence its log is

$$\ell(\alpha, p) = \sum_{t=1}^{T} \log \left[\frac{\alpha^p}{\Gamma(p)} x_t^{p-1} \exp(-\alpha x_t) \right] = \sum_{t=1}^{T} \ell_t \tag{21.4}$$

where

$$\ell_t = p \cdot \log(\alpha x_t) - \gamma(p) - \log x_t - \alpha x_t$$

and $\gamma(\cdot)$ is the log of the gamma function. In order to estimate the parameters α and p via ML, we need to maximize (21.4) with respect to them. The corresponding gretl code snippet is

```
scalar alpha = 1
scalar p = 1

mle logl =  p*ln(alpha * x) - lngamma(p) - ln(x) - alpha * x
  params alpha p
end mle
```

The first two statements

```
alpha = 1
p = 1
```

are necessary to ensure that the variables `alpha` and `p` exist before the computation of `logl` is attempted. Inside the `mle` block these variables (which could be either scalars, vectors or a combination of the two — see below for an example) are identified as the parameters that should be adjusted to maximize the likelihood via the `params` keyword. Their values will be changed by the execution of the `mle` command; upon successful completion, they will be replaced by the ML estimates. The starting value is 1 for both; this is arbitrary and does not matter much in this example (more on this later).

The above code can be made more readable, and marginally more efficient, by defining a variable to hold $\alpha \cdot x_t$. This command can be embedded in the `mle` block as follows:

```
mle logl =  p*ln(ax) - lngamma(p) - ln(x) - ax
  series ax = alpha*x
  params alpha p
end mle
```

The variable `ax` is not added to the `params` list, of course, since it is just an auxiliary variable to facilitate the calculations. You can insert as many such auxiliary lines as you require before the `params` line, with the restriction that they must contain either (a) commands to generate series, scalars or matrices or (b) print commands (which may be used to aid in debugging).

In a simple example like this, the choice of the starting values is almost inconsequential; the algorithm is likely to converge no matter what the starting values are. However, consistent method-of-moments estimators of p and α can be simply recovered from the sample mean m and variance V: since it can be shown that

$$E(x_t) = p/\alpha \qquad V(x_t) = p/\alpha^2$$

it follows that the following estimators

$$\begin{aligned} \bar{\alpha} &= m/V \\ \bar{p} &= m \cdot \bar{\alpha} \end{aligned}$$

are consistent, and therefore suitable to be used as starting point for the algorithm. The gretl script code then becomes

```
scalar m = mean(x)
scalar alpha = m/var(x)
scalar p = m*alpha

mle logl =  p*ln(ax) - lngamma(p) - ln(x) - ax
  series ax = alpha*x
  params alpha p
end mle
```

Another thing to note is that sometimes parameters are constrained within certain boundaries: in this case, for example, both α and p must be positive numbers. Gretl does not check for this: it is the user's responsibility to ensure that the function is always evaluated at an admissible point in the parameter space during the iterative search for the maximum. An effective technique is to define a variable for checking that the parameters are admissible and setting the log-likelihood as undefined if the check fails. An example, which uses the conditional assignment operator, follows:

```
scalar m = mean(x)
scalar alpha = m/var(x)
scalar p = m*alpha

mle logl = check ? p*ln(ax) - lngamma(p) - ln(x) - ax : NA
  series ax = alpha*x
  scalar check = (alpha>0) && (p>0)
  params alpha p
end mle
```

21.3 Stochastic frontier cost function

When modeling a cost function, it is sometimes worthwhile to incorporate explicitly into the statistical model the notion that firms may be inefficient, so that the observed cost deviates from the theoretical figure not only because of unobserved heterogeneity between firms, but also because two firms could be operating at a different efficiency level, despite being identical under all other respects. In this case we may write

$$C_i = C_i^* + u_i + v_i$$

where C_i is some variable cost indicator, C_i^* is its "theoretical" value, u_i is a zero-mean disturbance term and v_i is the inefficiency term, which is supposed to be nonnegative by its very nature.

A linear specification for C_i^* is often chosen. For example, the Cobb–Douglas cost function arises when C_i^* is a linear function of the logarithms of the input prices and the output quantities.

The *stochastic frontier* model is a linear model of the form $y_i = x_i\beta + \varepsilon_i$ in which the error term ε_i is the sum of u_i and v_i. A common postulate is that $u_i \sim N(0, \sigma_u^2)$ and $v_i \sim |N(0, \sigma_v^2)|$. If independence between u_i and v_i is also assumed, then it is possible to show that the density function of ε_i has the form:

$$f(\varepsilon_i) = \sqrt{\frac{2}{\pi}}\Phi\left(\frac{\lambda\varepsilon_i}{\sigma}\right)\frac{1}{\sigma}\phi\left(\frac{\varepsilon_i}{\sigma}\right) \tag{21.5}$$

where $\Phi(\cdot)$ and $\phi(\cdot)$ are, respectively, the distribution and density function of the standard normal, $\sigma = \sqrt{\sigma_u^2 + \sigma_v^2}$ and $\lambda = \frac{\sigma_u}{\sigma_v}$.

As a consequence, the log-likelihood for one observation takes the form (apart form an irrelevant constant)

$$\ell_t = \log\Phi\left(\frac{\lambda\varepsilon_i}{\sigma}\right) - \left[\log(\sigma) + \frac{\varepsilon_i^2}{2\sigma^2}\right]$$

Therefore, a Cobb–Douglas cost function with stochastic frontier is the model described by the following equations:

$$\begin{aligned}
\log C_i &= \log C_i^* + \varepsilon_i \\
\log C_i^* &= c + \sum_{j=1}^{m} \beta_j \log y_{ij} + \sum_{j=1}^{n} \alpha_j \log p_{ij} \\
\varepsilon_i &= u_i + v_i \\
u_i &\sim N(0, \sigma_u^2) \\
v_i &\sim \left| N(0, \sigma_v^2) \right|
\end{aligned}$$

In most cases, one wants to ensure that the homogeneity of the cost function with respect to the prices holds by construction. Since this requirement is equivalent to $\sum_{j=1}^{n} \alpha_j = 1$, the above equation for C_i^* can be rewritten as

$$\log C_i - \log p_{in} = c + \sum_{j=1}^{m} \beta_j \log y_{ij} + \sum_{j=2}^{n} \alpha_j (\log p_{ij} - \log p_{in}) + \varepsilon_i \tag{21.6}$$

The above equation could be estimated by OLS, but it would suffer from two drawbacks: first, the OLS estimator for the intercept c is inconsistent because the disturbance term has a non-zero expected value; second, the OLS estimators for the other parameters are consistent, but inefficient in view of the non-normality of ε_i. Both issues can be addressed by estimating (21.6) by maximum likelihood. Nevertheless, OLS estimation is a quick and convenient way to provide starting values for the MLE algorithm.

Example 21.1 shows how to implement the model described so far. The `banks91` file contains part of the data used in Lucchetti, Papi and Zazzaro (2001).

The script in example 21.1 is relatively easy to modify to show how one can use vectors (that is, 1-dimensional matrices) for storing the parameters to optimize: example 21.2 holds essentially the same script in which the parameters of the cost function are stored together in a vector. Of course, this makes also possible to use variable lists and other refinements which make the code more compact and readable.

21.4 GARCH models

GARCH models are handled by gretl via a native function. However, it is instructive to see how they can be estimated through the `mle` command.[2]

The following equations provide the simplest example of a GARCH(1,1) model:

$$\begin{aligned}
y_t &= \mu + \varepsilon_t \\
\varepsilon_t &= u_t \cdot \sigma_t \\
u_t &\sim N(0, 1) \\
h_t &= \omega + \alpha \varepsilon_{t-1}^2 + \beta h_{t-1}.
\end{aligned}$$

Since the variance of y_t depends on past values, writing down the log-likelihood function is not simply a matter of summing the log densities for individual observations. As is common in time series models, y_t cannot be considered independent of the other observations in our sample, and consequently the density function for the whole sample (the joint density for all observations) is not just the product of the marginal densities.

Maximum likelihood estimation, in these cases, is achieved by considering *conditional* densities, so what we maximize is a conditional likelihood function. If we define the information set at time t as

$$F_t = \{y_t, y_{t-1}, \ldots\},$$

[2] The `gig` addon, which handles other variants of conditionally heteroskedastic models, uses `mle` as its internal engine.

Example 21.1: Estimation of stochastic frontier cost function (with scalar parameters)

```
open banks91.gdt

# transformations
series cost = ln(VC)
series q1 = ln(Q1)
series q2 = ln(Q2)
series p1 = ln(P1)
series p2 = ln(P2)
series p3 = ln(P3)

# Cobb-Douglas cost function with homogeneity restrictions
# (for initialization)
genr rcost = cost - p1
genr rp2 = p2 - p1
genr rp3 = p3 - p1

ols rcost const q1 q2 rp2 rp3

# Cobb-Douglas cost function with homogeneity restrictions
# and inefficiency

scalar b0 = $coeff(const)
scalar b1 = $coeff(q1)
scalar b2 = $coeff(q2)
scalar b3 = $coeff(rp2)
scalar b4 = $coeff(rp3)

scalar su = 0.1
scalar sv = 0.1

mle logl = ln(cnorm(e*lambda/ss)) - (ln(ss) + 0.5*(e/ss)^2)
  scalar ss = sqrt(su^2 + sv^2)
  scalar lambda = su/sv
  series e = rcost - b0*const - b1*q1 - b2*q2 - b3*rp2 - b4*rp3
  params b0 b1 b2 b3 b4 su sv
end mle
```

Example 21.2: Estimation of stochastic frontier cost function (with matrix parameters)

```
open banks91.gdt

# transformations
series cost = ln(VC)
series q1 = ln(Q1)
series q2 = ln(Q2)
series p1 = ln(P1)
series p2 = ln(P2)
series p3 = ln(P3)

# Cobb-Douglas cost function with homogeneity restrictions
# (for initialization)
genr rcost = cost - p1
genr rp2 = p2 - p1
genr rp3 = p3 - p1
list X = const q1 q2 rp2 rp3

ols rcost X
X = const q1 q2 rp2 rp3
# Cobb-Douglas cost function with homogeneity restrictions
# and inefficiency

matrix b = $coeff
scalar su = 0.1
scalar sv = 0.1

mle logl = ln(cnorm(e*lambda/ss)) - (ln(ss) + 0.5*(e/ss)^2)
  scalar ss = sqrt(su^2 + sv^2)
  scalar lambda = su/sv
  series e = rcost - lincomb(X, b)
  params b su sv
end mle
```

then the density of y_t conditional on F_{t-1} is normal:

$$y_t|F_{t-1} \sim N[\mu, h_t].$$

By means of the properties of conditional distributions, the joint density can be factorized as follows

$$f(y_t, y_{t-1}, \ldots) = \left[\prod_{t=1}^{T} f(y_t|F_{t-1})\right] \cdot f(y_0)$$

If we treat y_0 as fixed, then the term $f(y_0)$ does not depend on the unknown parameters, and therefore the conditional log-likelihood can then be written as the sum of the individual contributions as

$$\ell(\mu, \omega, \alpha, \beta) = \sum_{t=1}^{T} \ell_t \tag{21.7}$$

where

$$\ell_t = \log\left[\frac{1}{\sqrt{h_t}}\phi\left(\frac{y_t - \mu}{\sqrt{h_t}}\right)\right] = -\frac{1}{2}\left[\log(h_t) + \frac{(y_t - \mu)^2}{h_t}\right]$$

The following script shows a simple application of this technique, which uses the data file `djclose`; it is one of the example dataset supplied with gretl and contains daily data from the Dow Jones stock index.

```
open djclose

series y = 100*ldiff(djclose)

scalar mu = 0.0
scalar omega = 1
scalar alpha = 0.4
scalar beta = 0.0

mle ll = -0.5*(log(h) + (e^2)/h)
  series e = y - mu
  series h = var(y)
  series h = omega + alpha*(e(-1))^2 + beta*h(-1)
  params mu omega alpha beta
end mle
```

21.5 Analytical derivatives

Computation of the score vector is essential for the working of the BFGS method. In all the previous examples, no explicit formula for the computation of the score was given, so the algorithm was fed numerically evaluated gradients. Numerical computation of the score for the i-th parameter is performed via a finite approximation of the derivative, namely

$$\frac{\partial \ell(\theta_1, \ldots, \theta_n)}{\partial \theta_i} \simeq \frac{\ell(\theta_1, \ldots, \theta_i + h, \ldots, \theta_n) - \ell(\theta_1, \ldots, \theta_i - h, \ldots, \theta_n)}{2h}$$

where h is a small number.

In many situations, this is rather efficient and accurate. A better approximation to the true derivative may be obtained by forcing `mle` to use a technique known as *Richardson Extrapolation*, which gives extremely precise results, but is considerably more CPU-intensive. This feature may be turned on by using the `set` command as in

```
set bfgs_richardson on
```

However, one might want to avoid the approximation and specify an exact function for the derivatives. As an example, consider the following script:

```
nulldata 1000

genr x1 = normal()
genr x2 = normal()
genr x3 = normal()

genr ystar = x1 + x2 + x3 + normal()
genr y = (ystar > 0)

scalar b0 = 0
scalar b1 = 0
scalar b2 = 0
scalar b3 = 0

mle logl = y*ln(P) + (1-y)*ln(1-P)
  series ndx = b0 + b1*x1 + b2*x2 + b3*x3
  series P = cnorm(ndx)
  params b0 b1 b2 b3
end mle --verbose
```

Here, 1000 data points are artificially generated for an ordinary probit model:[3] y_t is a binary variable, which takes the value 1 if $y_t^* = \beta_1 x_{1t} + \beta_2 x_{2t} + \beta_3 x_{3t} + \varepsilon_t > 0$ and 0 otherwise. Therefore, $y_t = 1$ with probability $\Phi(\beta_1 x_{1t} + \beta_2 x_{2t} + \beta_3 x_{3t}) = \pi_t$. The probability function for one observation can be written as

$$P(y_t) = \pi_t^{y_t}(1 - \pi_t)^{1-y_t}$$

Since the observations are independent and identically distributed, the log-likelihood is simply the sum of the individual contributions. Hence

$$\ell = \sum_{t=1}^{T} y_t \log(\pi_t) + (1 - y_t)\log(1 - \pi_t)$$

The `--verbose` switch at the end of the `end mle` statement produces a detailed account of the iterations done by the BFGS algorithm.

In this case, numerical differentiation works rather well; nevertheless, computation of the analytical score is straightforward, since the derivative $\frac{\partial \ell}{\partial \beta_i}$ can be written as

$$\frac{\partial \ell}{\partial \beta_i} = \frac{\partial \ell}{\partial \pi_t} \cdot \frac{\partial \pi_t}{\partial \beta_i}$$

via the chain rule, and it is easy to see that

$$\begin{aligned} \frac{\partial \ell}{\partial \pi_t} &= \frac{y_t}{\pi_t} - \frac{1 - y_t}{1 - \pi_t} \\ \frac{\partial \pi_t}{\partial \beta_i} &= \phi(\beta_1 x_{1t} + \beta_2 x_{2t} + \beta_3 x_{3t}) \cdot x_{it} \end{aligned}$$

The `mle` block in the above script can therefore be modified as follows:

```
mle logl = y*ln(P) + (1-y)*ln(1-P)
  series ndx = b0 + b1*x1 + b2*x2 + b3*x3
  series P = cnorm(ndx)
  series m = dnorm(ndx)*(y/P - (1-y)/(1-P))
  deriv b0 = m
  deriv b1 = m*x1
  deriv b2 = m*x2
  deriv b3 = m*x3
end mle --verbose
```

[3]Again, gretl does provide a native `probit` command (see section 32.1), but a probit model makes for a nice example here.

Note that the params statement has been replaced by a series of deriv statements; these have the double function of identifying the parameters over which to optimize and providing an analytical expression for their respective score elements.

21.6 Debugging ML scripts

We have discussed above the main sorts of statements that are permitted within an mle block, namely

- auxiliary commands to generate helper variables;
- deriv statements to specify the gradient with respect to each of the parameters; and
- a params statement to identify the parameters in case analytical derivatives are not given.

For the purpose of debugging ML estimators one additional sort of statement is allowed: you can print the value of a relevant variable at each step of the iteration. This facility is more restricted then the regular print command. The command word print should be followed by the name of just one variable (a scalar, series or matrix).

In the last example above a key variable named m was generated, forming the basis for the analytical derivatives. To track the progress of this variable one could add a print statement within the ML block, as in

```
series m = dnorm(ndx)*(y/P - (1-y)/(1-P))
print m
```

21.7 Using functions

The mle command allows you to estimate models that gretl does not provide natively: in some cases, it may be a good idea to wrap up the mle block in a user-defined function (see Chapter 13), so as to extend gretl's capabilities in a modular and flexible way.

As an example, we will take a simple case of a model that gretl does not yet provide natively: the zero-inflated Poisson model, or ZIP for short.[4] In this model, we assume that we observe a mixed population: for some individuals, the variable y_t is (conditionally on a vector of exogenous covariates x_t) distributed as a Poisson random variate; for some others, y_t is identically 0. The trouble is, we don't know which category a given individual belongs to.

For instance, suppose we have a sample of women, and the variable y_t represents the number of children that woman t has. There may be a certain proportion, α, of women for whom $y_t = 0$ with certainty (maybe out of a personal choice, or due to physical impossibility). But there may be other women for whom $y_t = 0$ just as a matter of chance — they haven't happened to have any children at the time of observation.

In formulae:

$$\begin{aligned} P(y_t = k|x_t) &= \alpha d_t + (1-\alpha)\left[e^{-\mu_t}\frac{\mu_t^{y_t}}{y_t!}\right] \\ \mu_t &= \exp(x_t\beta) \\ d_t &= \begin{cases} 1 & \text{for} \quad y_t = 0 \\ 0 & \text{for} \quad y_t > 0 \end{cases} \end{aligned}$$

Writing a mle block for this model is not difficult:

```
mle ll = logprob
  series xb = exp(b0 + b1 * x)
```

[4]The actual ZIP model is in fact a bit more general than the one presented here. The specialized version discussed in this section was chosen for the sake of simplicity. For futher details, see Greene (2003).

```
    series d = (y=0)
    series poiprob = exp(-xb) * xb^y / gamma(y+1)
    series logprob = (alpha>0) && (alpha<1) ? \
      log(alpha*d + (1-alpha)*poiprob) : NA
    params alpha b0 b1
  end mle -v
```

However, the code above has to be modified each time we change our specification by, say, adding an explanatory variable. Using functions, we can simplify this task considerably and eventually be able to write something easy like

```
  list X = const x
  zip(y, X)
```

Example 21.3: Zero-inflated Poisson Model — user-level function

```
/*
  user-level function: estimate the model and print out
  the results
*/
function void zip(series y, list X)
    matrix coef_stde = zip_estimate(y, X)
    printf "\nZero-inflated Poisson model:\n"
    string parnames = "alpha,"
    string parnames += varname(X)
    modprint coef_stde parnames
end function
```

Let's see how this can be done. First we need to define a function called `zip()` that will take two arguments: a dependent variable `y` and a list of explanatory variables `X`. An example of such function can be seen in script 21.3. By inspecting the function code, you can see that the actual estimation does not happen here: rather, the `zip()` function merely uses the built-in `modprint` command to print out the results coming from another user-written function, namely `zip_estimate()`.

The function `zip_estimate()` is not meant to be executed directly; it just contains the number-crunching part of the job, whose results are then picked up by the end function `zip()`. In turn, `zip_estimate()` calls other user-written functions to perform other tasks. The whole set of "internal" functions is shown in the panel 21.4.

All the functions shown in 21.3 and 21.4 can be stored in a separate `inp` file and executed once, at the beginning of our job, by means of the `include` command. Assuming the name of this script file is `zip_est.inp`, the following is an example script which (a) includes the script file, (b) generates a simulated dataset, and (c) performs the estimation of a ZIP model on the artificial data.

```
  set echo off
  set messages off

  # include the user-written functions
  include zip_est.inp

  # generate the artificial data
  nulldata 1000
  set seed 732237
  scalar truep = 0.2
  scalar b0 = 0.2
  scalar b1 = 0.5
```

Example 21.4: Zero-inflated Poisson Model — internal functions

```
/* compute log probabilities for the plain Poisson model */
function series ln_poi_prob(series y, list X, matrix beta)
    series xb = lincomb(X, beta)
    return -exp(xb) + y*xb - lngamma(y+1)
end function

/* compute log probabilities for the zero-inflated Poisson model */
function series ln_zip_prob(series y, list X, matrix beta, scalar p0)
    # check if the probability is in [0,1]; otherwise, return NA
    if (p0>1) || (p0<0)
        series ret = NA
    else
        series ret = ln_poi_prob(y, X, beta) + ln(1-p0)
        series ret = (y=0) ? ln(p0 + exp(ret)) : ret
    endif
    return ret
end function

/* do the actual estimation (silently) */
function matrix zip_estimate(series y, list X)
    # initialize alpha to a "sensible" value: half the frequency
    # of zeros in the sample
    scalar alpha = mean(y=0)/2
    # initialize the coeffs (we assume the first explanatory
    # variable is the constant here)
    matrix coef = zeros(nelem(X), 1)
    coef[1] = mean(y) / (1-alpha)
    # do the actual ML estimation
    mle ll = ln_zip_prob(y, X, coef, alpha)
        params alpha coef
    end mle --hessian --quiet
    return $coeff ~ $stderr
end function
```

```
series x = normal()
series y = (uniform()<truep) ? 0 : genpois(exp(b0 + b1*x))
list X = const x

# estimate the zero-inflated Poisson model
zip(y, X)
```

The results are as follows:

```
Zero-inflated Poisson model:

             coefficient   std. error   z-stat   p-value
  -----------------------------------------------------
  alpha       0.203069     0.0238035     8.531   1.45e-17 ***
  const       0.257014     0.0417129     6.161   7.21e-10 ***
  x           0.466657     0.0321235    14.53    8.17e-48 ***
```

A further step may then be creating a function package for accessing your new `zip()` function via gretl's graphical interface. For details on how to do this, see section 13.5.

21.8 Advanced use of `mle`: functions, analytical derivatives, algorithm choice

All the techniques decribed in the previous sections may be combined, and `mle` can be used for solving non-standard estimation problems (provided, of course, that one chooses maximum likelihood as the preferred inference method).

The strategy that, as of this writing, has proven most successful in designing scripts for this purpose is:

- Modularize your code as much as possible.
- Use analytical derivatives whenever possible.
- Choose your optimization method wisely.

In the rest of this section, we will expand on the probit example of section 21.5 to give the reader an idea of what a "heavy-duty" application of `mle` looks like. Most of the code fragments come from `mle-advanced.inp`, which is one of the sample scripts supplied with the standard installation of gretl (see under *File > Script files > Practice File*).

BFGS with and without analytical derivatives

The example in section 21.5 can be made more general by using matrices and user-written functions. Consider the following code fragment:

```
list X = const x1 x2 x3
matrix b = zeros(nelem(X),1)

mle logl = y*ln(P) + (1-y)*ln(1-P)
    series ndx = lincomb(X, b)
    series P = cnorm(ndx)
    params b
end mle
```

In this context, the fact that the model we are estimating has four explanatory variables is totally incidental: the code is written in such a way that we could change the content of the list X without having to make any other modification. This was made possible by:

1. gathering the parameters to estimate into a single vector b rather than using separate scalars;

2. using the `nelem()` function to initialize b, so that its dimension is kept track of automatically;

3. using the `lincomb()` function to compute the index function.

A parallel enhancement could be achieved in the case of analytically computed derivatives: since b is now a vector, `mle` expects the argument to the `deriv` keyword to be a matrix, in which each column is the partial derivative to the corresponding element of b. It is useful to re-write the score for the i-th observation as

$$\frac{\partial \ell_i}{\partial \beta} = m_i \mathbf{x}_i' \tag{21.8}$$

where m_i is the "signed Mills' ratio", that is

$$m_i = y_i \frac{\phi(\mathbf{x}_i'\beta)}{\Phi(\mathbf{x}_i'\beta)} - (1 - y_i)\frac{\phi(\mathbf{x}_i'\beta)}{1 - \Phi(\mathbf{x}_i'\beta)},$$

which was computed in section 21.5 via

```
series P = cnorm(ndx)
series m = dnorm(ndx)*(y/P - (1-y)/(1-P))
```

Here, we will code it in a somewhat terser way as

```
series m = y ? invmills(-ndx) : -invmills(ndx)
```

and make use of the conditional assignment operator and of the specialized function `invmills()` for efficiency. Building the score matrix is now easily achieved via

```
mle logl = y*ln(P) + (1-y)*ln(1-P)
    series ndx = lincomb(X, b)
    series P = cnorm(ndx)
    series m = y ? invmills(-ndx) : -invmills(ndx)
    matrix mX = {X}
    deriv b = mX .* {m}
end mle
```

in which the `{}` operator was used to turn series and lists into matrices (see chapter 15). However, proceeding in this way for more complex models than probit may imply inserting into the `mle` block a long series of instructions; the example above merely happens to be short because the score matrix for the probit model is very easy to write in matrix form.

A better solution is writing a user-level function to compute the score and using that inside the `mle` block, as in

```
function matrix score(matrix b, series y, list X)
    series ndx = lincomb(X, b)
    series m = y ? invmills(-ndx) : -invmills(ndx)
    return {m} .* {X}
end function

[...]

mle logl = y*ln(P) + (1-y)*ln(1-P)
    series ndx = lincomb(X, b)
    series P = cnorm(ndx)
    deriv b = score(b, y, X)
end mle
```

In this way, no matter how complex the computation of the score is, the `mle` block remains nicely compact.

Newton's method and the analytical Hessian

Since version 1.9.7, gretl offers the user the option of using Newton's method for maximizing the log-likelihood. In terms of the notation used in section 21.1, the direction for updating the inital parameter vector θ_0 is given by

$$d\left[\mathbf{g}(\theta_0)\right] = -\lambda \mathbf{H}(\theta_0)^{-1}\mathbf{g}(\theta_0), \tag{21.9}$$

where $\mathbf{H}(\theta)$ is the Hessian of the total loglikelihood computed at θ and $0 < \lambda < 1$ is a scalar called the *step length*.

The above expression makes a few points clear:

1. At each step, it must be possible to compute not only the score $\mathbf{g}(\theta)$, but also its derivative $\mathbf{H}(\theta)$;
2. the matrix $\mathbf{H}(\theta)$ should be nonsingular;
3. it is assumed that for some positive value of λ, $\ell(\theta_1) > \ell(\theta_0)$; in other words, that going in the direction $d\left[\mathbf{g}(\theta_0)\right]$ leads upwards for some step length.

The strength of Newton's method lies in the fact that, if the loglikelihood is globally concave, then (21.9) enjoys certain optimality properties and the number of iterations required to reach the maximum is often much smaller than it would be with other methods, such as BFGS. However, it may have some disadvantages: for a start, the Hessian $\mathbf{H}(\theta)$ may be difficult or very expensive to compute; moreover, the loglikelihood may not be globally concave, so for some values of θ, the matrix $\mathbf{H}(\theta)$ is not negative definite or perhaps even singular. Those cases are handled by gretl's implementation of Newton's algorithm by means of several heuristic techniques[5], but a number of adverse consequences may occur, which range from *longer* computation time for optimization to non-convergence of the algorithm.

As a consequence, using Newton's method is advisable only when the computation of the Hessian is not too CPU-intensive and the nature of the estimator is such that it is known in advance that the loglikelihood is globally concave. The probit models satisfies both requisites, so we will expand the preceding example to illustrate how to use Newton's method in gretl.

A first example may be given simply by issuing the command

```
set optimizer newton
```

before the `mle` block.[6] This will instruct gretl to use Newton's method instead of BFGS. If the `deriv` keyword is used, gretl will differentiate the score function numerically; otherwise, if the score has to be computed itself numerically, gretl will calculate $\mathbf{H}(\theta)$ by differentiating the loglikelihood numerically twice. The latter solution, though, is generally to be avoided, as may be extremely time-consuming and may yield imprecise results.

A much better option is to calculate the Hessian analytically and have gretl use its true value rather than a numerical approximation. In most cases, this is both much faster and numerically stable, but of course comes at the price of having to differentiate the loglikelihood twice to respect with the parameters and translate the resulting expressions into efficient hansl code.

Luckily, both tasks are relatively easy in the probit case: the matrix of second derivatives of ℓ_i may be written as

$$\frac{\partial^2 \ell_i}{\partial\beta\partial\beta'} = -m_i\left(m_i + \mathbf{x}_i'\beta\right)\mathbf{x}_i\mathbf{x}_i'$$

so the total Hessian is

$$\sum_{i=1}^{n}\frac{\partial^2 \ell_i}{\partial\beta\partial\beta'} = -X'\begin{bmatrix} w_1 & & & \\ & w_2 & & \\ & & \ddots & \\ & & & w_n \end{bmatrix} X \tag{21.10}$$

[5]The gist to it is that, if $\mathbf{H}$ is not negative definite, it is substituted by $k \cdot \mathrm{dg}(\mathbf{H}) + (1-k)\cdot\mathbf{H}$, where k is a suitable scalar; however, if you're interested in the precise details, you'll be much better off looking at the source code: the file you'll want to look at is `lib/src/gretl_bfgs.c`.

[6]To go back to BFGS, you use `set optimizer bfgs`.

where $w_i = m_i\left(m_i + \mathbf{x}_i'\beta\right)$. It can be shown that $w_i > 0$, so the Hessian is guaranteed to be negative definite in all sensible cases and the conditions are ideal for applying Newton's method.

A hansl translation of equation (21.10) may look like

```
function void Hess(matrix *H, matrix b, series y, list X)
    /* computes the negative Hessian for a Probit model */
    series ndx = lincomb(X, b)
    series m = y ? invmills(-ndx) : -invmills(ndx)
    series w = m*(m+ndx)
    matrix mX = {X}
    H = (mX .* {w})'mX
end function
```

There are two characteristics worth noting of the function above. For a start, it doesn't return anything: the result of the computation is simply stored in the matrix pointed at by the first argument of the function. Second, the result is not the Hessian proper, but rather its negative. This function becomes usable from within an `mle` block by the keyword `hessian`. The syntax is

```
mle ...
    ...
    hessian funcname(&mat_addr, ...)
end mle
```

In other words, the `hessian` keyword must be followed by the call to a function whose first argument is a matrix pointer which is supposed to be filled with the *negative* of the Hessian at θ.

Another feature worth noting is that gretl does not perform any numerical check on whether the function computes the Hessian correctly or not. On the one hand, this means that you can trick `mle` into using alternatives to the Hessian and thereby implement other optimization methods. For example, if you substitute in equation 21.9 the Hessian $\mathbf{H}$ with the negative of the OPG matrix $-\mathbf{G}'\mathbf{G}$, as defined in (21.2), you get the so-called BHHH optimization method (see Berndt *et al.* (1974)). Again, the sample file `mle-advanced.inp` provides an example. On the other hand, you may want to perform a check of your analytically-computed $\mathbf{H}$ matrix versus a numerical approximation.

If you have a function that computes the score, this is relatively simple to do by using the `fdjac` function, briefly described in section 31.4, which computes a numerical approximation to a derivative. In practice, you need a function computing $\mathbf{g}(\theta)$ as a row vector and then use `fdjac` to differentiate it numerically with respect to θ. The result can then be compared to your analytically-computed Hessian. The code fragment below shows an example of how this can be done in the probit case:

```
function matrix totalscore(matrix *b, series y, list X)
    /* computes the total score */
    return sumc(score(b, y, X))
end function

function void check(matrix b, series y, list X)
    /* compares the analytical Hessian to its numerical
    approximation obtained via fdjac */
    matrix aH
    Hess(&aH, b, y, X) # stores the analytical Hessian into aH

    matrix nH = fdjac(b, "totalscore(&b, y, X)")
    nH = 0.5*(nH + nH') # force symmetry

    printf "Numerical Hessian\n%16.6f\n", nH
    printf "Analytical Hessian (negative)\n%16.6f\n", aH
    printf "Check (should be zero)\n%16.6f\n", nH + aH
end function
```

Chapter 22

GMM estimation

22.1 Introduction and terminology

The Generalized Method of Moments (GMM) is a very powerful and general estimation method, which encompasses practically all the parametric estimation techniques used in econometrics. It was introduced in Hansen (1982) and Hansen and Singleton (1982); an excellent and thorough treatment is given in chapter 17 of Davidson and MacKinnon (1993).

The basic principle on which GMM is built is rather straightforward. Suppose we wish to estimate a scalar parameter θ based on a sample $x_1, x_2, \ldots, x_T$. Let θ_0 indicate the "true" value of θ. Theoretical considerations (either of statistical or economic nature) may suggest that a relationship like the following holds:

$$E\left[x_t - g(\theta)\right] = 0 \Leftrightarrow \theta = \theta_0, \tag{22.1}$$

with $g(\cdot)$ a continuous and invertible function. That is to say, there exists a function of the data and the parameter, with the property that it has expectation zero if and only if it is evaluated at the true parameter value. For example, economic models with rational expectations lead to expressions like (22.1) quite naturally.

If the sampling model for the x_ts is such that some version of the Law of Large Numbers holds, then

$$\bar{X} = \frac{1}{T}\sum_{t=1}^{T} x_t \xrightarrow{\text{p}} g(\theta_0);$$

hence, since $g(\cdot)$ is invertible, the statistic

$$\hat{\theta} = g^{-1}(\bar{X}) \xrightarrow{\text{p}} \theta_0,$$

so $\hat{\theta}$ is a consistent estimator of θ. A different way to obtain the same outcome is to choose, as an estimator of θ, the value that minimizes the objective function

$$F(\theta) = \left[\frac{1}{T}\sum_{t=1}^{T}(x_t - g(\theta))\right]^2 = \left[\bar{X} - g(\theta)\right]^2; \tag{22.2}$$

the minimum is trivially reached at $\hat{\theta} = g^{-1}(\bar{X})$, since the expression in square brackets equals 0.

The above reasoning can be generalized as follows: suppose θ is an n-vector and we have m relations like

$$E\left[f_i(x_t, \theta)\right] = 0 \quad \text{for } i = 1 \ldots m, \tag{22.3}$$

where $E[\cdot]$ is a conditional expectation on a set of p variables z_t, called the *instruments*. In the above simple example, $m = 1$ and $f(x_t, \theta) = x_t - g(\theta)$, and the only instrument used is $z_t = 1$. Then, it must also be true that

$$E\left[f_i(x_t, \theta) \cdot z_{j,t}\right] = E\left[f_{i,j,t}(\theta)\right] = 0 \quad \text{for } i = 1 \ldots m \quad \text{and } j = 1 \ldots p; \tag{22.4}$$

equation (22.4) is known as an *orthogonality condition*, or *moment condition*. The GMM estimator is defined as the minimum of the quadratic form

$$F(\theta, W) = \bar{\mathbf{f}}' W \bar{\mathbf{f}}, \tag{22.5}$$

where $\bar{\mathbf{f}}$ is a $(1 \times m \cdot p)$ vector holding the average of the orthogonality conditions and W is some symmetric, positive definite matrix, known as the *weights* matrix. A necessary condition for the minimum to exist is the order condition $n \leq m \cdot p$.

The statistic

$$\hat{\theta} = \operatorname*{Argmin}_{\theta} F(\theta, W) \tag{22.6}$$

is a consistent estimator of θ whatever the choice of W. However, to achieve maximum asymptotic efficiency W must be proportional to the inverse of the long-run covariance matrix of the orthogonality conditions; if W is not known, a consistent estimator will suffice.

These considerations lead to the following empirical strategy:

1. Choose a positive definite W and compute the *one-step* GMM estimator $\hat{\theta}_1$. Customary choices for W are $I_{m \cdot p}$ or $I_m \otimes (Z'Z)^{-1}$.

2. Use $\hat{\theta}_1$ to estimate $V(f_{i,j,t}(\theta))$ and use its inverse as the weights matrix. The resulting estimator $\hat{\theta}_2$ is called the *two-step* estimator.

3. Re-estimate $V(f_{i,j,t}(\theta))$ by means of $\hat{\theta}_2$ and obtain $\hat{\theta}_3$; iterate until convergence. Asymptotically, these extra steps are unnecessary, since the two-step estimator is consistent and efficient; however, the iterated estimator often has better small-sample properties and should be independent of the choice of W made at step 1.

In the special case when the number of parameters n is equal to the total number of orthogonality conditions $m \cdot p$, the GMM estimator $\hat{\theta}$ is the same for any choice of the weights matrix W, so the first step is sufficient; in this case, the objective function is 0 at the minimum.

If, on the contrary, $n < m \cdot p$, the second step (or successive iterations) is needed to achieve efficiency, and the estimator so obtained can be very different, in finite samples, from the one-step estimator. Moreover, the value of the objective function at the minimum, suitably scaled by the number of observations, yields *Hansen's J statistic*; this statistic can be interpreted as a test statistic that has a χ^2 distribution with $m \cdot p - n$ degrees of freedom under the null hypothesis of correct specification. See Davidson and MacKinnon (1993, section 17.6) for details.

In the following sections we will show how these ideas are implemented in gretl through some examples.

22.2 GMM as Method of Moments

This section draws from a kind contribution by Alecos Papadopoulos, whom we thank.

A very simple illustration of GMM can be given by dropping the "G", via an example of the time-honored statistical technique known as "method of moments"; let's see how to estimate the parameters of a gamma distribution, which we also used as an example for ML estimation in section 21.2.

Assume that we have an i.i.d. sample of size T from a gamma distribution. The gamma density can be parameterized in terms of the two parameters p (shape) and θ (scale), both real and positive.[1] In order to estimate them by the method of moments, we need two moment conditions so that we have two equations and two unknowns (in the GMM jargon, this amounts to exact identification). The two relations we need are

$$E(x_i) = p \cdot \theta \qquad V(x_i) = p \cdot \theta^2$$

These will become our moment conditions; substituting the finite sample analogues of the theoretical moments we have

$$\bar{X} = \hat{p} \cdot \hat{\theta} \tag{22.7}$$
$$V(x_i) = \hat{p} \cdot \hat{\theta}^2 \tag{22.8}$$

[1] In section 21.2 we used a slightly different, perhaps more common, parametrization, employing $\theta = 1/\alpha$. We are switching to the shape/scale parametrization here for the sake of convenience.

Of course, the two equations above are easy to solve analytically, giving $\hat{\theta} = \frac{\hat{V}}{\bar{X}}$ and $\hat{p} = \frac{\bar{X}}{\hat{\theta}}$, ($\hat{V}$ being the sample variance of x_i), but it's instructive to see how the gmm command will solve this system of equations numerically.

We feed gretl the necessary ingredients for GMM estimation in a command block, starting with gmm and ending with end gmm. Three elements are compulsory within a gmm block:

1. one or more orthog statements
2. one weights statement
3. one params statement

The three elements should be given in the stated order.

The orthog statements are used to specify the orthogonality conditions. They must follow the syntax

```
orthog x ; Z
```

where x may be a series, matrix or list of series and Z may also be a series, matrix or list. Note the structure of the orthogonality condition: it is assumed that the term to the left of the semicolon represents a quantity that depends on the estimated parameters (and so must be updated in the process of iterative estimation), while the term on the right is a constant function of the data.

The weights statement is used to specify the initial weighting matrix and its syntax is straightforward.

The params statement specifies the parameters with respect to which the GMM criterion should be minimized; it follows the same logic and rules as in the mle and nls commands.

The minimum is found through numerical minimization via BFGS (see chapters 31 and 21). The progress of the optimization procedure can be observed by appending the --verbose switch to the end gmm line.

Equations 22.7 and 22.8 are not yet in the "moment condition" form required by the gmm command. We need to transform them and arrive at something looking like $E(e_{j,i}z_{j,i}) = 0$, with $j = 1 \ldots 2$. Therefore, we need two corresponding observable variables e_1 and e_2 with corresponding instruments z_1 and z_2 and tell gretl that $\hat{E}(e_j z_j) = 0$ must be satisfied (where we used the $\hat{E}(\cdot)$ notation to indicate sample moments).

If we define the instrument as a series of ones, and set $e_{1,i} = x_i - p\theta$, then we can re-write the first moment condition as

$$\hat{E}[(x_i - p\theta) \cdot 1] = 0.$$

This is in the form required by the gmm command: in the required input statement "orthog e ; z", e will be the variable on the left (defined as a series) and z will the variable to the right of the semicolon. Since $z_{1,i} = 1$ for all i, we can use the built-in series const for that.

For the second moment condition we have, analogously,

$$\hat{E}\left\{\left[(x_i - \bar{X})^2 - p\theta^2\right] \cdot 1\right\} = 0,$$

so that by setting $e_{2,i} = (x_i - \bar{X})^2 - p\theta^2$ and $z_2 = z_1$ we can re-write the second moment condition as $\hat{E}[e_{2,i} \cdot 1] = 0$.

The weighting matrix, which is required by the gmm command, can be set to any 2×2 positive definite matrix, since under exact identification the choice does not matter and its dimension is determined by the number of orthogonality conditions. Therefore, we'll use the identity matrix.

Example code follows:

```
# create an empty data set
nulldata 200

# fix a random seed
set seed 2207092

#generate a gamma random variable with, say, shape p = 3 and scale theta = 2
series x = randgen(G, 3, 2)

#declare and set some initial value for parameters p and theta
scalar p = 1
scalar theta =1

#create the weight matrix as the identity matrix
matrix W = I(2)

#declare the series to be used in the orthogonality conditions
series e1 = 0
series e2 = 0

gmm
    scalar m = mean(x)
    series e1 = x - p*theta
    series e2 = (x - m)^2 - p*theta^2
    orthog e1 ; const
    orthog e2 ; const
    weights W
    params p theta
end gmm
```

The corresponding output is

```
Model 1: 1-step GMM, using observations 1-200

             estimate    std. error      z      p-value
  -----------------------------------------------------
  p          3.09165     0.346565     8.921   4.63e-19 ***
  theta      1.89983     0.224418     8.466   2.55e-17 ***

  GMM criterion: Q = 4.97341e-28 (TQ = 9.94682e-26)
```

If we want to use the unbiased estimator for the sample variance, we'd have to adjust the second moment condition by substituting

```
series e2 = (x - m)^2 - p*theta^2
```

with

```
scalar adj = $nobs / ($nobs - 1)
series e2 = adj * (x - m)^2 - p*theta^2
```

with the corresponding slight change in the output:

```
Model 1: 1-step GMM, using observations 1-200

             estimate    std. error      z      p-value
  -----------------------------------------------------
  p          3.07619     0.344832     8.921   4.63e-19 ***
  theta      1.90937     0.225546     8.466   2.55e-17 ***

  GMM criterion: Q = 2.80713e-28 (TQ = 5.61426e-26)
```

One can observe tiny improvements in the point estimates, since both moved a tad closer to the true values. This, however, is just a small-sample effect and not something you should expect in larger samples.

22.3 OLS as GMM

Let us now move to an example that is closer to econometrics proper: the linear model $y_t = x_t\beta + u_t$. Although most of us are used to read it as the sum of a hazily defined "systematic part" plus an equally hazy "disturbance", a more rigorous interpretation of this familiar expression comes from the *hypothesis* that the conditional mean $E(y_t|x_t)$ is linear and the *definition* of u_t as $y_t - E(y_t|x_t)$.

From the definition of u_t, it follows that $E(u_t|x_t) = 0$. The following orthogonality condition is therefore available:

$$E\left[f(\beta)\right] = 0, \tag{22.9}$$

where $f(\beta) = (y_t - x_t\beta)x_t$. The definitions given in section 22.1 therefore specialize here to:

- θ is β;
- the instrument is x_t;
- $f_{i,j,t}(\theta)$ is $(y_t - x_t\beta)x_t = u_t x_t$; the orthogonality condition is interpretable as the requirement that the regressors should be uncorrelated with the disturbances;
- W can be any symmetric positive definite matrix, since the number of parameters equals the number of orthogonality conditions. Let's say we choose I.
- The function $F(\theta, W)$ is in this case

  $$F(\theta, W) = \left[\frac{1}{T}\sum_{t=1}^{T}(\hat{u}_t x_t)\right]^2$$

  and it is easy to see why OLS and GMM coincide here: the GMM objective function has the same minimizer as the objective function of OLS, the residual sum of squares. Note, however, that the two functions are not equal to one another: at the minimum, $F(\theta, W) = 0$ while the minimized sum of squared residuals is zero only in the special case of a perfect linear fit.

The code snippet contained in Example 22.1 uses gretl's `gmm` command to make the above operational. In example 22.1, the series `e` holds the "residuals" and the series `x` holds the regressor. If `x` had been a list (a matrix), the `orthog` statement would have generated one orthogonality condition for each element (column) of `x`.

Example 22.1: OLS via GMM

```
/* initialize stuff */
series e = 0
scalar beta = 0
matrix W = I(1)

/* proceed with estimation */
gmm
  series e = y - x*beta
  orthog e ; x
  weights W
  params beta
end gmm
```

22.4 TSLS as GMM

Moving closer to the proper domain of GMM, we now consider two-stage least squares (TSLS) as a case of GMM.

TSLS is employed in the case where one wishes to estimate a linear model of the form $y_t = X_t\beta + u_t$, but where one or more of the variables in the matrix X are potentially endogenous—correlated with the error term, u. We proceed by identifying a set of instruments, Z_t, which are explanatory for the endogenous variables in X but which are plausibly uncorrelated with u. The classic two-stage procedure is (1) regress the endogenous elements of X on Z; then (2) estimate the equation of interest, with the endogenous elements of X replaced by their fitted values from (1).

An alternative perspective is given by GMM. We define the residual $\hat{u}_t$ as $y_t - X_t\hat{\beta}$, as usual. But instead of relying on $E(u|X) = 0$ as in OLS, we base estimation on the condition $E(u|Z) = 0$. In this case it is natural to base the initial weighting matrix on the covariance matrix of the instruments. Example 22.2 presents a model from Stock and Watson's *Introduction to Econometrics*. The demand for cigarettes is modeled as a linear function of the logs of price and income; income is treated as exogenous while price is taken to be endogenous and two measures of tax are used as instruments. Since we have two instruments and one endogenous variable the model is over-identified.

In the GMM context, this happens when you have more orthogonality conditions than parameters to estimate. If so, asymptotic efficiency gains can be expected by iterating the procedure once or more. This is accomplished ny specifying, after the `end gmm` statement, two mutually exclusive options: `--two-step` or `--iterate`, whose meaning should be obvious. Note that, when the problem is over-identified, the weights matrix will influence the solution you get from the 1- and 2- step procedure.

☞ In cases other than one-step estimation the specified weights matrix will be overwritten with the *final* weights on completion of the `gmm` command. If you wish to execute more than one GMM block with a common starting-point it is therefore necessary to reinitialize the weights matrix between runs.

Partial output from this script is shown in 22.3. The estimated standard errors from GMM are robust by default; if we supply the `--robust` option to the `tsls` command we get identical results.[2]

After the `end gmm` statement two mutually exclusive options can be specified: `--two-step` or `--iterate`, whose meaning should be obvious.

22.5 Covariance matrix options

The covariance matrix of the estimated parameters depends on the choice of W through

$$\hat{\Sigma} = (J'WJ)^{-1}J'W\Omega WJ(J'WJ)^{-1} \tag{22.10}$$

where J is a Jacobian term

$$J_{ij} = \frac{\partial \bar{f}_i}{\partial \theta_j}$$

and Ω is the long-run covariance matrix of the orthogonality conditions.

Gretl computes J by numeric differentiation (there is no provision for specifying a user-supplied analytical expression for J at the moment). As for Ω, a consistent estimate is needed. The simplest choice is the sample covariance matrix of the f_ts:

$$\hat{\Omega}_0(\theta) = \frac{1}{T}\sum_{t=1}^{T} f_t(\theta)f_t(\theta)' \tag{22.11}$$

This estimator is robust with respect to heteroskedasticity, but not with respect to autocorrelation. A heteroskedasticity- and autocorrelation-consistent (HAC) variant can be obtained using

[2]The data file used in this example is available in the Stock and Watson package for gretl. See http://gretl.sourceforge.net/gretl_data.html.

Example 22.2: TSLS via GMM

```
open cig_ch10.gdt
# real avg price including sales tax
genr ravgprs = avgprs / cpi
# real avg cig-specific tax
genr rtax = tax / cpi
# real average total tax
genr rtaxs = taxs / cpi
# real average sales tax
genr rtaxso = rtaxs - rtax
# logs of consumption, price, income
genr lpackpc = log(packpc)
genr lravgprs = log(ravgprs)
genr perinc = income / (pop*cpi)
genr lperinc = log(perinc)
# restrict sample to 1995 observations
smpl --restrict year=1995
# Equation (10.16) by tsls
list xlist = const lravgprs lperinc
list zlist = const rtaxso rtax lperinc
tsls lpackpc xlist ; zlist --robust

# setup for gmm
matrix Z = { zlist }
matrix W = inv(Z'Z)
series e = 0
scalar b0 = 1
scalar b1 = 1
scalar b2 = 1

gmm e = lpackpc - b0 - b1*lravgprs - b2*lperinc
  orthog e ; Z
  weights W
  params b0 b1 b2
end gmm
```

Example 22.3: TSLS via GMM: partial output

```
Model 1: TSLS estimates using the 48 observations 1-48
Dependent variable: lpackpc
Instruments: rtaxso rtax
Heteroskedasticity-robust standard errors, variant HC0

      VARIABLE        COEFFICIENT        STDERROR       T STAT    P-VALUE

    const              9.89496          0.928758       10.654   <0.00001 ***
    lravgprs          -1.27742          0.241684       -5.286   <0.00001 ***
    lperinc            0.280405         0.245828        1.141    0.25401

Model 2: 1-step GMM estimates using the 48 observations 1-48
e = lpackpc - b0 - b1*lravgprs - b2*lperinc

      PARAMETER        ESTIMATE          STDERROR       T STAT    P-VALUE

    b0                 9.89496          0.928758       10.654   <0.00001 ***
    b1                -1.27742          0.241684       -5.286   <0.00001 ***
    b2                 0.280405         0.245828        1.141    0.25401

    GMM criterion = 0.0110046
```

the Bartlett kernel or similar. A univariate version of this is used in the context of the `lrvar()` function—see equation (9.1). The multivariate version is set out in equation (22.12).

$$\hat{\Omega}_k(\theta) = \frac{1}{T}\sum_{t=k}^{T-k}\left[\sum_{i=-k}^{k} w_i f_t(\theta) f_{t-i}(\theta)'\right], \tag{22.12}$$

Gretl computes the HAC covariance matrix by default when a GMM model is estimated on time series data. You can control the kernel and the bandwidth (that is, the value of k in 22.12) using the `set` command. See chapter 17 for further discussion of HAC estimation. You can also ask gretl *not* to use the HAC version by saying

```
set force_hc on
```

22.6 A real example: the Consumption Based Asset Pricing Model

To illustrate gretl's implementation of GMM, we will replicate the example given in chapter 3 of Hall (2005). The model to estimate is a classic application of GMM, and provides an example of a case when orthogonality conditions do not stem from statistical considerations, but rather from economic theory.

A rational individual who must allocate his income between consumption and investment in a financial asset must in fact choose the consumption path of his whole lifetime, since investment translates into future consumption. It can be shown that an optimal consumption path should satisfy the following condition:

$$pU'(c_t) = \delta^k E\left[r_{t+k}U'(c_{t+k})|\mathcal{F}_t\right], \tag{22.13}$$

where p is the asset price, $U(\cdot)$ is the individual's utility function, δ is the individual's subjective discount rate and r_{t+k} is the asset's rate of return between time t and time $t+k$. $\mathcal{F}_t$ is the *information set* at time t; equation (22.13) says that the utility "lost" at time t by purchasing the asset instead of consumption goods must be matched by a corresponding increase in the (discounted) future utility of the consumption financed by the asset's return. Since the future is uncertain, the individual considers his expectation, conditional on what is known at the time when the choice is made.

We have said nothing about the nature of the asset, so equation (22.13) should hold whatever asset we consider; hence, it is possible to build a system of equations like (22.13) for each asset whose price we observe.

If we are willing to believe that

- the economy as a whole can be represented as a single gigantic and immortal representative individual, and
- the function $U(x) = \frac{x^{\alpha}-1}{\alpha}$ is a faithful representation of the individual's preferences,

then, setting $k = 1$, equation (22.13) implies the following for any asset j:

$$E\left[\delta \frac{r_{j,t+1}}{p_{j,t}} \left(\frac{C_{t+1}}{C_t}\right)^{\alpha-1} \middle| \mathcal{F}_t\right] = 1, \tag{22.14}$$

where C_t is aggregate consumption and α and δ are the risk aversion and discount rate of the representative individual. In this case, it is easy to see that the "deep" parameters α and δ can be estimated via GMM by using

$$e_t = \delta \frac{r_{j,t+1}}{p_{j,t}} \left(\frac{C_{t+1}}{C_t}\right)^{\alpha-1} - 1$$

as the moment condition, while any variable known at time t may serve as an instrument.

In the example code given in 22.4, we replicate selected portions of table 3.7 in Hall (2005). The variable `consrat` is defined as the ratio of monthly consecutive real per capita consumption (services and nondurables) for the US, and `ewr` is the return-price ratio of a fictitious asset constructed by averaging all the stocks in the NYSE. The instrument set contains the constant and two lags of each variable.

The command `set force_hc on` on the second line of the script has the sole purpose of replicating the given example: as mentioned above, it forces gretl to compute the long-run variance of the orthogonality conditions according to equation (22.11) rather than (22.12).

We run `gmm` four times: one-step estimation for each of two initial weights matrices, then iterative estimation starting from each set of initial weights. Since the number of orthogonality conditions (5) is greater than the number of estimated parameters (2), the choice of initial weights should make a difference, and indeed we see fairly substantial differences between the one-step estimates (Models 1 and 2). On the other hand, iteration reduces these differences almost to the vanishing point (Models 3 and 4).

Part of the output is given in 22.5. It should be noted that the J test leads to a rejection of the hypothesis of correct specification. This is perhaps not surprising given the heroic assumptions required to move from the microeconomic principle in equation (22.13) to the aggregate system that is actually estimated.

22.7 Caveats

A few words of warning are in order: despite its ingenuity, GMM is possibly the most fragile estimation method in econometrics. The number of non-obvious choices one has to make when using GMM is high, and in finite samples each of these can have dramatic consequences on the eventual output. Some of the factors that may affect the results are:

1. Orthogonality conditions can be written in more than one way: for example, if $E(x_t - \mu) = 0$, then $E(x_t/\mu - 1) = 0$ holds too. It is possible that a different specification of the moment conditions leads to different results.

2. As with all other numerical optimization algorithms, weird things may happen when the objective function is nearly flat in some directions or has multiple minima. BFGS is usually quite good, but there is no guarantee that it always delivers a sensible solution, if one at all.

Example 22.4: Estimation of the Consumption Based Asset Pricing Model

```
open hall.gdt
set force_hc on

scalar alpha = 0.5
scalar delta = 0.5
series e = 0

list inst = const consrat(-1) consrat(-2) ewr(-1) ewr(-2)

matrix V0 = 100000*I(nelem(inst))
matrix Z = { inst }
matrix V1 = $nobs*inv(Z'Z)

gmm e = delta*ewr*consrat^(alpha-1) - 1
  orthog e ; inst
  weights V0
  params alpha delta
end gmm

gmm e = delta*ewr*consrat^(alpha-1) - 1
  orthog e ; inst
  weights V1
  params alpha delta
end gmm

gmm e = delta*ewr*consrat^(alpha-1) - 1
  orthog e ; inst
  weights V0
  params alpha delta
end gmm --iterate

gmm e = delta*ewr*consrat^(alpha-1) - 1
  orthog e ; inst
  weights V1
  params alpha delta
end gmm --iterate
```

Example 22.5: Estimation of the Consumption Based Asset Pricing Model — output

```
Model 1: 1-step GMM estimates using the 465 observations 1959:04-1997:12
e = d*ewr*consrat^(alpha-1) - 1

      PARAMETER        ESTIMATE          STDERROR       T STAT    P-VALUE

  alpha                 -3.14475          6.84439       -0.459    0.64590
  d                      0.999215         0.0121044     82.549   <0.00001 ***

  GMM criterion = 2778.08

Model 2: 1-step GMM estimates using the 465 observations 1959:04-1997:12
e = d*ewr*consrat^(alpha-1) - 1

      PARAMETER        ESTIMATE          STDERROR       T STAT    P-VALUE

  alpha                  0.398194         2.26359        0.176    0.86036
  d                      0.993180         0.00439367   226.048   <0.00001 ***

  GMM criterion = 14.247

Model 3: Iterated GMM estimates using the 465 observations 1959:04-1997:12
e = d*ewr*consrat^(alpha-1) - 1

      PARAMETER        ESTIMATE          STDERROR       T STAT    P-VALUE

  alpha                 -0.344325         2.21458       -0.155    0.87644
  d                      0.991566         0.00423620   234.070   <0.00001 ***

  GMM criterion = 5491.78
  J test: Chi-square(3) = 11.8103 (p-value 0.0081)

Model 4: Iterated GMM estimates using the 465 observations 1959:04-1997:12
e = d*ewr*consrat^(alpha-1) - 1

      PARAMETER        ESTIMATE          STDERROR       T STAT    P-VALUE

  alpha                 -0.344315         2.21359       -0.156    0.87639
  d                      0.991566         0.00423469   234.153   <0.00001 ***

  GMM criterion = 5491.78
  J test: Chi-square(3) = 11.8103 (p-value 0.0081)
```

3. The 1-step and, to a lesser extent, the 2-step estimators may be sensitive to apparently trivial details, like the re-scaling of the instruments. Different choices for the initial weights matrix can also have noticeable consequences.

4. With time-series data, there is no hard rule on the appropriate number of lags to use when computing the long-run covariance matrix (see section 22.5). Our advice is to go by trial and error, since results may be greatly influenced by a poor choice. Future versions of gretl will include more options on covariance matrix estimation.

One of the consequences of this state of things is that replicating various well-known published studies may be extremely difficult. Any non-trivial result is virtually impossible to reproduce unless all details of the estimation procedure are carefully recorded.

Chapter 23

Model selection criteria

23.1 Introduction

In some contexts the econometrician chooses between alternative models based on a formal hypothesis test. For example, one might choose a more general model over a more restricted one if the restriction in question can be formulated as a testable null hypothesis, and the null is rejected on an appropriate test.

In other contexts one sometimes seeks a criterion for model selection that somehow measures the balance between goodness of fit or likelihood, on the one hand, and parsimony on the other. The balancing is necessary because the addition of extra variables to a model cannot reduce the degree of fit or likelihood, and is very likely to increase it somewhat even if the additional variables are not truly relevant to the data-generating process.

The best known such criterion, for linear models estimated via least squares, is the adjusted R^2,

$$\bar{R}^2 = 1 - \frac{\mathrm{SSR}/(n-k)}{\mathrm{TSS}/(n-1)}$$

where n is the number of observations in the sample, k denotes the number of parameters estimated, and SSR and TSS denote the sum of squared residuals and the total sum of squares for the dependent variable, respectively. Compared to the ordinary coefficient of determination or unadjusted R^2,

$$R^2 = 1 - \frac{\mathrm{SSR}}{\mathrm{TSS}}$$

the "adjusted" calculation penalizes the inclusion of additional parameters, other things equal.

23.2 Information criteria

A more general criterion in a similar spirit is Akaike's (1974) "Information Criterion" (AIC). The original formulation of this measure is

$$\mathrm{AIC} = -2\ell(\hat{\theta}) + 2k \tag{23.1}$$

where $\ell(\hat{\theta})$ represents the maximum loglikelihood as a function of the vector of parameter estimates, $\hat{\theta}$, and k (as above) denotes the number of "independently adjusted parameters within the model." In this formulation, with AIC negatively related to the likelihood and positively related to the number of parameters, the researcher seeks the minimum AIC.

The AIC can be confusing, in that several variants of the calculation are "in circulation." For example, Davidson and MacKinnon (2004) present a simplified version,

$$\mathrm{AIC} = \ell(\hat{\theta}) - k$$

which is just -2 times the original: in this case, obviously, one wants to maximize AIC.

In the case of models estimated by least squares, the loglikelihood can be written as

$$\ell(\hat{\theta}) = -\frac{n}{2}(1 + \log 2\pi - \log n) - \frac{n}{2}\log \mathrm{SSR} \tag{23.2}$$

Substituting (23.2) into (23.1) we get

$$\mathrm{AIC} = n(1 + \log 2\pi - \log n) + n \log \mathrm{SSR} + 2k$$

which can also be written as

$$\text{AIC} = n \log\left(\frac{\text{SSR}}{n}\right) + 2k + n(1 + \log 2\pi) \tag{23.3}$$

Some authors simplify the formula for the case of models estimated via least squares. For instance, William Greene writes

$$\text{AIC} = \log\left(\frac{\text{SSR}}{n}\right) + \frac{2k}{n} \tag{23.4}$$

This variant can be derived from (23.3) by dividing through by n and subtracting the constant $1 + \log 2\pi$. That is, writing AIC_G for the version given by Greene, we have

$$\text{AIC}_G = \frac{1}{n}\text{AIC} - (1 + \log 2\pi)$$

Finally, Ramanathan gives a further variant:

$$\text{AIC}_R = \left(\frac{\text{SSR}}{n}\right) e^{2k/n}$$

which is the exponential of the one given by Greene.

Gretl began by using the Ramanathan variant, but since version 1.3.1 the program has used the original Akaike formula (23.1), and more specifically (23.3) for models estimated via least squares.

Although the Akaike criterion is designed to favor parsimony, arguably it does not go far enough in that direction. For instance, if we have two nested models with $k - 1$ and k parameters respectively, and if the null hypothesis that parameter k equals 0 is true, in large samples the AIC will nonetheless tend to select the less parsimonious model about 16 percent of the time (see Davidson and MacKinnon, 2004, chapter 15).

An alternative to the AIC which avoids this problem is the Schwarz (1978) "Bayesian information criterion" (BIC). The BIC can be written (in line with Akaike's formulation of the AIC) as

$$\text{BIC} = -2\ell(\hat{\theta}) + k \log n$$

The multiplication of k by $\log n$ in the BIC means that the penalty for adding extra parameters grows with the sample size. This ensures that, asymptotically, one will not select a larger model over a correctly specified parsimonious model.

A further alternative to AIC, which again tends to select more parsimonious models than AIC, is the Hannan–Quinn criterion or HQC (Hannan and Quinn, 1979). Written consistently with the formulations above, this is

$$\text{HQC} = -2\ell(\hat{\theta}) + 2k \log \log n$$

The Hannan–Quinn calculation is based on the law of the iterated logarithm (note that the last term is the log of the log of the sample size). The authors argue that their procedure provides a "strongly consistent estimation procedure for the order of an autoregression", and that "compared to other strongly consistent procedures this procedure will underestimate the order to a lesser degree."

Gretl reports the AIC, BIC and HQC (calculated as explained above) for most sorts of models. The key point in interpreting these values is to know whether they are calculated such that smaller values are better, or such that larger values are better. In gretl, smaller values are better: one wants to minimize the chosen criterion.

Chapter 24

Time series filters

In addition to the usual application of lags and differences, gretl provides fractional differencing and various filters commonly used in macroeconomics for trend-cycle decomposition: notably the Hodrick–Prescott filter (Hodrick and Prescott, 1997), the Baxter–King bandpass filter (Baxter and King, 1999) and the Butterworth filter (Butterworth, 1930).

24.1 Fractional differencing

The concept of differencing a time series d times is pretty obvious when d is an integer; it may seem odd when d is fractional. However, this idea has a well-defined mathematical content: consider the function

$$f(z) = (1 - z)^{-d},$$

where z and d are real numbers. By taking a Taylor series expansion around $z = 0$, we see that

$$f(z) = 1 + dz + \frac{d(d+1)}{2} z^2 + \cdots$$

or, more compactly,

$$f(z) = 1 + \sum_{i=1}^{\infty} \psi_i z^i$$

with

$$\psi_k = \frac{\prod_{i=1}^{k}(d + i - 1)}{k!} = \psi_{k-1}\frac{d + k - 1}{k}$$

The same expansion can be used with the lag operator, so that if we defined

$$Y_t = (1 - L)^{0.5} X_t$$

this could be considered shorthand for

$$Y_t = X_t - 0.5X_{t-1} - 0.125X_{t-2} - 0.0625X_{t-3} - \cdots$$

In gretl this transformation can be accomplished by the syntax

```
genr Y = fracdiff(X,0.5)
```

24.2 The Hodrick–Prescott filter

This filter is accessed using the `hpfilt()` function, which takes as its first argument the name of the variable to be processed. (A further optional argument is explained below.)

A time series y_t may be decomposed into a trend or growth component g_t and a cyclical component c_t.

$$y_t = g_t + c_t, \quad t = 1, 2, \ldots, T$$

The Hodrick–Prescott filter effects such a decomposition by minimizing the following:

$$\sum_{t=1}^{T} (y_t - g_t)^2 + \lambda \sum_{t=2}^{T-1} \left((g_{t+1} - g_t) - (g_t - g_{t-1})\right)^2 .$$

The first term above is the sum of squared cyclical components $c_t = y_t - g_t$. The second term is a multiple λ of the sum of squares of the trend component's second differences. This second term penalizes variations in the growth rate of the trend component: the larger the value of λ, the higher is the penalty and hence the smoother the trend series.

Note that the `hpfilt` function in gretl produces the cyclical component, c_t, of the original series. If you want the smoothed trend you can subtract the cycle from the original:

```
genr ct = hpfilt(yt)
genr gt = yt - ct
```

Hodrick and Prescott (1997) suggest that a value of $\lambda = 1600$ is reasonable for quarterly data. The default value in gretl is 100 times the square of the data frequency (which, of course, yields 1600 for quarterly data). The value can be adjusted using an optional second argument to `hpfilt()`, as in

```
genr ct = hpfilt(yt, 1300)
```

24.3 The Baxter and King filter

This filter is accessed using the `bkfilt()` function, which again takes the name of the variable to be processed as its first argument. The operation of the filter can be controlled via three further optional argument.

Consider the spectral representation of a time series y_t:

$$y_t = \int_{-\pi}^{\pi} e^{i\omega} \mathrm{d}Z(\omega)$$

To extract the component of y_t that lies between the frequencies $\underline{\omega}$ and $\overline{\omega}$ one could apply a bandpass filter:

$$c_t^* = \int_{-\pi}^{\pi} F^*(\omega) e^{i\omega} \mathrm{d}Z(\omega)$$

where $F^*(\omega) = 1$ for $\underline{\omega} < |\omega| < \overline{\omega}$ and 0 elsewhere. This would imply, in the time domain, applying to the series a filter with an infinite number of coefficients, which is undesirable. The Baxter and King bandpass filter applies to y_t a finite polynomial in the lag operator $A(L)$:

$$c_t = A(L) y_t$$

where $A(L)$ is defined as

$$A(L) = \sum_{i=-k}^{k} a_i L^i$$

The coefficients a_i are chosen such that $F(\omega) = A(e^{i\omega})A(e^{-i\omega})$ is the best approximation to $F^*(\omega)$ for a given k. Clearly, the higher k the better the approximation is, but since $2k$ observations have to be discarded, a compromise is usually sought. Moreover, the filter has also other appealing theoretical properties, among which the property that $A(1) = 0$, so a series with a single unit root is made stationary by application of the filter.

In practice, the filter is normally used with monthly or quarterly data to extract the "business cycle" component, namely the component between 6 and 36 quarters. Usual choices for k are 8 or 12 (maybe higher for monthly series). The default values for the frequency bounds are 8 and 32, and the default value for the approximation order, k, is 8. You can adjust these values using the full form of `bkfilt()`, which is

`bkfilt(`*seriesname*`,` *f1*`,` *f2*`,` *k*`)`

where *f1* and *f2* represent the lower and upper frequency bounds respectively.

24.4 The Butterworth filter

The Butterworth filter (Butterworth, 1930) is an approximation to an "ideal" square-wave filter. The ideal filter divides the spectrum of a time series into a pass-band (frequencies less than some chosen $\omega^\star$ for a low-pass filter, or frequencies greater than $\omega^\star$ for high-pass) and a stop-band; the gain is 1 for the pass-band and 0 for the stop-band. The ideal filter is unattainable in practice since it would require an infinite number of coefficients, but the Butterworth filter offers a remarkably good approximation. This filter is derived and persuasively advocated by Pollock (2000).

For data y, the filtered sequence x is given by

$$x = y - \lambda\Sigma Q(M + \lambda Q'\Sigma Q)^{-1}Q'y \tag{24.1}$$

where

$$\Sigma = \{2I_T - (L_T + L_T^{-1})\}^{T-2} \quad \text{and} \quad M = \{2I_T + (L_T + L_T^{-1})\}^T$$

I_T denotes the identity matrix of order T; $L_T = [e_1, e_2, \ldots, e_{T-1}, 0]$ is the finite-sample matrix version of the lag operator; and Q is defined such that pre-multiplication of a T-vector of data by Q' of order $(T-2) \times T$ produces the second differences of the data. The matrix product

$$Q'\Sigma Q = \{2I_T - (L_T + L_T^{-1})\}^T$$

is a Toeplitz matrix.

The behavior of the Butterworth filter is governed by two parameters: the frequency cutoff $\omega^\star$ and an integer order, n, which determines the number of coefficients used. The λ that appears in (24.1) is $\tan(\omega^\star/2)^{-2n}$. Higher values of n produce a better approximation to the ideal filter in principle (i.e. a sharper cut between the pass-band and the stop-band) but there is a downside: with a greater number of coefficients numerical instability may be an issue, and the influence of the initial values in the sample may be exaggerated.

In gretl the Butterworth filter is implemented by the `bwfilt()` function,[1] which takes three arguments: the series to filter, the order n and the frequency cutoff, $\omega^\star$, expressed in degrees. The cutoff value must be greater than 0 and less than 180. This function operates as a low-pass filter; for the high-pass variant, subtract the filtered series from the original, as in

```
series bwcycle = y - bwfilt(y, 8, 67)
```

Pollock recommends that the parameters of the Butterworth filter be tuned to the data: one should examine the periodogram of the series in question (possibly after removal of a polynomial trend) in search of a "dead spot" of low power between the frequencies one wishes to exclude and the frequencies one wishes to retain. If $\omega^\star$ is placed in such a dead spot then the job of separation can be done with a relatively small n, hence avoiding numerical problems. By way of illustration, consider the periodogram for quarterly observations on new cars sales in the US,[2] 1975:1 to 1990:4 (the upper panel in Figure 24.1).

A seasonal pattern is clearly visible in the periodogram, centered at an angle of 90° or 4 periods. If we set $\omega^\star = 68°$ (or thereabouts) we should be able to excise the seasonality quite cleanly using $n = 8$. The result is shown in the lower panel of the Figure, along with the frequency response or gain plot for the chosen filter. Note the smooth and reasonably steep drop-off in gain centered on the nominal cutoff of $68° \approx 3\pi/8$.

The apparatus that supports this sort of analysis in the gretl GUI can be found under the Variable menu in the main window: the items Periodogram and Filter. In the periodogram dialog box you have the option of expressing the frequency axis in degrees, which is helpful when selecting a Butterworth filter; and in the Butterworth filter dialog you have the option of plotting the frequency response as well as the smoothed series and/or the residual or cycle.

---

[1] The code for this filter is based on D. S. G. Pollock's programs IDEOLOG and DETREND. The Pascal source code for the former is available from `http://www.le.ac.uk/users/dsgp1` and the C sources for the latter were kindly made available to us by the author.

[2] This is the variable QNC from the Ramanathan data file `data9-7`.

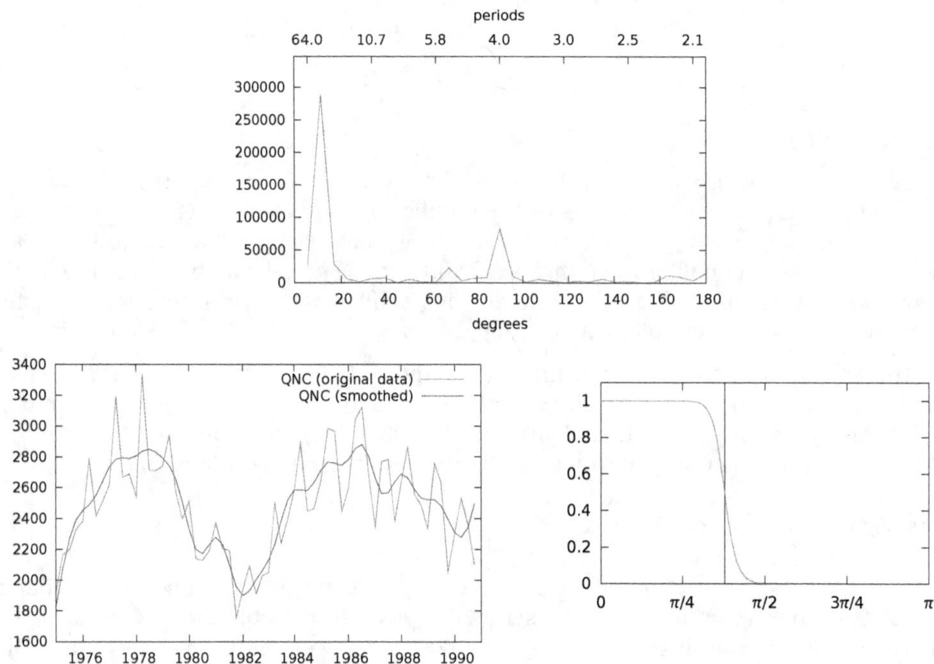

Figure 24.1: The Butterworth filter applied

Chapter 25

Univariate time series models

25.1 Introduction

Time series models are discussed in this chapter and the next two. Here we concentrate on ARIMA models, unit root tests, and GARCH. The following chapter deals with VARs, and chapter 27 with cointegration and error correction.

25.2 ARIMA models

Representation and syntax

The `arma` command performs estimation of AutoRegressive, Integrated, Moving Average (ARIMA) models. These are models that can be written in the form

$$\phi(L)y_t = \theta(L)\epsilon_t \tag{25.1}$$

where $\phi(L)$, and $\theta(L)$ are polynomials in the lag operator, L, defined such that $L^n x_t = x_{t-n}$, and ϵ_t is a white noise process. The exact content of y_t, of the AR polynomial $\phi()$, and of the MA polynomial $\theta()$, will be explained in the following.

Mean terms

The process y_t as written in equation (25.1) has, without further qualifications, mean zero. If the model is to be applied to real data, it is necessary to include some term to handle the possibility that y_t has non-zero mean. There are two possible ways to represent processes with nonzero mean: one is to define μ_t as the *unconditional* mean of y_t, namely the central value of its marginal distribution. Therefore, the series $\tilde{y}_t = y_t - \mu_t$ has mean 0, and the model (25.1) applies to $\tilde{y}_t$. In practice, assuming that μ_t is a linear function of some observable variables x_t, the model becomes

$$\phi(L)(y_t - x_t\beta) = \theta(L)\epsilon_t \tag{25.2}$$

This is sometimes known as a "regression model with ARMA errors"; its structure may be more apparent if we represent it using two equations:

$$\begin{aligned} y_t &= x_t\beta + u_t \\ \phi(L)u_t &= \theta(L)\epsilon_t \end{aligned}$$

The model just presented is also sometimes known as "ARMAX" (ARMA + eXogenous variables). It seems to us, however, that this label is more appropriately applied to a different model: another way to include a mean term in (25.1) is to base the representation on the *conditional* mean of y_t, that is the central value of the distribution of y_t *given its own past*. Assuming, again, that this can be represented as a linear combination of some observable variables z_t, the model would expand to

$$\phi(L)y_t = z_t\gamma + \theta(L)\epsilon_t \tag{25.3}$$

The formulation (25.3) has the advantage that γ can be immediately interpreted as the vector of marginal effects of the z_t variables on the conditional mean of y_t. And by adding lags of z_t to this specification one can estimate *Transfer Function models* (which generalize ARMA by adding the effects of exogenous variable distributed across time).

Gretl provides a way to estimate both forms. Models written as in (25.2) are estimated by maximum likelihood; models written as in (25.3) are estimated by conditional maximum likelihood. (For more on these options see the section on "Estimation" below.)

In the special case when $x_t = z_t = 1$ (that is, the models include a constant but no exogenous variables) the two specifications discussed above reduce to

$$\phi(L)(y_t - \mu) = \theta(L)\epsilon_t \tag{25.4}$$

and

$$\phi(L)y_t = \alpha + \theta(L)\epsilon_t \tag{25.5}$$

respectively. These formulations are essentially equivalent, but if they represent one and the same process μ and α are, fairly obviously, not numerically identical; rather

$$\alpha = \left(1 - \phi_1 - \ldots - \phi_p\right)\mu$$

The gretl syntax for estimating (25.4) is simply

```
arma p q ; y
```

The AR and MA lag orders, p and q, can be given either as numbers or as pre-defined scalars. The parameter μ can be dropped if necessary by appending the option `--nc` ("no constant") to the command. If estimation of (25.5) is needed, the switch `--conditional` must be appended to the command, as in

```
arma p q ; y --conditional
```

Generalizing this principle to the estimation of (25.2) or (25.3), you get that

```
arma p q ; y const x1 x2
```

would estimate the following model:

$$y_t - x_t\beta = \phi_1\left(y_{t-1} - x_{t-1}\beta\right) + \ldots + \phi_p\left(y_{t-p} - x_{t-p}\beta\right) + \epsilon_t + \theta_1\epsilon_{t-1} + \ldots + \theta_q\epsilon_{t-q}$$

where in this instance $x_t\beta = \beta_0 + x_{t,1}\beta_1 + x_{t,2}\beta_2$. Appending the `--conditional` switch, as in

```
arma p q ; y const x1 x2 --conditional
```

would estimate the following model:

$$y_t = x_t\gamma + \phi_1 y_{t-1} + \ldots + \phi_p y_{t-p} + \epsilon_t + \theta_1\epsilon_{t-1} + \ldots + \theta_q\epsilon_{t-q}$$

Ideally, the issue broached above could be made moot by writing a more general specification that nests the alternatives; that is

$$\phi(L)\left(y_t - x_t\beta\right) = z_t\gamma + \theta(L)\epsilon_t; \tag{25.6}$$

we would like to generalize the `arma` command so that the user could specify, for any estimation method, whether certain exogenous variables should be treated as x_ts or z_ts, but we're not yet at that point (and neither are most other software packages).

Seasonal models

A more flexible lag structure is desirable when analyzing time series that display strong seasonal patterns. Model (25.1) can be expanded to

$$\phi(L)\Phi(L^s)y_t = \theta(L)\Theta(L^s)\epsilon_t. \tag{25.7}$$

For such cases, a fuller form of the syntax is available, namely,

```
arma p q ; P Q ; y
```

where p and q represent the non-seasonal AR and MA orders, and P and Q the seasonal orders. For example,

```
arma 1 1 ; 1 1 ; y
```

would be used to estimate the following model:

$$(1 - \phi L)(1 - \Phi L^s)(y_t - \mu) = (1 + \theta L)(1 + \Theta L^s)\epsilon_t$$

If y_t is a quarterly series (and therefore $s = 4$), the above equation can be written more explicitly as

$$y_t - \mu = \phi(y_{t-1} - \mu) + \Phi(y_{t-4} - \mu) - (\phi \cdot \Phi)(y_{t-5} - \mu) + \epsilon_t + \theta\epsilon_{t-1} + \Theta\epsilon_{t-4} + (\theta \cdot \Theta)\epsilon_{t-5}$$

Such a model is known as a "multiplicative seasonal ARMA model".

Gaps in the lag structure

The standard way to specify an ARMA model in gretl is via the AR and MA orders, p and q respectively. In this case all lags from 1 to the given order are included. In some cases one may wish to include only certain specific AR and/or MA lags. This can be done in either of two ways.

- One can construct a matrix containing the desired lags (positive integer values) and supply the name of this matrix in place of p or q.
- One can give a comma-separated list of lags, enclosed in braces, in place of p or q.

The following code illustrates these options:

```
matrix pvec = {1,4}
arma pvec 1 ; y
arma {1,4} 1 ; y
```

Both forms above specify an ARMA model in which AR lags 1 and 4 are used (but not 2 and 3).

This facility is available only for the non-seasonal component of the ARMA specification.

Differencing and ARIMA

The above discussion presupposes that the time series y_t has already been subjected to all the transformations deemed necessary for ensuring stationarity (see also section 25.3). Differencing is the most common of these transformations, and gretl provides a mechanism to include this step into the arma command: the syntax

```
arma p d q ; y
```

would estimate an ARMA(p, q) model on $\Delta^d y_t$. It is functionally equivalent to

```
series tmp = y
loop i=1..d
  tmp = diff(tmp)
endloop
arma p q ; tmp
```

except with regard to forecasting after estimation (see below).

When the series y_t is differenced before performing the analysis the model is known as ARIMA ("I" for Integrated); for this reason, gretl provides the arima command as an alias for arma.

Seasonal differencing is handled similarly, with the syntax

```
arma p d q ; P D Q ; y
```

where D is the order for seasonal differencing. Thus, the command

```
arma 1 0 0 ; 1 1 1 ; y
```

would produce the same parameter estimates as

```
genr dsy = sdiff(y)
arma 1 0 ; 1 1 ; dsy
```

where we use the `sdiff` function to create a seasonal difference (e.g. for quarterly data, $y_t - y_{t-4}$).

In specifying an ARIMA model with exogenous regressors we face a choice which relates back to the discussion of the variant models (25.2) and (25.3) above. If we choose model (25.2), the "regression model with ARMA errors", how should this be extended to the case of ARIMA? The issue is whether or not the differencing that is applied to the dependent variable should also be applied to the regressors. Consider the simplest case, ARIMA with non-seasonal differencing of order 1. We may estimate either

$$\phi(L)(1-L)(y_t - X_t\beta) = \theta(L)\epsilon_t \qquad (25.8)$$

or

$$\phi(L)\left((1-L)y_t - X_t\beta\right) = \theta(L)\epsilon_t \qquad (25.9)$$

The first of these formulations can be described as a regression model with ARIMA errors, while the second preserves the levels of the X variables. As of gretl version 1.8.6, the default model is (25.8), in which differencing is applied to both y_t and X_t. However, when using the default estimation method (native exact ML, see below), the option `--y-diff-only` may be given, in which case gretl estimates (25.9).[1]

Estimation

The default estimation method for ARMA models is exact maximum likelihood estimation (under the assumption that the error term is normally distributed), using the Kalman filter in conjunction with the BFGS maximization algorithm. The gradient of the log-likelihood with respect to the parameter estimates is approximated numerically. This method produces results that are directly comparable with many other software packages. The constant, and any exogenous variables, are treated as in equation (25.2). The covariance matrix for the parameters is computed using a numerical approximation to the Hessian at convergence.

The alternative method, invoked with the `--conditional` switch, is conditional maximum likelihood (CML), also known as "conditional sum of squares" (see Hamilton, 1994, p. 132). This method was exemplified in the script 12.3, and only a brief description will be given here. Given a sample of size T, the CML method minimizes the sum of squared one-step-ahead prediction errors generated by the model for the observations $t_0, \ldots, T$. The starting point t_0 depends on the orders of the AR polynomials in the model. The numerical maximization method used is BHHH, and the covariance matrix is computed using a Gauss-Newton regression.

The CML method is nearly equivalent to maximum likelihood under the hypothesis of normality; the difference is that the first $(t_0 - 1)$ observations are considered fixed and only enter the likelihood function as conditioning variables. As a consequence, the two methods are asymptotically equivalent under standard conditions—except for the fact, discussed above, that our CML implementation treats the constant and exogenous variables as per equation (25.3).

The two methods can be compared as in the following example

```
open data10-1
arma 1 1 ; r
arma 1 1 ; r --conditional
```

[1] Prior to gretl 1.8.6, the default model was (25.9). We changed this for the sake of consistency with other software.

which produces the estimates shown in Table 25.1. As you can see, the estimates of ϕ and θ are quite similar. The reported constants differ widely, as expected—see the discussion following equations (25.4) and (25.5). However, dividing the CML constant by $1 - \phi$ we get 7.38, which is not far from the ML estimate of 6.93.

Table 25.1: ML and CML estimates

Parameter	ML		CML	
μ	6.93042	(0.923882)	1.07322	(0.488661)
ϕ	0.855360	(0.0511842)	0.852772	(0.0450252)
θ	0.588056	(0.0986096)	0.591838	(0.0456662)

Convergence and initialization

The numerical methods used to maximize the likelihood for ARMA models are not guaranteed to converge. Whether or not convergence is achieved, and whether or not the true maximum of the likelihood function is attained, may depend on the starting values for the parameters. Gretl employs one of the following two initialization mechanisms, depending on the specification of the model and the estimation method chosen.

1. Estimate a pure AR model by Least Squares (nonlinear least squares if the model requires it, otherwise OLS). Set the AR parameter values based on this regression and set the MA parameters to a small positive value (0.0001).
2. The Hannan–Rissanen method: First estimate an autoregressive model by OLS and save the residuals. Then in a second OLS pass add appropriate lags of the first-round residuals to the model, to obtain estimates of the MA parameters.

To see the details of the ARMA estimation procedure, add the `--verbose` option to the command. This prints a notice of the initialization method used, as well as the parameter values and log-likelihood at each iteration.

Besides the built-in initialization mechanisms, the user has the option of specifying a set of starting values manually. This is done via the `set` command: the first argument should be the keyword `initvals` and the second should be the name of a pre-specified matrix containing starting values. For example

```
matrix start = { 0, 0.85, 0.34 }
set initvals start
arma 1 1 ; y
```

The specified matrix should have just as many parameters as the model: in the example above there are three parameters, since the model implicitly includes a constant. The constant, if present, is always given first; otherwise the order in which the parameters are expected is the same as the order of specification in the `arma` or `arima` command. In the example the constant is set to zero, ϕ_1 to 0.85, and θ_1 to 0.34.

You can get gretl to revert to automatic initialization via the command `set initvals auto`.

Two variants of the BFGS algorithm are available in gretl. In general we recommend the default variant, which is based on an implementation by Nash (1990), but for some problems the alternative, limited-memory version (L-BFGS-B, see Byrd *et al.*, 1995) may increase the chances of convergence on the ML solution. This can be selected via the `--lbfgs` option to the `arma` command.

Estimation via X-12-ARIMA

As an alternative to estimating ARMA models using "native" code, gretl offers the option of using the external program X-12-ARIMA. This is the seasonal adjustment software produced and maintained by the U.S. Census Bureau; it is used for all official seasonal adjustments at the Bureau.

Gretl includes a module which interfaces with X-12-ARIMA: it translates `arma` commands using the syntax outlined above into a form recognized by X-12-ARIMA, executes the program, and retrieves the results for viewing and further analysis within gretl. To use this facility you have to install X-12-ARIMA separately. Packages for both MS Windows and GNU/Linux are available from the gretl website, `http://gretl.sourceforge.net/`.

To invoke X-12-ARIMA as the estimation engine, append the flag `--x-12-arima`, as in

```
arma p q ; y --x-12-arima
```

As with native estimation, the default is to use exact ML but there is the option of using conditional ML with the `--conditional` flag. However, please note that when X-12-ARIMA is used in conditional ML mode, the comments above regarding the variant treatments of the mean of the process y_t *do not apply*. That is, when you use X-12-ARIMA the model that is estimated is (25.2), regardless of whether estimation is by exact ML or conditional ML. In addition, the treatment of exogenous regressors in the context of ARIMA differencing is always that shown in equation (25.8).

Forecasting

ARMA models are often used for forecasting purposes. The autoregressive component, in particular, offers the possibility of forecasting a process "out of sample" over a substantial time horizon.

Gretl supports forecasting on the basis of ARMA models using the method set out by Box and Jenkins (1976).[2] The Box and Jenkins algorithm produces a set of integrated AR coefficients which take into account any differencing of the dependent variable (seasonal and/or non-seasonal) in the ARIMA context, thus making it possible to generate a forecast for the level of the original variable. By contrast, if you first difference a series manually and then apply ARMA to the differenced series, forecasts will be for the differenced series, not the level. This point is illustrated in Example 25.1. The parameter estimates are identical for the two models. The forecasts differ but are mutually consistent: the variable `fcdiff` emulates the ARMA forecast (static, one step ahead within the sample range, and dynamic out of sample).

25.3 Unit root tests

The ADF test

The Augmented Dickey–Fuller (ADF) test is, as implemented in gretl, the t-statistic on φ in the following regression:

$$\Delta y_t = \mu_t + \varphi y_{t-1} + \sum_{i=1}^{p} \gamma_i \Delta y_{t-i} + \epsilon_t . \tag{25.10}$$

This test statistic is probably the best-known and most widely used unit root test. It is a one-sided test whose null hypothesis is $\varphi = 0$ versus the alternative $\varphi < 0$ (and hence large negative values of the test statistic lead to the rejection of the null). Under the null, y_t must be differenced at least once to achieve stationarity; under the alternative, y_t is already stationary and no differencing is required.

One peculiar aspect of this test is that its limit distribution is non-standard under the null hypothesis: moreover, the shape of the distribution, and consequently the critical values for the test, depends on the form of the μ_t term. A full analysis of the various cases is inappropriate here: Hamilton (1994) contains an excellent discussion, but any recent time series textbook covers this topic. Suffice it to say that gretl allows the user to choose the specification for μ_t among four different alternatives:

[2] See in particular their "Program 4" on p. 505ff.

Example 25.1: ARIMA forecasting

```
open greene18_2.gdt
# log of quarterly U.S. nominal GNP, 1950:1 to 1983:4
genr y = log(Y)
# and its first difference
genr dy = diff(y)
# reserve 2 years for out-of-sample forecast
smpl ; 1981:4
# Estimate using ARIMA
arima 1 1 1 ; y
# forecast over full period
smpl --full
fcast fc1
# Return to sub-sample and run ARMA on the first difference of y
smpl ; 1981:4
arma 1 1 ; dy
smpl --full
fcast fc2
genr fcdiff = (t<=1982:1)? (fc1 - y(-1)) : (fc1 - fc1(-1))
# compare the forecasts over the later period
smpl 1981:1 1983:4
print y fc1 fc2 fcdiff --byobs
```

The output from the last command is:

	y	fc1	fc2	fcdiff
1981:1	7.964086	7.940930	0.02668	0.02668
1981:2	7.978654	7.997576	0.03349	0.03349
1981:3	8.009463	7.997503	0.01885	0.01885
1981:4	8.015625	8.033695	0.02423	0.02423
1982:1	8.014997	8.029698	0.01407	0.01407
1982:2	8.026562	8.046037	0.01634	0.01634
1982:3	8.032717	8.063636	0.01760	0.01760
1982:4	8.042249	8.081935	0.01830	0.01830
1983:1	8.062685	8.100623	0.01869	0.01869
1983:2	8.091627	8.119528	0.01891	0.01891
1983:3	8.115700	8.138554	0.01903	0.01903
1983:4	8.140811	8.157646	0.01909	0.01909

μ_t	command option
0	`--nc`
μ_0	`--c`
$\mu_0 + \mu_1 t$	`--ct`
$\mu_0 + \mu_1 t + \mu_1 t^2$	`--ctt`

These option flags are not mutually exclusive; when they are used together the statistic will be reported separately for each selected case. By default, gretl uses the combination `--c --ct`. For each case, approximate p-values are calculated by means of the algorithm developed in MacKinnon (1996).

The gretl command used to perform the test is `adf`; for example

```
adf 4 x1
```

would compute the test statistic as the t-statistic for φ in equation 25.10 with $p = 4$ in the two cases $\mu_t = \mu_0$ and $\mu_t = \mu_0 + \mu_1 t$.

The number of lags (p in equation 25.10) should be chosen as to ensure that (25.10) is a parametrization flexible enough to represent adequately the short-run persistence of Δy_t. Setting p too low results in size distortions in the test, whereas setting p too high leads to low power. As a convenience to the user, the parameter p can be automatically determined. Setting p to a negative number triggers a sequential procedure that starts with p lags and decrements p until the t-statistic for the parameter γ_p exceeds 1.645 in absolute value.

The ADF-GLS test

Elliott, Rothenberg and Stock (1996) proposed a variant of the ADF test which involves an alternative method of handling the parameters pertaining to the deterministic term μ_t: these are estimated first via Generalized Least Squares, and in a second stage an ADF regression is performed using the GLS residuals. This variant offers greater power than the regular ADF test for the cases $\mu_t = \mu_0$ and $\mu_t = \mu_0 + \mu_1 t$.

The ADF-GLS test is available in gretl via the `--gls` option to the `adf` command. When this option is selected the `--nc` and `--ctt` options become unavailable, and only one case can be selected at a time; by default the constant-only model is used but a trend can be added using the `--ct` flag. When a trend is present in this test MacKinnon-type p-values are not available; instead we show critical values from Table 1 in Elliott *et al.* (1996).

The KPSS test

The KPSS test (Kwiatkowski, Phillips, Schmidt and Shin, 1992) is a unit root test in which the null hypothesis is opposite to that in the ADF test: under the null, the series in question is stationary; the alternative is that the series is $I(1)$.

The basic intuition behind this test statistic is very simple: if y_t can be written as $y_t = \mu + u_t$, where u_t is some zero-mean stationary process, then not only does the sample average of the y_ts provide a consistent estimator of μ, but the long-run variance of u_t is a well-defined, finite number. Neither of these properties hold under the alternative.

The test itself is based on the following statistic:

$$\eta = \frac{\sum_{i=1}^{T} S_t^2}{T^2 \bar{\sigma}^2} \tag{25.11}$$

where $S_t = \sum_{s=1}^{t} e_s$ and $\bar{\sigma}^2$ is an estimate of the long-run variance of $e_t = (y_t - \bar{y})$. Under the null, this statistic has a well-defined (nonstandard) asymptotic distribution, which is free of nuisance parameters and has been tabulated by simulation. Under the alternative, the statistic diverges.

As a consequence, it is possible to construct a one-sided test based on η, where H_0 is rejected if η is bigger than the appropriate critical value; gretl provides the 90, 95 and 99 percent quantiles. The critical values are computed via the method presented by Sephton (1995), which offers greater accuracy than the values tabulated in Kwiatkowski *et al.* (1992).

Usage example:

```
kpss m y
```

where m is an integer representing the bandwidth or window size used in the formula for estimating the long run variance:

$$\bar{\sigma}^2 = \sum_{i=-m}^{m} \left(1 - \frac{|i|}{m+1}\right) \hat{\gamma}_i$$

The $\hat{\gamma}_i$ terms denote the empirical autocovariances of e_t from order $-m$ through m. For this estimator to be consistent, m must be large enough to accommodate the short-run persistence of e_t, but not too large compared to the sample size T. If the supplied m is non-positive a default value is computed, namely the integer part of $4\left(\frac{T}{100}\right)^{1/4}$.

The above concept can be generalized to the case where y_t is thought to be stationary around a deterministic trend. In this case, formula (25.11) remains unchanged, but the series e_t is defined as the residuals from an OLS regression of y_t on a constant and a linear trend. This second form of the test is obtained by appending the `--trend` option to the `kpss` command:

```
kpss n y --trend
```

Note that in this case the asymptotic distribution of the test is different and the critical values reported by gretl differ accordingly.

Panel unit root tests

The most commonly used unit root tests for panel data involve a generalization of the ADF procedure, in which the joint null hypothesis is that a given times series is non-stationary for all individuals in the panel.

In this context the ADF regression (25.10) can be rewritten as

$$\Delta y_{it} = \mu_{it} + \varphi_i y_{i,t-1} + \sum_{j=1}^{p_i} \gamma_{ij} \Delta y_{i,t-j} + \epsilon_{it} \tag{25.12}$$

The model (25.12) allows for maximal heterogeneity across the individuals in the panel: the parameters of the deterministic term, the autoregressive coefficient φ, and the lag order p are all specific to the individual, indexed by i.

One possible modification of this model is to impose the assumption that $\varphi_i = \varphi$ for all i; that is, the individual time series share a common autoregressive root (although they may differ in respect of other statistical properties). The choice of whether or not to impose this assumption has an important bearing on the hypotheses under test. Under model (25.12) the joint null is $\varphi_i = 0$ for all i, meaning that all the individual time series are non-stationary, and the alternative (simply the negation of the null) is that *at least one* individual time series is stationary. When a common φ is assumed, the null is that $\varphi = 0$ and the alternative is that $\varphi < 0$. The null still says that all the individual series are non-stationary, but the alternative now says that they are *all* stationary. The choice of model should take this point into account, as well as the gain in power from forming a pooled estimate of φ and, of course, the plausibility of assuming a common AR(1) coefficient.[3]

In gretl, the formulation (25.12) is used automatically when the `adf` command is used on panel data. The joint test statistic is formed using the method of Im, Pesaran and Shin (2003). In this context the behavior of `adf` differs from regular time-series data: only one case of the deterministic term is handled per invocation of the command; the default is that μ_{it} includes

[3] If the assumption of a common φ seems excessively restrictive, bear in mind that we routinely assume common slope coefficients when estimating panel models, even if this is unlikely to be literally true.

just a constant but the `--nc` and `--ct` flags can be used to suppress the constant or to include a trend, respectively; and the quadratic trend option `--ctt` is not available.

The alternative that imposes a common value of φ is implemented via the `levinlin` command. The test statistic is computed as per Levin, Lin and Chu (2002). As with the `adf` command, the first argument is the lag order and the second is the name of the series to test; and the default case for the deterministic component is a constant only. The options `--nc` and `--ct` have the same effect as with `adf`. One refinement is that the lag order may be given in either of two forms: if a scalar is given, this is taken to represent a common value of p for all individuals, but you may instead provide a vector holding a set of p_i values, hence allowing the order of autocorrelation of the series to differ by individual. So, for example, given

```
levinlin 2 y
levinlin {2,2,3,3,4,4} y
```

the first command runs a joint ADF test with a common lag order of 2, while the second (which assumes a panel with six individuals) allows for differing short-run dynamics. The first argument to `levinlin` can be given as a set of comma-separated integers enclosed in braces, as shown above, or as the name of an appropriately dimensioned pre-defined matrix (see chapter 15).

Besides variants of the ADF test, the KPSS test also can be used with panel data via the `kpss` command. In this case the test (of the null hypothesis that the given time series is *stationary* for all individuals) is implemented using the method of Choi (2001). This is an application of *meta-analysis*, the statistical technique whereby an overall or composite p-value for the test of a given null hypothesis can be computed from the p-values of a set of separate tests. Unfortunately, in the case of the KPSS test we are limited by the unavailability of precise p-values, although if an individual test statistic falls between the 10 percent and 1 percent critical values we are able to interpolate with a fair degree of confidence. This gives rise to four cases.

1. All the individual KPSS test statistics fall between the 10 percent and 1 percent critical values: the Choi method gives us a plausible composite p-value.
2. Some of the KPSS test statistics exceed the 1 percent value and none fall short of the 10 percent value: we can give an upper bound for the composite p-value by setting the unknown p-values to 0.01.
3. Some of the KPSS test statistics fall short of the 10 percent critical value but none exceed the 1 percent value: we can give a lower bound to the composite p-value by setting the unknown p-values to 0.10.
4. None of the above conditions are satisfied: the Choi method fails to produce any result for the composite KPSS test.

25.4 Cointegration tests

The generally recommended test for cointegration is the Johansen test, which is discussed in detail in chapter 27. In this context we offer a few remarks on the cointegration test of Engle and Granger (1987), which builds on the ADF test discussed above (section 25.3).

For the Engle–Granger test, the procedure is:

1. Test each series for a unit root using an ADF test.
2. Run a "cointegrating regression" via OLS. For this we select one of the potentially cointegrated variables as dependent, and include the other potentially cointegrated variables as regressors.
3. Perform an ADF test on the residuals from the cointegrating regression.

The idea is that cointegration is supported if (a) the null of non-stationarity is *not* rejected for each of the series individually, in step 1, while (b) the null *is* rejected for the residuals at step 3. That is, each of the individual series is $I(1)$ but some linear combination of the series is $I(0)$.

This test is implemented in gretl by the `coint` command, which requires an integer lag order (for the ADF tests) followed by a list of variables to be tested, the first of which will be taken as dependent in the cointegrating regression. Please see the online help for `coint`, or the *Gretl Command Reference*, for further details.

25.5 ARCH and GARCH

Heteroskedasticity means a non-constant variance of the error term in a regression model. Autoregressive Conditional Heteroskedasticity (ARCH) is a phenomenon specific to time series models, whereby the variance of the error displays autoregressive behavior; for instance, the time series exhibits successive periods where the error variance is relatively large, and successive periods where it is relatively small. This sort of behavior is reckoned to be common in asset markets: an unsettling piece of news can lead to a period of increased volatility in the market.

An ARCH error process of order q can be represented as

$$u_t = \sigma_t \varepsilon_t; \qquad \sigma_t^2 \equiv \mathrm{E}(u_t^2|\Omega_{t-1}) = \alpha_0 + \sum_{i=1}^{q} \alpha_i u_{t-i}^2$$

where the ε_ts are independently and identically distributed (iid) with mean zero and variance 1, and where σ_t is taken to be the positive square root of σ_t^2. Ω_{t-1} denotes the information set as of time $t-1$ and σ_t^2 is the conditional variance: that is, the variance conditional on information dated $t-1$ and earlier.

It is important to notice the difference between ARCH and an ordinary autoregressive error process. The simplest (first-order) case of the latter can be written as

$$u_t = \rho u_{t-1} + \varepsilon_t; \qquad -1 < \rho < 1$$

where the ε_ts are independently and identically distributed with mean zero and variance σ^2. With an AR(1) error, if ρ is positive then a positive value of u_t will tend to be followed by a positive u_{t+1}. With an ARCH error process, a disturbance u_t of large absolute value will tend to be followed by further large absolute values, but with no presumption that the successive values will be of the same sign. ARCH in asset prices is a "stylized fact" and is consistent with market efficiency; on the other hand autoregressive behavior of asset prices would violate market efficiency.

One can test for ARCH of order q in the following way:

1. Estimate the model of interest via OLS and save the squared residuals, $\hat{u}_t^2$.
2. Perform an auxiliary regression in which the current squared residual is regressed on a constant and q lags of itself.
3. Find the TR^2 value (sample size times unadjusted R^2) for the auxiliary regression.
4. Refer the TR^2 value to the χ^2 distribution with q degrees of freedom, and if the p-value is "small enough" reject the null hypothesis of homoskedasticity in favor of the alternative of ARCH(q).

This test is implemented in gretl via the `modtest` command with the `--arch` option, which must follow estimation of a time-series model by OLS (either a single-equation model or a VAR). For example,

```
ols y 0 x
modtest 4 --arch
```

This example specifies an ARCH order of $q = 4$; if the order argument is omitted, q is set equal to the periodicity of the data. In the graphical interface, the ARCH test is accessible from the "Tests" menu in the model window (again, for single-equation OLS or VARs).

GARCH

The simple ARCH(q) process is useful for introducing the general concept of conditional heteroskedasticity in time series, but it has been found to be insufficient in empirical work. The dynamics of the error variance permitted by ARCH(q) are not rich enough to represent the patterns found in financial data. The generalized ARCH or GARCH model is now more widely used.

The representation of the variance of a process in the GARCH model is somewhat (but not exactly) analogous to the ARMA representation of the level of a time series. The variance at time t is allowed to depend on both past values of the variance and past values of the realized squared disturbance, as shown in the following system of equations:

$$y_t = X_t\beta + u_t \tag{25.13}$$

$$u_t = \sigma_t\varepsilon_t \tag{25.14}$$

$$\sigma_t^2 = \alpha_0 + \sum_{i=1}^{q} \alpha_i u_{t-i}^2 + \sum_{j=1}^{p} \delta_j \sigma_{t-j}^2 \tag{25.15}$$

As above, ε_t is an iid sequence with unit variance. X_t is a matrix of regressors (or in the simplest case, just a vector of 1s allowing for a non-zero mean of y_t). Note that if $p = 0$, GARCH collapses to ARCH(q): the generalization is embodied in the δ_j terms that multiply previous values of the error variance.

In principle the underlying innovation, ε_t, could follow any suitable probability distribution, and besides the obvious candidate of the normal or Gaussian distribution the Student's t distribution has been used in this context. Currently gretl only handles the case where ε_t is assumed to be Gaussian. However, when the `--robust` option to the `garch` command is given, the estimator gretl uses for the covariance matrix can be considered Quasi-Maximum Likelihood even with non-normal disturbances. See below for more on the options regarding the GARCH covariance matrix.

Example:

```
garch p q ; y const x
```

where `p` ≥ 0 and `q` > 0 denote the respective lag orders as shown in equation (25.15). These values can be supplied in numerical form or as the names of pre-defined scalar variables.

GARCH estimation

Estimation of the parameters of a GARCH model is by no means a straightforward task. (Consider equation 25.15: the conditional variance at any point in time, σ_t^2, depends on the conditional variance in earlier periods, but σ_t^2 is not observed, and must be inferred by some sort of Maximum Likelihood procedure.) By default gretl uses native code that employs the BFGS maximizer; you also have the option (activated by the `--fcp` command-line switch) of using the method proposed by Fiorentini *et al.* (1996),[4] which was adopted as a benchmark in the study of GARCH results by McCullough and Renfro (1998). It employs analytical first and second derivatives of the log-likelihood, and uses a mixed-gradient algorithm, exploiting the information matrix in the early iterations and then switching to the Hessian in the neighborhood of the maximum likelihood. (This progress can be observed if you append the `--verbose` option to gretl's `garch` command.)

Several options are available for computing the covariance matrix of the parameter estimates in connection with the `garch` command. At a first level, one can choose between a "standard" and a "robust" estimator. By default, the Hessian is used unless the `--robust` option is given, in which case the QML estimator is used. A finer choice is available via the `set` command, as shown in Table 25.2.

It is not uncommon, when one estimates a GARCH model for an arbitrary time series, to find that the iterative calculation of the estimates fails to converge. For the GARCH model to make

[4]The algorithm is based on Fortran code deposited in the archive of the *Journal of Applied Econometrics* by the authors, and is used by kind permission of Professor Fiorentini.

Table 25.2: Options for the GARCH covariance matrix

command	*effect*
`set garch_vcv hessian`	Use the Hessian
`set garch_vcv im`	Use the Information Matrix
`set garch_vcv op`	Use the Outer Product of the Gradient
`set garch_vcv qml`	QML estimator
`set garch_vcv bw`	Bollerslev–Wooldridge "sandwich" estimator

sense, there are strong restrictions on the admissible parameter values, and it is not always the case that there exists a set of values inside the admissible parameter space for which the likelihood is maximized.

The restrictions in question can be explained by reference to the simplest (and much the most common) instance of the GARCH model, where $p = q = 1$. In the GARCH(1, 1) model the conditional variance is

$$\sigma_t^2 = \alpha_0 + \alpha_1 u_{t-1}^2 + \delta_1 \sigma_{t-1}^2 \tag{25.16}$$

Taking the unconditional expectation of (25.16) we get

$$\sigma^2 = \alpha_0 + \alpha_1 \sigma^2 + \delta_1 \sigma^2$$

so that

$$\sigma^2 = \frac{\alpha_0}{1 - \alpha_1 - \delta_1}$$

For this unconditional variance to exist, we require that $\alpha_1 + \delta_1 < 1$, and for it to be positive we require that $\alpha_0 > 0$.

A common reason for non-convergence of GARCH estimates (that is, a common reason for the non-existence of α_i and δ_i values that satisfy the above requirements and at the same time maximize the likelihood of the data) is misspecification of the model. It is important to realize that GARCH, in itself, allows *only* for time-varying volatility in the data. If the *mean* of the series in question is not constant, or if the error process is not only heteroskedastic but also autoregressive, it is necessary to take this into account when formulating an appropriate model. For example, it may be necessary to take the first difference of the variable in question and/or to add suitable regressors, X_t, as in (25.13).

Chapter 26

Vector Autoregressions

Gretl provides a standard set of procedures for dealing with the multivariate time-series models known as VARs (*Vector AutoRegression*). More general models—such as VARMAs, nonlinear models or multivariate GARCH models—are not provided as of now, although it is entirely possible to estimate them by writing custom procedures in the gretl scripting language. In this chapter, we will briefly review gretl's VAR toolbox.

26.1 Notation

A VAR is a structure whose aim is to model the time persistence of a vector of n time series, y_t, via a multivariate autoregression, as in

$$y_t = A_1 y_{t-1} + A_2 y_{t-2} + \cdots + A_p y_{t-p} + Bx_t + \epsilon_t \tag{26.1}$$

The number of lags p is called the *order* of the VAR. The vector x_t, if present, contains a set of exogenous variables, often including a constant, possibly with a time trend and seasonal dummies. The vector ϵ_t is typically assumed to be a vector white noise, with covariance matrix Σ.

Equation (26.1) can be written more compactly as

$$A(L)y_t = Bx_t + \epsilon_t \tag{26.2}$$

where $A(L)$ is a matrix polynomial in the lag operator, or as

$$\begin{bmatrix} y_t \\ y_{t-1} \\ \cdots \\ y_{t-p-1} \end{bmatrix} = \mathbf{A} \begin{bmatrix} y_{t-1} \\ y_{t-2} \\ \cdots \\ y_{t-p} \end{bmatrix} + \begin{bmatrix} B \\ 0 \\ \cdots \\ 0 \end{bmatrix} x_t + \begin{bmatrix} \epsilon_t \\ 0 \\ \cdots \\ 0 \end{bmatrix} \tag{26.3}$$

The matrix $\mathbf{A}$ is known as the "companion matrix" and equals

$$\mathbf{A} = \begin{bmatrix} A_1 & A_2 & \cdots & A_p \\ I & 0 & \cdots & 0 \\ 0 & I & \cdots & 0 \\ \vdots & \vdots & \ddots & \vdots \end{bmatrix}$$

Equation (26.3) is known as the "companion form" of the VAR.

Another representation of interest is the so-called "VMA representation", which is written in terms of an infinite series of matrices Θ_i defined as

$$\Theta_i = \frac{\partial y_t}{\partial \epsilon_{t-i}} \tag{26.4}$$

The Θ_i matrices may be derived by recursive substitution in equation (26.1): for example, assuming for simplicity that $B = 0$ and $p = 1$, equation (26.1) would become

$$y_t = Ay_{t-1} + \epsilon_t$$

which could be rewritten as

$$y_t = A^{n+1} y_{t-n-1} + \epsilon_t + A\epsilon_{t-1} + A^2\epsilon_{t-2} + \cdots + A^n \epsilon_{t-n}$$

In this case $\Theta_i = A^i$. In general, it is possible to compute Θ_i as the $n \times n$ north-west block of the i-th power of the companion matrix $\mathbf{A}$ (so Θ_0 is always an identity matrix).

The VAR is said to be *stable* if all the eigenvalues of the companion matrix $\mathbf{A}$ are smaller than 1 in absolute value, or equivalently, if the matrix polynomial $A(L)$ in equation (26.2) is such that $|A(z)| = 0$ implies $|z| > 1$. If this is the case, $\lim_{n\to\infty} \Theta_n = 0$ and the vector y_t is stationary; as a consequence, the equation

$$y_t - E(y_t) = \sum_{i=0}^{\infty} \Theta_i \epsilon_{t-i} \tag{26.5}$$

is a legitimate Wold representation.

If the VAR is not stable, then the inferential procedures that are called for become somewhat more specialized, except for some simple cases. In particular, if the number of eigenvalues of $\mathbf{A}$ with modulus 1 is between 1 and $n-1$, the canonical tool to deal with these models is the cointegrated VAR model, discussed in chapter 27.

26.2 Estimation

The gretl command for estimating a VAR is `var` which, in the command line interface, is invoked in the following manner:

```
[ modelname <- ] var p Ylist [; Xlist]
```

where `p` is a scalar (the VAR order) and `Ylist` is a list of variables specifying the content of y_t. The optional `Xlist` argument can be used to specify a set of exogenous variables. If this argument is omitted, the vector x_t is taken to contain a constant (only); if present, it must be separated from `Ylist` by a semicolon. Note, however, that a few common choices can be obtained in a simpler way: the options `--trend` and `--seasonals` call for inclusion of a linear trend and a set of seasonal dummies respectively. In addition the `--nc` option (no constant) can be used to suppress the standard inclusion of a constant.

The "<-" construct can be used to store the model under a name (see section 3.2), if so desired. To estimate a VAR using the graphical interface, choose "Time Series, Vector Autoregression", under the Model menu.

The parameters in eq. (26.1) are typically free from restrictions, which implies that multivariate OLS provides a consistent and asymptotically efficient estimator of all the parameters.[1] Given the simplicity of OLS, this is what every software package, including gretl, uses; example script 26.1 exemplifies the fact that the `var` command gives you exactly the output you would have from a battery of OLS regressions. The advantage of using the dedicated command is that, after estimation is done, it makes it much easier to access certain quantities and manage certain tasks. For example, the `$coeff` accessor returns the estimated coefficients as a matrix with n columns and `$sigma` returns an estimate of the matrix Σ, the covariance matrix of ϵ_t.

Moreover, for each variable in the system an F test is automatically performed, in which the null hypothesis is that no lags of variable j are significant in the equation for variable i. This is commonly known as a **Granger causality** test.

Periodicity	horizon
Quarterly	20 (5 years)
Monthly	24 (2 years)
Daily	3 weeks
All other cases	10

Table 26.1: VMA horizon as a function of the dataset periodicity

In addition, two accessors become available for the companion matrix (`$compan`) and the VMA representation (`$vma`). The latter deserves a detailed description: since the VMA representation

[1]In fact, under normality of ϵ_t OLS is indeed the conditional ML estimator. You may want to use other methods if you need to estimate a VAR in which some parameters are constrained.

Example 26.1: Estimation of a VAR via OLS

Input:

```
open sw_ch14.gdt
genr infl = 400*sdiff(log(PUNEW))

scalar p = 2
list X = LHUR infl
list Xlag = lags(p,X)

loop foreach i X
    ols $i const Xlag
endloop

var p X
```

Output (selected portions):

```
Model 1: OLS, using observations 1960:3-1999:4 (T = 158)
Dependent variable: LHUR

             coefficient   std. error   t-ratio   p-value
  --------------------------------------------------------
  const        0.113673    0.0875210     1.299    0.1960
  LHUR_1       1.54297     0.0680518    22.67     8.78e-51 ***
  LHUR_2      -0.583104    0.0645879    -9.028    7.00e-16 ***
  infl_1       0.0219040   0.00874581    2.505    0.0133   **
  infl_2      -0.0148408   0.00920536   -1.612    0.1090

Mean dependent var   6.019198   S.D. dependent var   1.502549
Sum squared resid    8.654176   S.E. of regression   0.237830

...

VAR system, lag order 2
OLS estimates, observations 1960:3-1999:4 (T = 158)
Log-likelihood = -322.73663
Determinant of covariance matrix = 0.20382769
AIC = 4.2119
BIC = 4.4057
HQC = 4.2906
Portmanteau test: LB(39) = 226.984, df = 148 [0.0000]

Equation 1: LHUR

             coefficient   std. error   t-ratio   p-value
  --------------------------------------------------------
  const        0.113673    0.0875210     1.299    0.1960
  LHUR_1       1.54297     0.0680518    22.67     8.78e-51 ***
  LHUR_2      -0.583104    0.0645879    -9.028    7.00e-16 ***
  infl_1       0.0219040   0.00874581    2.505    0.0133   **
  infl_2      -0.0148408   0.00920536   -1.612    0.1090

Mean dependent var   6.019198   S.D. dependent var   1.502549
Sum squared resid    8.654176   S.E. of regression   0.237830
```

(26.5) is of infinite order, gretl defines a *horizon* up to which the Θ_i matrices are computed automatically. By default, this is a function of the periodicity of the data (see table 26.1), but it can be set by the user to any desired value via the `set` command with the `horizon` parameter, as in

```
set horizon 30
```

Calling the horizon h, the `$vma` accessor returns an $(h + 1) \times n^2$ matrix, in which the $(i + 1)$-th row is the vectorized form of Θ_i.

VAR lag-order selection

In order to help the user choose the most appropriate VAR order, gretl provides a special variant of the `var` command:

`var` *p* *Ylist* `[;` *Xlist*`] --lagselect`

When the `--lagselect` option is given, estimation is performed for all lags up to *p* and a table is printed: it displays, for each order, a Likelihood Ratio test for the order p versus $p - 1$, plus an array of information criteria (see chapter 23). For each information criterion in the table, a star indicates what appears to be the "best" choice. The same output can be obtained through the graphical interface via the "Time Series, VAR lag selection" entry under the Model menu.

Example 26.2: VAR lag selection via Information Criteria

Input:

```
open denmark
list Y = 1 2 3 4
var 4 Y --lagselect
var 6 Y --lagselect
```

Output (selected portions):

```
VAR system, maximum lag order 4

The asterisks below indicate the best (that is, minimized) values
of the respective information criteria, AIC = Akaike criterion,
BIC = Schwarz Bayesian criterion and HQC = Hannan-Quinn criterion.

lags        loglik    p(LR)        AIC          BIC          HQC

   1     609.15315            -23.104045   -22.346466*  -22.814552
   2     631.70153  0.00013  -23.360844*  -21.997203   -22.839757*
   3     642.38574  0.16478  -23.152382   -21.182677   -22.399699
   4     653.22564  0.15383  -22.950025   -20.374257   -21.965748

VAR system, maximum lag order 6

The asterisks below indicate the best (that is, minimized) values
of the respective information criteria, AIC = Akaike criterion,
BIC = Schwarz Bayesian criterion and HQC = Hannan-Quinn criterion.

lags        loglik    p(LR)        AIC          BIC          HQC

   1     594.38410            -23.444249   -22.672078*  -23.151288*
   2     615.43480  0.00038  -23.650400*  -22.260491   -23.123070
   3     624.97613  0.26440  -23.386781   -21.379135   -22.625083
   4     636.03766  0.13926  -23.185210   -20.559827   -22.189144
   5     658.36014  0.00016  -23.443271   -20.200150   -22.212836
   6     669.88472  0.11243  -23.260601   -19.399743   -21.795797
```

Warning: in finite samples the choice of the maximum lag, p, may affect the outcome of the procedure. *This is not a bug*, but rather an unavoidable side effect of the way these comparisons should be made. If your sample contains T observations and you invoke the lag selection procedure with maximum order p, gretl examines all VARs of order ranging form 1 to p, estimated on a uniform sample of $T - p$ observations. In other words, the comparison procedure does not use all the available data when estimating VARs of order less than p, so as to ensure that all the models in the comparison are estimated on the same data range. Choosing a different value of p may therefore alter the results, although this is unlikely to happen if your sample size is reasonably large.

An example of this unpleasant phenomenon is given in example script 26.2. As can be seen, according to the Hannan-Quinn criterion, order 2 seems preferable to order 1 if the maximum tested order is 4, but the situation is reversed if the maximum tested order is 6.

26.3 Structural VARs

Gretl does not currently provide a native implementation for the general class of models known as "Structural VARs"; however, it provides an implementation of the Cholesky decomposition-based approach, the classic and most popular SVAR variant.

IRF and FEVD

Assume that the disturbance in equation (26.1) can be thought of as a linear function of a vector of *structural shocks* u_t, which are assumed to have unit variance and to be mutually unncorrelated, so $V(u_t) = I$. If $\epsilon_t = Ku_t$, it follows that $\Sigma = V(\epsilon_t) = KK'$.

The main object of interest in this setting is the sequence of matrices

$$C_k = \frac{\partial y_t}{\partial u_{t-i}} = \Theta_k K, \qquad (26.6)$$

known as the structural VMA representation. From the C_k matrices defined in equation (26.6) two quantities of interest may be derived: the **Impulse Response Function** (IRF) and the **Forecast Error Variance Decomposition** (FEVD).

The IRF of variable i to shock j is simply the sequence of the elements in row i and column j of the C_k matrices. In symbols:

$$\mathcal{I}_{i,j,k} = \frac{\partial y_{i,t}}{\partial u_{j,t-k}}$$

As a rule, Impulse Response Functions are plotted as a function of k, and are interpreted as the effect that a shock has on an observable variable through time. Of course, what we observe are the estimated IRFs, so it is natural to endow them with confidence intervals: following common practice, gretl computes the confidence intervals by using the bootstrap;[2] details are given later in this section.

Another quantity of interest that may be computed from the structural VMA representation is the Forecast Error Variance Decomposition (FEVD). The forecast error variance after h steps is given by

$$\Omega_h = \sum_{k=0}^{h} C_k C_k'$$

hence the variance for variable i is

$$\omega_i^2 = [\Omega_h]_{i,i} = \sum_{k=0}^{h} \mathrm{diag}(C_k C_k')_i = \sum_{k=0}^{h} \sum_{l=1}^{n} ({}_k c_{i.l})^2$$

[2]It is possible, in principle, to compute analytical confidence intervals via an asymptotic approximation, but this is not a very popular choice: asymptotic formulae are known to often give a very poor approximation of the finite-sample properties.

where ${}_k c_{i.l}$ is, trivially, the i, l element of C_k. As a consequence, the share of uncertainty on variable i that can be attributed to the j-th shock after h periods equals

$$\mathcal{VD}_{i,j,h} = \frac{\sum_{k=0}^{h} ({}_k c_{i.j})^2}{\sum_{k=0}^{h} \sum_{l=1}^{n} ({}_k c_{i.l})^2}.$$

This makes it possible to quantify which shocks are most important to determine a certain variable in the short and/or in the long run.

Triangularization

The formula 26.6 takes K as known, while of course it has to be estimated. The estimation problem has been the subject of an enormous body of literature we will not even attempt to summarize here: see for example (Lütkepohl, 2005, chapter 9).

Suffice it to say that the most popular choice dates back to Sims (1980), and consists in assuming that K is lower triangular, so its estimate is simply the Cholesky decomposition of the estimate of Σ. The main consequence of this choice is that the ordering of variables within the vector y_t becomes meaningful: since K is also the matrix of Impulse Response Functions at lag 0, the triangularity assumption means that the first variable in the ordering responds instantaneously only to shock number 1, the second one only to shocks 1 and 2, and so forth. For this reason, each variable is thought to "own" one shock: variable 1 owns shock number 1, and so on.

In this sort of exercise, therefore, the ordering of the y variables is important, and the applied literature has developed the "most exogenous first" mantra—where, in this setting, "exogenous" really means "instantaneously insensitive to structural shocks".[3] To put it differently, if variable `foo` comes before variable `bar` in the Y list, it follows that the shock owned by `foo` affects `bar` instantaneously, but not vice versa.

Impulse Response Functions and the FEVD can be printed out via the command line interface by using the `--impulse-response` and `--variance-decomp` options, respectively. If you need to store them into matrices, you can compute the structural VMA and proceed from there. For example, the following code snippet shows you how to compute a matrix containing the IRFs:

```
open denmark
list Y = 1 2 3 4
scalar n = nelem(Y)
var 2 Y --quiet --impulse

matrix K = cholesky($sigma)
matrix V = $vma
matrix IRF = V * (K ** I(n))
print IRF
```

in which the equality

$$\text{vec}(C_k) = \text{vec}(\Theta_k K) = (K' \otimes I)\text{vec}(\Theta_k)$$

was used.

FIXME: show all the nice stuff we have under the GUI.

IRF bootstrap

FIXME: todo

[3]The word "exogenous" has caught on in this context, but it's a rather unfortunate choice: for a start, each shock impacts on every variable after one lag, so nothing is really exogenous here. A better choice of words would probably have been something like "sturdy", but it's too late now.

Chapter 27

Cointegration and Vector Error Correction Models

27.1 Introduction

The twin concepts of cointegration and error correction have drawn a good deal of attention in macroeconometrics over recent years. The attraction of the Vector Error Correction Model (VECM) is that it allows the researcher to embed a representation of economic equilibrium relationships within a relatively rich time-series specification. This approach overcomes the old dichotomy between (a) structural models that faithfully represented macroeconomic theory but failed to fit the data, and (b) time-series models that were accurately tailored to the data but difficult if not impossible to interpret in economic terms.

The basic idea of cointegration relates closely to the concept of unit roots (see section 25.3). Suppose we have a set of macroeconomic variables of interest, and we find we cannot reject the hypothesis that some of these variables, considered individually, are non-stationary. Specifically, suppose we judge that a subset of the variables are individually integrated of order 1, or I(1). That is, while they are non-stationary in their levels, their first differences are stationary. Given the statistical problems associated with the analysis of non-stationary data (for example, the threat of spurious regression), the traditional approach in this case was to take first differences of all the variables before proceeding with the analysis.

But this can result in the loss of important information. It may be that while the variables in question are I(1) when taken individually, there exists a linear combination of the variables that is stationary without differencing, or I(0). (There could be more than one such linear combination.) That is, while the ensemble of variables may be "free to wander" over time, nonetheless the variables are "tied together" in certain ways. And it may be possible to interpret these ties, or *cointegrating vectors*, as representing equilibrium conditions.

For example, suppose we find some or all of the following variables are I(1): money stock, M, the price level, P, the nominal interest rate, R, and output, Y. According to standard theories of the demand for money, we would nonetheless expect there to be an equilibrium relationship between real balances, interest rate and output; for example

$$m - p = \gamma_0 + \gamma_1 y + \gamma_2 r \qquad \gamma_1 > 0, \gamma_2 < 0$$

where lower-case variable names denote logs. In equilibrium, then,

$$m - p - \gamma_1 y - \gamma_2 r = \gamma_0$$

Realistically, we should not expect this condition to be satisfied each period. We need to allow for the possibility of short-run disequilibrium. But if the system moves back towards equilibrium following a disturbance, it follows that the vector $x = (m, p, y, r)'$ is bound by a cointegrating vector $\beta' = (\beta_1, \beta_2, \beta_3, \beta_4)$, such that $\beta' x$ is stationary (with a mean of γ_0). Furthermore, if equilibrium is correctly characterized by the simple model above, we have $\beta_2 = -\beta_1$, $\beta_3 < 0$ and $\beta_4 > 0$. These things are testable within the context of cointegration analysis.

There are typically three steps in this sort of analysis:

1. Test to determine the number of cointegrating vectors, the *cointegrating rank* of the system.

2. Estimate a VECM with the appropriate rank, but subject to no further restrictions.

3. Probe the interpretation of the cointegrating vectors as equilibrium conditions by means of restrictions on the elements of these vectors.

The following sections expand on each of these points, giving further econometric details and explaining how to implement the analysis using gretl.

27.2 Vector Error Correction Models as representation of a cointegrated system

Consider a VAR of order p with a deterministic part given by μ_t (typically, a polynomial in time). One can write the n-variate process y_t as

$$y_t = \mu_t + A_1 y_{t-1} + A_2 y_{t-2} + \cdots + A_p y_{t-p} + \epsilon_t \tag{27.1}$$

But since $y_{t-i} \equiv y_{t-1} - (\Delta y_{t-1} + \Delta y_{t-2} + \cdots + \Delta y_{t-i+1})$, we can re-write the above as

$$\Delta y_t = \mu_t + \Pi y_{t-1} + \sum_{i=1}^{p-1} \Gamma_i \Delta y_{t-i} + \epsilon_t, \tag{27.2}$$

where $\Pi = \sum_{i=1}^{p} A_i - I$ and $\Gamma_i = -\sum_{j=i+1}^{p} A_j$. This is the VECM representation of (27.1).

The interpretation of (27.2) depends crucially on r, the rank of the matrix Π.

- If $r = 0$, the processes are all I(1) and not cointegrated.
- If $r = n$, then Π is invertible and the processes are all I(0).
- Cointegration occurs in between, when $0 < r < n$ and Π can be written as $\alpha\beta'$. In this case, y_t is I(1), but the combination $z_t = \beta' y_t$ is I(0). If, for example, $r = 1$ and the first element of β was -1, then one could write $z_t = -y_{1,t} + \beta_2 y_{2,t} + \cdots + \beta_n y_{n,t}$, which is equivalent to saying that

  $$y_{1_t} = \beta_2 y_{2,t} + \cdots + \beta_n y_{n,t} - z_t$$

  is a long-run equilibrium relationship: the deviations z_t may not be 0 but they are stationary. In this case, (27.2) can be written as

  $$\Delta y_t = \mu_t + \alpha\beta' y_{t-1} + \sum_{i=1}^{p-1} \Gamma_i \Delta y_{t-i} + \epsilon_t. \tag{27.3}$$

  If β were known, then z_t would be observable and all the remaining parameters could be estimated via OLS. In practice, the procedure estimates β first and then the rest.

The rank of Π is investigated by computing the eigenvalues of a closely related matrix whose rank is the same as Π: however, this matrix is by construction symmetric and positive semidefinite. As a consequence, all its eigenvalues are real and non-negative, and tests on the rank of Π can therefore be carried out by testing how many eigenvalues are 0.

If all the eigenvalues are significantly different from 0, then all the processes are stationary. If, on the contrary, there is at least one zero eigenvalue, then the y_t process is integrated, although some linear combination $\beta' y_t$ might be stationary. At the other extreme, if no eigenvalues are significantly different from 0, then not only is the process y_t non-stationary, but the same holds for any linear combination $\beta' y_t$; in other words, no cointegration occurs.

Estimation typically proceeds in two stages: first, a sequence of tests is run to determine r, the cointegration rank. Then, for a given rank the parameters in equation (27.3) are estimated. The two commands that gretl offers for estimating these systems are `coint2` and `vecm`, respectively.

The syntax for `coint2` is

```
coint2 p ylist [ ; xlist [ ; zlist ] ]
```

where p is the number of lags in (27.1); `ylist` is a list containing the y_t variables; `xlist` is an optional list of exogenous variables; and `zlist` is another optional list of exogenous variables whose effects are assumed to be confined to the cointegrating relationships.

The syntax for `vecm` is

```
vecm p r ylist [ ; xlist [ ; zlist ] ]
```

where p is the number of lags in (27.1); r is the cointegration rank; and the lists `ylist`, `xlist` and `zlist` have the same interpretation as in `coint2`.

Both commands can be given specific options to handle the treatment of the deterministic component μ_t. These are discussed in the following section.

27.3 Interpretation of the deterministic components

Statistical inference in the context of a cointegrated system depends on the hypotheses one is willing to make on the deterministic terms, which leads to the famous "five cases."

In equation (27.2), the term μ_t is usually understood to take the form

$$\mu_t = \mu_0 + \mu_1 \cdot t.$$

In order to have the model mimic as closely as possible the features of the observed data, there is a preliminary question to settle. Do the data appear to follow a deterministic trend? If so, is it linear or quadratic?

Once this is established, one should impose restrictions on μ_0 and μ_1 that are consistent with this judgement. For example, suppose that the data do not exhibit a discernible trend. This means that Δy_t is on average zero, so it is reasonable to assume that its expected value is also zero. Write equation (27.2) as

$$\Gamma(L)\Delta y_t = \mu_0 + \mu_1 \cdot t + \alpha z_{t-1} + \epsilon_t, \tag{27.4}$$

where $z_t = \beta' y_t$ is assumed to be stationary and therefore to possess finite moments. Taking unconditional expectations, we get

$$0 = \mu_0 + \mu_1 \cdot t + \alpha m_z.$$

Since the left-hand side does not depend on t, the restriction $\mu_1 = 0$ is a safe bet. As for μ_0, there are just two ways to make the above expression true: either $\mu_0 = 0$ with $m_z = 0$, or μ_0 equals $-\alpha m_z$. The latter possibility is less restrictive in that the vector μ_0 may be non-zero, but is constrained to be a linear combination of the columns of α. In that case, μ_0 can be written as $\alpha \cdot c$, and one may write (27.4) as

$$\Gamma(L)\Delta y_t = \alpha \begin{bmatrix} \beta' & c \end{bmatrix} \begin{bmatrix} y_{t-1} \\ 1 \end{bmatrix} + \epsilon_t.$$

The long-run relationship therefore contains an intercept. This type of restriction is usually written

$$\alpha'_\perp \mu_0 = 0,$$

where $\alpha_\perp$ is the left null space of the matrix α.

An intuitive understanding of the issue can be gained by means of a simple example. Consider a series x_t which behaves as follows

$$x_t = m + x_{t-1} + \varepsilon_t$$

where m is a real number and ε_t is a white noise process: x_t is then a random walk with drift m. In the special case $m = 0$, the drift disappears and x_t is a pure random walk.

Consider now another process y_t, defined by

$$y_t = k + x_t + u_t$$

where, again, k is a real number and u_t is a white noise process. Since u_t is stationary by definition, x_t and y_t cointegrate: that is, their difference

$$z_t = y_t - x_t = k + u_t$$

is a stationary process. For $k = 0$, z_t is simple zero-mean white noise, whereas for $k \neq 0$ the process z_t is white noise with a non-zero mean.

After some simple substitutions, the two equations above can be represented jointly as a VAR(1) system

$$\begin{bmatrix} y_t \\ x_t \end{bmatrix} = \begin{bmatrix} k+m \\ m \end{bmatrix} + \begin{bmatrix} 0 & 1 \\ 0 & 1 \end{bmatrix} \begin{bmatrix} y_{t-1} \\ x_{t-1} \end{bmatrix} + \begin{bmatrix} u_t + \varepsilon_t \\ \varepsilon_t \end{bmatrix}$$

or in VECM form

$$\begin{aligned} \begin{bmatrix} \Delta y_t \\ \Delta x_t \end{bmatrix} &= \begin{bmatrix} k+m \\ m \end{bmatrix} + \begin{bmatrix} -1 & 1 \\ 0 & 0 \end{bmatrix} \begin{bmatrix} y_{t-1} \\ x_{t-1} \end{bmatrix} + \begin{bmatrix} u_t + \varepsilon_t \\ \varepsilon_t \end{bmatrix} = \\ &= \begin{bmatrix} k+m \\ m \end{bmatrix} + \begin{bmatrix} -1 \\ 0 \end{bmatrix} \begin{bmatrix} 1 & -1 \end{bmatrix} \begin{bmatrix} y_{t-1} \\ x_{t-1} \end{bmatrix} + \begin{bmatrix} u_t + \varepsilon_t \\ \varepsilon_t \end{bmatrix} = \\ &= \mu_0 + \alpha\beta' \begin{bmatrix} y_{t-1} \\ x_{t-1} \end{bmatrix} + \eta_t = \mu_0 + \alpha z_{t-1} + \eta_t, \end{aligned}$$

where β is the cointegration vector and α is the "loadings" or "adjustments" vector.

We are now ready to consider three possible cases:

1. $m \neq 0$: In this case x_t is trended, as we just saw; it follows that y_t also follows a linear trend because on average it keeps at a fixed distance k from x_t. The vector μ_0 is unrestricted.

2. $m = 0$ and $k \neq 0$: In this case, x_t is not trended and as a consequence neither is y_t. However, the mean distance between y_t and x_t is non-zero. The vector μ_0 is given by

$$\mu_0 = \begin{bmatrix} k \\ 0 \end{bmatrix}$$

   which is not null and therefore the VECM shown above does have a constant term. The constant, however, is subject to the restriction that its second element must be 0. More generally, μ_0 is a multiple of the vector α. Note that the VECM could also be written as

$$\begin{bmatrix} \Delta y_t \\ \Delta x_t \end{bmatrix} = \begin{bmatrix} -1 \\ 0 \end{bmatrix} \begin{bmatrix} 1 & -1 & -k \end{bmatrix} \begin{bmatrix} y_{t-1} \\ x_{t-1} \\ 1 \end{bmatrix} + \begin{bmatrix} u_t + \varepsilon_t \\ \varepsilon_t \end{bmatrix}$$

   which incorporates the intercept into the cointegration vector. This is known as the "restricted constant" case.

3. $m = 0$ and $k = 0$: This case is the most restrictive: clearly, neither x_t nor y_t are trended, and the mean distance between them is zero. The vector μ_0 is also 0, which explains why this case is referred to as "no constant."

In most cases, the choice between these three possibilities is based on a mix of empirical observation and economic reasoning. If the variables under consideration seem to follow a linear trend then we should not place any restriction on the intercept. Otherwise, the question arises of whether it makes sense to specify a cointegration relationship which includes a non-zero intercept. One example where this is appropriate is the relationship between two interest rates: generally these are not trended, but the VAR might still have an intercept because the difference between the two (the "interest rate spread") might be stationary around a non-zero mean (for example, because of a risk or liquidity premium).

The previous example can be generalized in three directions:

1. If a VAR of order greater than 1 is considered, the algebra gets more convoluted but the conclusions are identical.

2. If the VAR includes more than two endogenous variables the cointegration rank r can be greater than 1. In this case, α is a matrix with r columns, and the case with restricted constant entails the restriction that μ_0 should be some linear combination of the columns of α.

3. If a linear trend is included in the model, the deterministic part of the VAR becomes $\mu_0 + \mu_1 t$. The reasoning is practically the same as above except that the focus now centers on μ_1 rather than μ_0. The counterpart to the "restricted constant" case discussed above is a "restricted trend" case, such that the cointegration relationships include a trend but the first differences of the variables in question do not. In the case of an unrestricted trend, the trend appears in both the cointegration relationships and the first differences, which corresponds to the presence of a quadratic trend in the variables themselves (in levels).

In order to accommodate the five cases, gretl provides the following options to the `coint2` and `vecm` commands:

μ_t	*option flag*	*description*
0	`--nc`	no constant
$\mu_0, \alpha'_{\perp}\mu_0 = 0$	`--rc`	restricted constant
μ_0	`--uc`	unrestricted constant
$\mu_0 + \mu_1 t, \alpha'_{\perp}\mu_1 = 0$	`--crt`	constant + restricted trend
$\mu_0 + \mu_1 t$	`--ct`	constant + unrestricted trend

Note that for this command the above options are mutually exclusive. In addition, you have the option of using the `--seasonal` options, for augmenting μ_t with centered seasonal dummies. In each case, p-values are computed via the approximations devised by Doornik (1998).

27.4 The Johansen cointegration tests

The two Johansen tests for cointegration are used to establish the rank of β, or in other words the number of cointegrating vectors. These are the "λ-max" test, for hypotheses on individual eigenvalues, and the "trace" test, for joint hypotheses. Suppose that the eigenvalues λ_i are sorted from largest to smallest. The null hypothesis for the λ-max test on the i-th eigenvalue is that $\lambda_i = 0$. The corresponding trace test, instead, considers the hypothesis $\lambda_j = 0$ for all $j \geq i$.

The gretl command `coint2` performs these two tests. The corresponding menu entry in the GUI is "Model, Time Series, Cointegration Test, Johansen".

As in the ADF test, the asymptotic distribution of the tests varies with the deterministic component μ_t one includes in the VAR (see section 27.3 above). The following code uses the `denmark` data file, supplied with gretl, to replicate Johansen's example found in his 1995 book.

```
open denmark
coint2 2 LRM LRY IBO IDE --rc --seasonal
```

In this case, the vector y_t in equation (27.2) comprises the four variables `LRM`, `LRY`, `IBO`, `IDE`. The number of lags equals p in (27.2) (that is, the number of lags of the model written in VAR form). Part of the output is reported below:

```
Johansen test:
Number of equations = 4
Lag order = 2
Estimation period: 1974:3 - 1987:3 (T = 53)

Case 2: Restricted constant
Rank Eigenvalue Trace test p-value   Lmax test  p-value
```

```
0    0.43317      49.144 [0.1284]      30.087 [0.0286]
1    0.17758      19.057 [0.7833]      10.362 [0.8017]
2    0.11279      8.6950 [0.7645]      6.3427 [0.7483]
3   0.043411      2.3522 [0.7088]      2.3522 [0.7076]
```

Both the trace and λ-max tests accept the null hypothesis that the smallest eigenvalue is 0 (see the last row of the table), so we may conclude that the series are in fact non-stationary. However, some linear combination may be I(0), since the λ-max test rejects the hypothesis that the rank of Π is 0 (though the trace test gives less clear-cut evidence for this, with a p-value of 0.1284).

27.5 Identification of the cointegration vectors

The core problem in the estimation of equation (27.2) is to find an estimate of Π that has by construction rank r, so it can be written as $\Pi = \alpha\beta'$, where β is the matrix containing the cointegration vectors and α contains the "adjustment" or "loading" coefficients whereby the endogenous variables respond to deviation from equilibrium in the previous period.

Without further specification, the problem has multiple solutions (in fact, infinitely many). The parameters α and β are under-identified: if all columns of β are cointegration vectors, then any arbitrary linear combinations of those columns is a cointegration vector too. To put it differently, if $\Pi = \alpha_0\beta_0'$ for specific matrices α_0 and β_0, then Π also equals $(\alpha_0 Q)(Q^{-1}\beta_0')$ for any conformable non-singular matrix Q. In order to find a unique solution, it is therefore necessary to impose some restrictions on α and/or β. It can be shown that the minimum number of restrictions that is necessary to guarantee identification is r^2. Normalizing one coefficient per column to 1 (or -1, according to taste) is a trivial first step, which also helps in that the remaining coefficients can be interpreted as the parameters in the equilibrium relations, but this only suffices when $r = 1$.

The method that gretl uses by default is known as the "Phillips normalization", or "triangular representation".[1] The starting point is writing β in partitioned form as in

$$\beta = \begin{bmatrix} \beta_1 \\ \beta_2 \end{bmatrix},$$

where β_1 is an $r \times r$ matrix and β_2 is $(n-r) \times r$. Assuming that β_1 has full rank, β can be post-multiplied by β_1^{-1}, giving

$$\hat{\beta} = \begin{bmatrix} I \\ \beta_2\beta_1^{-1} \end{bmatrix} = \begin{bmatrix} I \\ -B \end{bmatrix},$$

The coefficients that gretl produces are $\hat{\beta}$, with B known as the matrix of unrestricted coefficients. In terms of the underlying equilibrium relationship, the Phillips normalization expresses the system of r equilibrium relations as

$$\begin{aligned} y_{1,t} &= b_{1,r+1}y_{r+1,t} + \ldots + b_{1,n}y_{n,t} \\ y_{2,t} &= b_{2,r+1}y_{r+1,t} + \ldots + b_{2,n}y_{n,t} \\ &\vdots \\ y_{r,t} &= b_{r,r+1}y_{r+1,t} + \ldots + b_{r,n}y_{r,t} \end{aligned}$$

where the first r variables are expressed as functions of the remaining $n - r$.

Although the triangular representation ensures that the statistical problem of estimating β is solved, the resulting equilibrium relationships may be difficult to interpret. In this case, the

[1] For comparison with other studies, you may wish to normalize β differently. Using the `set` command you can do `set vecm_norm diag` to select a normalization that simply scales the columns of the original β such that $\beta_{ij} = 1$ for $i = j$ and $i \leq r$, as used in the empirical section of Boswijk and Doornik (2004). Another alternative is `set vecm_norm first`, which scales β such that the elements on the first row equal 1. To suppress normalization altogether, use `set vecm_norm none`. (To return to the default: `set vecm_norm phillips`.)

user may want to achieve identification by specifying manually the system of r^2 constraints that gretl will use to produce an estimate of β.

As an example, consider the money demand system presented in section 9.6 of Verbeek (2004). The variables used are m (the log of real money stock M1), infl (inflation), cpr (the commercial paper rate), y (log of real GDP) and tbr (the Treasury bill rate).[2]

Estimation of β can be performed via the commands

```
open money.gdt
smpl 1954:1 1994:4
vecm 6 2 m infl cpr y tbr --rc
```

and the relevant portion of the output reads

```
Maximum likelihood estimates, observations 1954:1-1994:4 (T = 164)
Cointegration rank = 2
Case 2: Restricted constant

beta (cointegrating vectors, standard errors in parentheses)

m             1.0000        0.0000
            (0.0000)      (0.0000)
infl          0.0000        1.0000
            (0.0000)      (0.0000)
cpr          0.56108       -24.367
           (0.10638)      (4.2113)
y           -0.40446      -0.91166
           (0.10277)      (4.0683)
tbr         -0.54293        24.786
           (0.10962)      (4.3394)
const        -3.7483        16.751
           (0.78082)      (30.909)
```

Interpretation of the coefficients of the cointegration matrix β would be easier if a meaning could be attached to each of its columns. This is possible by hypothesizing the existence of two long-run relationships: a money demand equation

$$\mathtt{m} = c_1 + \beta_1 \mathtt{infl} + \beta_2 \mathtt{y} + \beta_3 \mathtt{tbr}$$

and a risk premium equation

$$\mathtt{cpr} = c_2 + \beta_4 \mathtt{infl} + \beta_5 \mathtt{y} + \beta_6 \mathtt{tbr}$$

which imply that the cointegration matrix can be normalized as

$$\beta = \begin{bmatrix} -1 & 0 \\ \beta_1 & \beta_4 \\ 0 & -1 \\ \beta_2 & \beta_5 \\ \beta_3 & \beta_6 \\ c_1 & c_2 \end{bmatrix}$$

This renormalization can be accomplished by means of the restrict command, to be given after the vecm command or, in the graphical interface, by selecting the "Test, Linear Restrictions" menu entry. The syntax for entering the restrictions should be fairly obvious:[3]

[2]This data set is available in the verbeek data package; see http://gretl.sourceforge.net/gretl_data.html.

[3]Note that in this context we are bending the usual matrix indexation convention, using the leading index to refer to the *column* of β (the particular cointegrating vector). This is standard practice in the literature, and defensible insofar as it is the columns of β (the cointegrating relations or equilibrium errors) that are of primary interest.

```
restrict
  b[1,1] = -1
  b[1,3] = 0
  b[2,1] = 0
  b[2,3] = -1
end restrict
```

which produces

```
Cointegrating vectors (standard errors in parentheses)

m          -1.0000        0.0000
          (0.0000)       (0.0000)
infl     -0.023026      0.041039
       (0.0054666)     (0.027790)
cpr         0.0000       -1.0000
          (0.0000)       (0.0000)
y          0.42545     -0.037414
        (0.033718)      (0.17140)
tbr      -0.027790        1.0172
       (0.0045445)     (0.023102)
const       3.3625       0.68744
         (0.25318)       (1.2870)
```

27.6 Over-identifying restrictions

One purpose of imposing restrictions on a VECM system is simply to achieve identification. If these restrictions are simply normalizations, they are not testable and should have no effect on the maximized likelihood. In addition, however, one may wish to formulate constraints on β and/or α that derive from the economic theory underlying the equilibrium relationships; substantive restrictions of this sort are then testable via a likelihood-ratio statistic.

Gretl is capable of testing general linear restrictions of the form

$$R_b \text{vec}(\beta) = q \tag{27.5}$$

and/or

$$R_a \text{vec}(\alpha) = 0 \tag{27.6}$$

Note that the β restriction may be non-homogeneous ($q \neq 0$) but the α restriction must be homogeneous. Nonlinear restrictions are not supported, and neither are restrictions that cross between β and α. In the case where $r > 1$ such restrictions may be in common across all the columns of β (or α) or may be specific to certain columns of these matrices. This is the case discussed in Boswijk (1995) and Boswijk and Doornik (2004), section 4.4.

The restrictions (27.5) and (27.6) may be written in explicit form as

$$\text{vec}(\beta) = H\phi + h_0 \tag{27.7}$$

and

$$\text{vec}(\alpha') = G\psi \tag{27.8}$$

respectively, where ϕ and ψ are the free parameter vectors associated with β and α respectively. We may refer to the free parameters collectively as θ (the column vector formed by concatenating ϕ and ψ). Gretl uses this representation internally when testing the restrictions.

If the list of restrictions that is passed to the `restrict` command contains more constraints than necessary to achieve identification, then an LR test is performed; moreover, the `restrict` command can be given the `--full` switch, in which case full estimates for the restricted system are printed (including the Γ_i terms), and the system thus restricted becomes the "current model" for the purposes of further tests. Thus you are able to carry out cumulative tests, as in Chapter 7 of Johansen (1995).

Syntax

The full syntax for specifying the restriction is an extension of that exemplified in the previous section. Inside a `restrict...end restrict` block, valid statements are of the form

parameter linear combination = *scalar*

where a parameter linear combination involves a weighted sum of individual elements of β or α (but not both in the same combination); the scalar on the right-hand side must be 0 for combinations involving α, but can be any real number for combinations involving β. Below, we give a few examples of valid restrictions:

```
b[1,1] = 1.618
b[1,4] + 2*b[2,5] = 0
a[1,3] = 0
a[1,1] - a[1,2] = 0
```

Special syntax is used when a certain constraint should be applied to all columns of β: in this case, one index is given for each b term, and the square brackets are dropped. Hence, the following syntax

```
restrict
  b1 + b2 = 0
end restrict
```

corresponds to

$$\beta = \begin{bmatrix} \beta_{11} & \beta_{21} \\ -\beta_{11} & -\beta_{21} \\ \beta_{13} & \beta_{23} \\ \beta_{14} & \beta_{24} \end{bmatrix}$$

The same convention is used for α: when only one index is given for an a term the restriction is presumed to apply to all r columns of α, or in other words the variable associated with the given row of α is weakly exogenous. For instance, the formulation

```
restrict
  a3 = 0
  a4 = 0
end restrict
```

specifies that variables 3 and 4 do not respond to the deviation from equilibrium in the previous period.[4]

A variant on the single-index syntax for common restrictions on α and β is available: you can replace the index number with the name of the corresponding variable, in square brackets. For example, instead of `a3 = 0` one could write `a[cpr] = 0`, if the third variable in the system is named `cpr`.

Finally, a short-cut is available for setting up complex restrictions (but currently only in relation to β): you can specify R_b and q, as in $R_b\text{vec}(\beta) = q$, by giving the names of previously defined matrices. For example,

```
matrix I4 = I(4)
matrix vR = I4**(I4~zeros(4,1))
matrix vq = mshape(I4,16,1)
restrict
  R = vR
  q = vq
end restrict
```

which manually imposes Phillips normalization on the β estimates for a system with cointegrating rank 4.

[4]Note that when two indices are given in a restriction on α the indexation is consistent with that for β restrictions: the leading index denotes the cointegrating vector and the trailing index the equation number.

An example

Brand and Cassola (2004) propose a money demand system for the Euro area, in which they postulate three long-run equilibrium relationships:

money demand	$m = \beta_l l + \beta_y y$
Fisher equation	$\pi = \phi l$
Expectation theory of interest rates	$l = s$

where m is real money demand, l and s are long- and short-term interest rates, y is output and π is inflation.[5] (The names for these variables in the gretl data file are `m_p`, `rl`, `rs`, `y` and `infl`, respectively.)

The cointegration rank assumed by the authors is 3 and there are 5 variables, giving 15 elements in the β matrix. $3 \times 3 = 9$ restrictions are required for identification, and a just-identified system would have $15 - 9 = 6$ free parameters. However, the postulated long-run relationships feature only three free parameters, so the over-identification rank is 3.

Example 27.1 replicates Table 4 on page 824 of the Brand and Cassola article.[6] Note that we use the `$lnl` accessor after the `vecm` command to store the unrestricted log-likelihood and the `$rlnl` accessor after `restrict` for its restricted counterpart.

The example continues in script 27.2, where we perform further testing to check whether (a) the income elasticity in the money demand equation is 1 ($\beta_y = 1$) and (b) the Fisher relation is homogeneous ($\phi = 1$). Since the `--full` switch was given to the initial `restrict` command, additional restrictions can be applied without having to repeat the previous ones. (The second script contains a few `printf` commands, which are not strictly necessary, to format the output nicely.) It turns out that both of the additional hypotheses are rejected by the data, with p-values of 0.002 and 0.004.

Another type of test that is commonly performed is the "weak exogeneity" test. In this context, a variable is said to be weakly exogenous if all coefficients on the corresponding row in the α matrix are zero. If this is the case, that variable does not adjust to deviations from any of the long-run equilibria and can be considered an autonomous driving force of the whole system.

The code in Example 27.3 performs this test for each variable in turn, thus replicating the first column of Table 6 on page 825 of Brand and Cassola (2004). The results show that weak exogeneity might perhaps be accepted for the long-term interest rate and real GDP (p-values 0.07 and 0.08 respectively).

Identification and testability

One point regarding VECM restrictions that can be confusing at first is that identification (does the restriction identify the system?) and testability (is the restriction testable?) are quite separate matters. Restrictions can be identifying but not testable; less obviously, they can be testable but not identifying.

This can be seen quite easily in relation to a rank-1 system. The restriction $\beta_1 = 1$ is identifying (it pins down the scale of β) but, being a pure scaling, it is not testable. On the other hand, the restriction $\beta_1 + \beta_2 = 0$ is testable—the system with this requirement imposed will almost certainly have a lower maximized likelihood—but it is not identifying; it still leaves open the scale of β.

We said above that the number of restrictions must equal at least r^2, where r is the cointegrating rank, for identification. This is a necessary and not a sufficient condition. In fact, when $r > 1$ it can be quite tricky to assess whether a given set of restrictions is identifying. Gretl

[5] A traditional formulation of the Fisher equation would reverse the roles of the variables in the second equation, but this detail is immaterial in the present context; moreover, the expectation theory of interest rates implies that the third equilibrium relationship should include a constant for the liquidity premium. However, since in this example the system is estimated with the constant term unrestricted, the liquidity premium gets merged in the system intercept and disappears from z_t.

[6] Modulo what appear to be a few typos in the article.

Example 27.1: Estimation of a money demand system with constraints on β

Input:

```
open brand_cassola.gdt

# perform a few transformations
m_p = m_p*100
y = y*100
infl = infl/4
rs = rs/4
rl = rl/4

# replicate table 4, page 824
vecm 2 3 m_p infl rl rs y -q
genr ll0 = $lnl

restrict --full
  b[1,1] = 1
  b[1,2] = 0
  b[1,4] = 0
  b[2,1] = 0
  b[2,2] = 1
  b[2,4] = 0
  b[2,5] = 0
  b[3,1] = 0
  b[3,2] = 0
  b[3,3] = 1
  b[3,4] = -1
  b[3,5] = 0
end restrict
genr ll1 = $rlnl
```

Partial output:

```
Unrestricted loglikelihood (lu) = 116.60268
Restricted loglikelihood (lr) = 115.86451
2 * (lu - lr) = 1.47635
P(Chi-Square(3) > 1.47635) = 0.68774

beta (cointegrating vectors, standard errors in parentheses)

m_p           1.0000        0.0000        0.0000
            (0.0000)      (0.0000)      (0.0000)
infl          0.0000        1.0000        0.0000
            (0.0000)      (0.0000)      (0.0000)
rl            1.6108      -0.67100        1.0000
           (0.62752)    (0.049482)      (0.0000)
rs            0.0000        0.0000       -1.0000
            (0.0000)      (0.0000)      (0.0000)
y            -1.3304        0.0000        0.0000
          (0.030533)      (0.0000)      (0.0000)
```

Example 27.2: Further testing of money demand system

Input:

```
restrict
  b[1,5] = -1
end restrict
genr ll_uie = $rlnl

restrict
  b[2,3] = -1
end restrict
genr ll_hfh = $rlnl

# replicate table 5, page 824
printf "Testing zero restrictions in cointegration space:\n"
printf "  LR-test, rank = 3: chi^2(3) = %6.4f [%6.4f]\n", 2*(ll0-ll1), \
        pvalue(X, 3, 2*(ll0-ll1))

printf "Unit income elasticity: LR-test, rank = 3:\n"
printf "  chi^2(4) = %g [%6.4f]\n", 2*(ll0-ll_uie), \
        pvalue(X, 4, 2*(ll0-ll_uie))

printf "Homogeneity in the Fisher hypothesis:\n"
printf "  LR-test, rank = 3: chi^2(4) = %6.3f [%6.4f]\n", 2*(ll0-ll_hfh), \
        pvalue(X, 4, 2*(ll0-ll_hfh))
```

Output:

```
Testing zero restrictions in cointegration space:
  LR-test, rank = 3: chi^2(3) = 1.4763 [0.6877]
Unit income elasticity: LR-test, rank = 3:
  chi^2(4) = 17.2071 [0.0018]
Homogeneity in the Fisher hypothesis:
  LR-test, rank = 3: chi^2(4) = 15.547 [0.0037]
```

Example 27.3: Testing for weak exogeneity

Input:

```
restrict
  a1 = 0
end restrict
ts_m = 2*(ll0 - $rlnl)

restrict
  a2 = 0
end restrict
ts_p = 2*(ll0 - $rlnl)

restrict
  a3 = 0
end restrict
ts_l = 2*(ll0 - $rlnl)

restrict
  a4 = 0
end restrict
ts_s = 2*(ll0 - $rlnl)

restrict
  a5 = 0
end restrict
ts_y = 2*(ll0 - $rlnl)

loop foreach i m p l s y --quiet
  printf "\Delta $i\t%6.3f [%6.4f]\n", ts_$i, pvalue(X, 6, ts_$i)
endloop
```

Output (variable, LR test, p-value):

```
\Delta m        18.111 [0.0060]
\Delta p        21.067 [0.0018]
\Delta l        11.819 [0.0661]
\Delta s        16.000 [0.0138]
\Delta y        11.335 [0.0786]
```

uses the method suggested by Doornik (1995), where identification is assessed via the rank of the information matrix.

It can be shown that for restrictions of the sort (27.7) and (27.8) the information matrix has the same rank as the Jacobian matrix

$$\mathcal{J}(\theta) = \left[(I_p \otimes \beta)G : (\alpha \otimes I_{p_1})H \right]$$

A sufficient condition for identification is that the rank of $\mathcal{J}(\theta)$ equals the number of free parameters. The rank of this matrix is evaluated by examination of its singular values at a randomly selected point in the parameter space. For practical purposes we treat this condition as if it were both necessary and sufficient; that is, we disregard the special cases where identification could be achieved without this condition being met.[7]

27.7 Numerical solution methods

In general, the ML estimator for the restricted VECM problem has no closed form solution, hence the maximum must be found via numerical methods.[8] In some cases convergence may be difficult, and gretl provides several choices to solve the problem.

Switching and LBFGS

Two maximization methods are available in gretl. The default is the switching algorithm set out in Boswijk and Doornik (2004). The alternative is a limited-memory variant of the BFGS algorithm (LBFGS), using analytical derivatives. This is invoked using the `--lbfgs` flag with the `restrict` command.

The switching algorithm works by explicitly maximizing the likelihood at each iteration, with respect to $\hat{\phi}$, $\hat{\psi}$ and $\hat{\Omega}$ (the covariance matrix of the residuals) in turn. This method shares a feature with the basic Johansen eigenvalues procedure, namely, it can handle a set of restrictions that does not fully identify the parameters.

LBFGS, on the other hand, requires that the model be fully identified. When using LBFGS, therefore, you may have to supplement the restrictions of interest with normalizations that serve to identify the parameters. For example, one might use all or part of the Phillips normalization (see section 27.5).

Neither the switching algorithm nor LBFGS is guaranteed to find the global ML solution.[9] The optimizer may end up at a local maximum (or, in the case of the switching algorithm, at a saddle point).

The solution (or lack thereof) may be sensitive to the initial value selected for θ. By default, gretl selects a starting point using a deterministic method based on Boswijk (1995), but two further options are available: the initialization may be adjusted using simulated annealing, or the user may supply an explicit initial value for θ.

The default initialization method is:

1. Calculate the unrestricted ML $\hat{\beta}$ using the Johansen procedure.

2. If the restriction on β is non-homogeneous, use the method proposed by Boswijk:

$$\phi_0 = -[(I_r \otimes \hat{\beta}_\perp)' H]^+ (I_r \otimes \hat{\beta}_\perp)' h_0 \tag{27.9}$$

where $\hat{\beta}_\perp' \hat{\beta} = 0$ and A^+ denotes the Moore–Penrose inverse of A. Otherwise

$$\phi_0 = (H'H)^{-1} H' \text{vec}(\hat{\beta}) \tag{27.10}$$

[7] See Boswijk and Doornik (2004), pp. 447–8 for discussion of this point.

[8] The exception is restrictions that are homogeneous, common to all β or all α (in case $r > 1$), and involve either β only or α only. Such restrictions are handled via the modified eigenvalues method set out by Johansen (1995). We solve directly for the ML estimator, without any need for iterative methods.

[9] In developing gretl's VECM-testing facilities we have considered a fair number of "tricky cases" from various sources. We'd like to thank Luca Fanelli of the University of Bologna and Sven Schreiber of Goethe University Frankfurt for their help in devising torture-tests for gretl's VECM code.

3. $\text{vec}(\beta_0) = H\phi_0 + h_0$.

4. Calculate the unrestricted ML $\hat{\alpha}$ conditional on β_0, as per Johansen:

$$\hat{\alpha} = S_{01}\beta_0(\beta_0' S_{11}\beta_0)^{-1} \tag{27.11}$$

5. If α is restricted by $\text{vec}(\alpha') = G\psi$, then $\psi_0 = (G'G)^{-1}G' \,\text{vec}(\hat{\alpha}')$ and $\text{vec}(\alpha_0') = G\psi_0$.

Alternative initialization methods

As mentioned above, gretl offers the option of adjusting the initialization using **simulated annealing**. This is invoked by adding the `--jitter` option to the `restrict` command.

The basic idea is this: we start at a certain point in the parameter space, and for each of n iterations (currently $n = 4096$) we randomly select a new point within a certain radius of the previous one, and determine the likelihood at the new point. If the likelihood is higher, we jump to the new point; otherwise, we jump with probability P (and remain at the previous point with probability $1-P$). As the iterations proceed, the system gradually "cools"—that is, the radius of the random perturbation is reduced, as is the probability of making a jump when the likelihood fails to increase.

In the course of this procedure many points in the parameter space are evaluated, starting with the point arrived at by the deterministic method, which we'll call θ_0. One of these points will be "best" in the sense of yielding the highest likelihood: call it θ^*. This point may or may not have a greater likelihood than θ_0. And the procedure has an end point, θ_n, which may or may not be "best".

The rule followed by gretl in selecting an initial value for θ based on simulated annealing is this: use θ^* if $\theta^* > \theta_0$, otherwise use θ_n. That is, if we get an improvement in the likelihood via annealing, we make full use of this; on the other hand, if we fail to get an improvement we nonetheless allow the annealing to randomize the starting point. Experiments indicated that the latter effect can be helpful.

Besides annealing, a further alternative is **manual initialization**. This is done by passing a predefined vector to the `set` command with parameter `initvals`, as in

```
set initvals myvec
```

The details depend on whether the switching algorithm or LBFGS is used. For the switching algorithm, there are two options for specifying the initial values. The more user-friendly one (for most people, we suppose) is to specify a matrix that contains $\text{vec}(\beta)$ followed by $\text{vec}(\alpha)$. For example:

```
open denmark.gdt
vecm 2 1 LRM LRY IBO IDE --rc --seasonals

matrix BA = {1, -1, 6, -6, -6, -0.2, 0.1, 0.02, 0.03}
set initvals BA
restrict
  b[1] = 1
  b[1] + b[2] = 0
  b[3] + b[4] = 0
end restrict
```

In this example—from Johansen (1995)—the cointegration rank is 1 and there are 4 variables. However, the model includes a restricted constant (the `--rc` flag) so that β has 5 elements. The α matrix has 4 elements, one per equation. So the matrix `BA` may be read as

$$(\beta_1, \beta_2, \beta_3, \beta_4, \beta_5, \alpha_1, \alpha_2, \alpha_3, \alpha_4)$$

The other option, which is compulsory when using LBFGS, is to specify the initial values in terms of the free parameters, ϕ and ψ. Getting this right is somewhat less obvious. As mentioned above, the implicit-form restriction $R\text{vec}(\beta) = q$ has explicit form $\text{vec}(\beta) = H\phi + h_0$, where $H = R_{\perp}$, the right nullspace of R. The vector ϕ is shorter, by the number of restrictions, than $\text{vec}(\beta)$. The savvy user will then see what needs to be done. The other point to take into account is that if α is unrestricted, the *effective* length of ψ is 0, since it is then optimal to compute α using Johansen's formula, conditional on β (equation 27.11 above). The example above could be rewritten as:

```
open denmark.gdt
vecm 2 1 LRM LRY IBO IDE --rc --seasonals

matrix phi = {-8, -6}
set initvals phi
restrict --lbfgs
  b[1] = 1
  b[1] + b[2] = 0
  b[3] + b[4] = 0
end restrict
```

In this more economical formulation the initializer specifies only the two free parameters in ϕ (5 elements in β minus 3 restrictions). There is no call to give values for ψ since α is unrestricted.

Scale removal

Consider a simpler version of the restriction discussed in the previous section, namely,

```
restrict
  b[1] = 1
  b[1] + b[2] = 0
end restrict
```

This restriction comprises a substantive, testable requirement—that β_1 and β_2 sum to zero—and a normalization or scaling, $\beta_1 = 1$. The question arises, might it be easier and more reliable to maximize the likelihood without imposing $\beta_1 = 1$?[10] If so, we could record this normalization, remove it for the purpose of maximizing the likelihood, then reimpose it by scaling the result.

Unfortunately it is not possible to say in advance whether "scale removal" of this sort will give better results, for any particular estimation problem. However, this does seem to be the case more often than not. Gretl therefore performs scale removal where feasible, unless you

- explicitly forbid this, by giving the `--no-scaling` option flag to the restrict command; or
- provide a specific vector of initial values; or
- select the LBFGS algorithm for maximization.

Scale removal is deemed infeasible if there are any cross-column restrictions on β, or any non-homogeneous restrictions involving more than one element of β.

In addition, experimentation has suggested to us that scale removal is inadvisable if the system is just identified with the normalization(s) included, so we do not do it in that case. By "just identified" we mean that the system would not be identified if any of the restrictions were removed. On that criterion the above example is not just identified, since the removal of the second restriction would not affect identification; and gretl would in fact perform scale removal in this case unless the user specified otherwise.

[10]As a numerical matter, that is. In principle this should make no difference.

Chapter 28

Multivariate models

By a multivariate model we mean one that includes more than one dependent variable. Certain specific types of multivariate model for time-series data are discussed elsewhere: chapter 26 deals with VARs and chapter 27 with VECMs. Here we discuss two general sorts of multivariate model, implemented in gretl via the `system` command: SUR systems (Seemingly Unrelated Regressions), in which all the regressors are taken to be exogenous and interest centers on the covariance of the error term across equations; and simultaneous systems, in which some regressors are assumed to be endogenous.

In this chapter we give an account of the syntax and use of the `system` command and its companions, `restrict` and `estimate`; we also explain the options and accessors available in connection with multivariate models.

28.1 The system command

The specification of a multivariate system takes the form of a block of statements, starting with `system` and ending with `end system`. Once a system is specified it can estimated via various methods, using the `estimate` command, with or without restrictions, which may be imposed via the `restrict` command.

Starting a system block

The first line of a `system` block may be augmented in either (or both) of two ways:

- An estimation method is specified for the system. This is done by following `system` with an expression of the form `method=`*estimator*, where *estimator* must be one of `ols` (Ordinary Least Squares), `tsls` (Two-Stage Least Squares), `sur` (Seemingly Unrelated Regressions), `3sls` (Three-Stage Least Squares), `liml` (Limited Information Maximum Likelihood) or `fiml` (Full Information Maximum Likelihood). Two examples:

  ```
  system method=sur
  system method=fiml
  ```

  OLS, TSLS and LIML are, of course, single-equation methods rather than true system estimators; they are included to facilitate comparisons.

- The system is assigned a name. This is done by giving the name first, followed by a back-arrow, "<-", followed by `system`. If the name contains spaces it must be enclosed in double-quotes. Here are two examples:

  ```
  sys1 <- system
  "System 1" <- system
  ```

  Note, however, that this naming method is not available within a user-defined function, only in the main body of a gretl script.

If the initial `system` line is augmented in the first way, the effect is that the system is estimated as soon as its definition is completed, using the specified method. The effect of the second option is that the system can then be referenced by the assigned name for the purposes of the `restrict` and `estimate` commands; in the gretl GUI an additional effect is that an icon for the system is added to the "Session view".

These two possibilities can be combined, as in

```
mysys <- system method=3sls
```

In this example the system is estimated immediately via Three-Stage Least Squares, and is also available for subsequent use under the name mysys.

If the system is not named via the back-arrow mechanism, it is still available for subsequent use via restrict and estimate; in this case you should use the generic name $system to refer to the last-defined multivariate system.

The body of a system block

The most basic element in the body of a system block is the equation statement, which is used to specify each equation within the system. This takes the same form as the regression specification for single-equation estimators, namely a list of series with the dependent variable given first, followed by the regressors, with the series given either by name or by ID number (order in the dataset). A system block must contain at least two equation statements, and for systems without endogenous regressors these statements are all that is required. So, for example, a minimal SUR specification might look like this:

```
system method=sur
  equation y1 const x1
  equation y2 const x2
end system
```

For simultaneous systems it is necessary to determine which regressors are endogenous and which exogenous. By default all regressors are treated as exogenous, except that any variable that appears as the dependent variable in one equation is automatically treated as endogeous if it appears as a regressor elsewhere. However, an explicit list of endogenous regressors may be supplied following the equations lines: this takes the form of the keyword endog followed by the names or ID numbers of the relevant regressors.

When estimation is via TSLS or 3SLS it is possible to specify a particular set of instruments for each equation. This is done by giving the equation lists in the format used with the tsls command: first the dependent variable, then the regressors, then a semicolon, then the instruments, as in

```
system method=3sls
  equation y1 const x11 x12 ; const x11 z1
  equation y2 const x21 x22 ; const x21 z2
end system
```

An alternative way of specifying instruments is to insert an extra line starting with instr, followed by the list of variables acting as instruments. This is especially useful for specifying the system with the equations keyword, see the following subsection. As in tsls, any regressors that are not also listed as instruments are treated as endogenous, so in the example above x11 and x21 are treated as exogenous while x21 and x22 are endogenous, and instrumented by z1 and z2 respectively.

One more sort of statement is allowed in a system block: that is, the keyword identity followed by an equation that defines an accounting relationship, rather then a stochastic one, between variables. For example,

```
identity Y = C + I + G + X
```

There can be more than one identity in a system block. But note that these statements are specific to estimation via FIML; they are ignored for other estimators.

Equation systems within functions

It is also possible to define a multivariate system in a programmatic way. This is useful if the precise specification of the system depends on some input parameters that are not known in advance, but are given when the script is actually run.[1]

The relevant syntax is given by the `equations` keyword (note the plural), which replaces the block of `equation` lines in the standard form. An `equations` line requires two list arguments. The first list must contain all series on the left-hand side of the system; thus the number of elements in this first list determines the number of equations in the system. The second list is a "list of lists", which is a special variant of the list data type. That is, for each equation of the system you must provide a list of right-hand side variables, and the lists for all equations must be joined by assigning them to another list object; in that assignment, they must be separated by a semicolon. Here is an example for a two-equation system:

```
list syslist = xlist1 ; xlist2
```

Therefore, specifying a system generically in this way just involves building the necessary list arguments, as shown in the following example:

```
open denmark
list LHS = LRM LRY
list RHS1 = const LRM(-1) IBO(-1) IDE(-1)
list RHS2 = const LRY(-1) IBO(-1)
list RHS = RHS1 ; RHS2
system method=ols
     equations LHS RHS
end system
```

As mentioned above, the option of assigning a specific name to a system is not available within functions, but the generic identifier `$system` can be used to similar effect. The following example shows how one can define a system, estimate it via two methods, apply a restriction, then re-estimate it subject to the restriction.

```
function void anonsys(series x, series y)
     system
         equation x const
         equation y const
     end system
     estimate $system method=ols
     estimate $system method=sur
     restrict $system
         b[1,1] - b[2,1] = 0
     end restrict
     estimate $system method=ols
end function
```

28.2 Restriction and estimation

The behavior of the `restrict` command is a little different for multivariate systems as compared with single-equation models.

In the single-equation case, `restrict` refers to the last-estimated model, and once the command is completed the restriction is tested. In the multivariate case, you must give the name of the system to which the restriction is to be applied (or `$system` to refer to the last-defined system), and the effect of the command is just to attach the restriction to the system; testing is not done until the next `estimate` command is given. In addition, in the system case the default is to produce full estimates of the restricted model; if you are not interested in the full estimates and just want the test statistic you can append the `--quiet` option to `estimate`.

[1]This feature was added in version 1.9.7 of gretl.

A given system restriction remains in force until it is replaced or removed. To return a system to its unrestricted state you can give an empty restrict block, as in

```
restrict sysname
end restrict
```

As illustrated above, you can use the `method` tag to specify an estimation method with the `estimate` command. If the system has already been estimated you can omit this tag and the previous method is used again.

The `estimate` command is the main locus for options regarding the details of estimation. The available options are as follows:

- If the estimation method is SUR or 3SLS and the `--iterate` flag is given, the estimator will be iterated. In the case of SUR, if the procedure converges the results are maximum likelihood estimates. Iteration of three-stage least squares, however, does not in general converge on the full-information maximum likelihood results. This flag is ignored for other estimators.
- If the equation-by-equation estimators OLS or TSLS are chosen, the default is to apply a degrees of freedom correction when calculating standard errors. This can be suppressed using the `--no-df-corr` flag. This flag has no effect with the other estimators; no degrees of freedom correction is applied in any case.
- By default, the formula used in calculating the elements of the cross-equation covariance matrix is

$$\hat{\sigma}_{ij} = \frac{\hat{u}_i'\hat{u}_j}{T}$$

where T is the sample size and $\hat{u}_i$ is the vector of residuals from equation i. But if the `--geomean` flag is given, a degrees of freedom correction is applied: the formula is

$$\hat{\sigma}_{ij} = \frac{\hat{u}_i'\hat{u}_j}{\sqrt{(T-k_i)(T-k_j)}}$$

where k_i denotes the number of independent parameters in equation i.

- If an iterative method is specified, the `--verbose` option calls for printing of the details of the iterations.
- When the system estimator is SUR or 3SLS the cross-equation covariance matrix is initially estimated via OLS or TSLS, respectively. In the case of a system subject to restrictions the question arises: should the initial single-equation estimator be restricted or unrestricted? The default is the former, but the `--unrestrict-init` flag can be used to select unrestricted initialization. (Note that this is unlikely to make much difference if the `--iterate` option is given.)

28.3 System accessors

After system estimation various matrices may be retrieved for further analysis. Let g denote the number of equations in the system and let K denote the total number of estimated parameters ($K = \sum_i k_i$). The accessors `$uhat` and `$yhat` get $T \times g$ matrices holding the residuals and fitted values respectively. The accessor `$coeff` gets the stacked K-vector of parameter estimates; `$vcv` gets the $K \times K$ variance matrix of the parameter estimates; and `$sigma` gets the $g \times g$ cross-equation covariance matrix, $\hat{\Sigma}$.

A test statistic for the hypothesis that Σ is diagonal can be retrieved as `$diagtest` and its p-value as `$diagpval`. This is the Breusch–Pagan test except when the estimator is (unrestricted) iterated SUR, in which case it's a Likelihood Ratio test. The Breusch–Pagan test is computed as

$$\mathrm{LM} = T \sum_{i=2}^{g} \sum_{j=1}^{i-1} r_{ij}^2$$

where $r_{ij} = \hat{\sigma}_{ij}/\sqrt{\hat{\sigma}_{ii}\hat{\sigma}_{jj}}$; the LR test is

$$\mathrm{LR} = T\left(\sum_{i=1}^{g} \log \hat{\sigma}_i^2 - \log|\hat{\Sigma}|\right)$$

where $\hat{\sigma}_i^2$ is $\hat{u}_i'\hat{u}_i/T$ from the individual OLS regressions. In both cases the test statistic is distributed asymptotically as χ^2 with $g(g-1)/2$ degrees of freedom.

Structural and reduced forms

Systems of simultaneous systems can be represented in structural form as

$$\Gamma y_t = A_1 y_{t-1} + A_2 y_{t-2} + \cdots + A_p y_{t-p} + B x_t + \epsilon_t$$

where y_t represents the vector of endogenous variables in period t, x_t denotes the vector of exogenous variables, and p is the maximum lag of the endogenous regressors. The structural-form matrices can be retrieved as `$sysGamma`, `$sysA` and `$sysB` respectively. If y_t is $m \times 1$ and x_t is $n \times 1$, then Γ is $m \times m$ and B is $m \times n$. If the system contains no lags of the endogenous variables then the A matrix is not defined, otherwise A is the horizontal concatenation of $A_1, \ldots, A_p$, and is therefore $m \times mp$.

From the structural form it is straightforward to obtain the reduced form, namely,

$$y_t = \Gamma^{-1}\left(\sum_{i=1}^{p} A_i y_{t-i}\right) + \Gamma^{-1} B x_t + v_t$$

where $v_t \equiv \Gamma^{-1}\epsilon_t$. The reduced form is used by gretl to generate forecasts in response to the `fcast` command. This means that—in contrast to single-equation estimation—the values produced via `fcast` for a static, within-sample forecast will in general differ from the fitted values retrieved via `$yhat`. The fitted values for equation i represent the expectation of y_{ti} conditional on the contemporaneous values of all the regressors, while the `fcast` values are conditional on the exogenous and predetermined variables only.

The above account has to be qualified for the case where a system is set up for estimation via TSLS or 3SLS using a specific list of instruments per equation, as described in section 28.1. In that case it is possible to include more endogenous regressors than explicit equations (although, of course, there must be sufficient instruments to achieve identification). In such systems endogenous regressors that have no associated explicit equation are treated "as if" exogenous when constructing the structural-form matrices. This means that forecasts are conditional on the observed values of the "extra" endogenous regressors rather than solely on the values of the exogenous and predetermined variables.

Chapter 29

Forecasting

29.1 Introduction

In some econometric contexts forecasting is the prime objective: one wants estimates of the future values of certain variables to reduce the uncertainty attaching to current decision making. In other contexts where real-time forecasting is not the focus prediction may nonetheless be an important moment in the analysis. For example, out-of-sample prediction can provide a useful check on the validity of an econometric model. In other cases we are interested in questions of "what if": for example, how might macroeconomic outcomes have differed over a certain period if a different policy had been pursued? In the latter cases "prediction" need not be a matter of actually projecting into the future but in any case it involves generating fitted values from a given model. The term "postdiction" might be more accurate but it is not commonly used; we tend to talk of prediction even when there is no true forecast in view.

This chapter offers an overview of the methods available within gretl for forecasting or prediction (whether forward in time or not) and explicates some of the finer points of the relevant commands.

29.2 Saving and inspecting fitted values

In the simplest case, the "predictions" of interest are just the (within sample) fitted values from an econometric model. For the single-equation linear model, $y_t = X_t\beta + u_t$, these are $\hat{y}_t = X_t\hat{\beta}$.

In command-line mode, the $\hat{y}$ series can be retrieved, after estimating a model, using the accessor `$yhat`, as in

```
series yh = $yhat
```

If the model in question takes the form of a system of equations, `$yhat` returns a matrix, each column of which contains the fitted values for a particular dependent variable. To extract the fitted series for, e.g., the dependent variable in the second equation, do

```
matrix Yh = $yhat
series yh2 = Yh[,2]
```

Having obtained a series of fitted values, you can use the `fcstats` function to produce a vector of statistics that characterize the accuracy of the predictions (see section 29.4 below).

The gretl GUI offers several ways of accessing and examining within-sample predictions. In the model display window the Save menu contains an item for saving fitted values, the Graphs menu allows plotting of fitted versus actual values, and the Analysis menu offers a display of actual, fitted and residual values.

29.3 The `fcast` command

The `fcast` command generates predictions based on the last estimated model. Several questions arise here: How to control the range over which predictions are generated? How to control the forecasting method (where a choice is available)? How to control the printing and/or saving of the results? Basic answers can be found in the *Gretl Command Reference*; we add some more details here.

The forecast range

The range defaults to the currently defined sample range. If this remains unchanged following estimation of the model in question, the forecast will be "within sample" and (with some qualifications noted below) it will essentially duplicate the information available via the retrieval of fitted values (see section 29.2 above).

A common situation is that a model is estimated over a given sample and then forecasts are wanted for a subsequent out-of-sample range. The simplest way to accomplish this is via the `--out-of-sample` option to `fcast`. For example, assuming we have a quarterly time-series dataset containing observations from 1980:1 to 2008:4, four of which are to be reserved for forecasting:

```
# reserve the last 4 observations
smpl 1980:1 2007:4
ols y 0 xlist
fcast --out-of-sample
```

This will generate a forecast from 2008:1 to 2008:4.

There are two other ways of adjusting the forecast range, offering finer control:

- Use the `smpl` command to adjust the sample range prior to invoking `fcast`.
- Use the optional *startobs* and *endobs* arguments to `fcast` (which should come right after the command word). These values set the forecast range independently of the sample range.

What if one wants to generate a true forecast that goes beyond the available data? In that case one can use the `dataset` command with the `addobs` parameter to add extra observations before forecasting. For example:

```
# use the entire dataset, which ends in 2008:4
ols y 0 xlist
dataset addobs 4
fcast 2009:1 2009:4
```

But this will work as stated only if the set of regressors in `xlist` does not contain any stochastic regressors other than lags of y. The `dataset addobs` command attempts to detect and extrapolate certain common deterministic variables (e.g., time trend, periodic dummy variables). In addition, lagged values of the dependent variable can be supported via a dynamic forecast (see below for discussion of the static/dynamic distinction). But "future" values of any other included regressors must be supplied before such a forecast is possible. Note that specific values in a series can be set directly by date, for example: `x1[2009:1] = 120.5`. Or, if the assumption of no change in the regressors is warranted, one can do something like this:

```
loop t=2009:1..2009:4
    loop foreach i xlist
        $i[t] = $i[2008:4]
    endloop
endloop
```

Static, dynamic and rolling forecasts

The distinction between static and dynamic forecasts applies only to dynamic models, i.e., those that feature one or more lags of the dependent variable. The simplest case is the AR(1) model,

$$y_t = \alpha_0 + \alpha_1 y_{t-1} + \epsilon_t \tag{29.1}$$

In some cases the presence of a lagged dependent variable is implicit in the dynamics of the error term, for example

$$\begin{aligned} y_t &= \beta + u_t \\ u_t &= \rho u_{t-1} + \epsilon_t \end{aligned}$$

which implies that

$$y_t = (1 - \rho)\beta + \rho y_{t-1} + \epsilon_t$$

Suppose we want to forecast y for period s using a dynamic model, say (29.1) for example. If we have data on y available for period $s - 1$ we could form a fitted value in the usual way: $\hat{y}_s = \hat{\alpha}_0 + \hat{\alpha}_1 y_{s-1}$. But suppose that data are available only up to $s - 2$. In that case we can apply the chain rule of forecasting:

$$\hat{y}_{s-1} = \hat{\alpha}_0 + \hat{\alpha}_1 y_{s-2}$$
$$\hat{y}_s = \hat{\alpha}_0 + \hat{\alpha}_1 \hat{y}_{s-1}$$

This is what is called a dynamic forecast. A static forecast, on the other hand, is simply a fitted value (even if it happens to be computed out-of-sample).

Printing and saving forecasts

To be written.

29.4 Univariate forecast evaluation statistics

Let y_t be the value of a variable of interest at time t and let f_t be a forecast of y_t. We define the forecast error as $e_t = y_t - f_t$. Given a series of T observations and associated forecasts we can construct several measures of the overall accuracy of the forecasts. Some commonly used measures are the Mean Error (ME), Mean Squared Error (MSE), Root Mean Squared Error (RMSE), Mean Absolute Error (MAE), Mean Percentage Error (MPE) and Mean Absolute Percentage Error (MAPE). These are defined as follows.

$$\text{ME} = \frac{1}{T}\sum_{t=1}^{T} e_t \qquad \text{MSE} = \frac{1}{T}\sum_{t=1}^{T} e_t^2 \qquad \text{RMSE} = \sqrt{\frac{1}{T}\sum_{t=1}^{T} e_t^2} \qquad \text{MAE} = \frac{1}{T}\sum_{t=1}^{T} |e_t|$$

$$\text{MPE} = \frac{1}{T}\sum_{t=1}^{T} 100\frac{e_t}{y_t} \qquad \text{MAPE} = \frac{1}{T}\sum_{t=1}^{T} 100\frac{|e_t|}{y_t}$$

A further relevant statistic is Theil's U (Theil, 1966), defined as the positive square root of

$$U^2 = \frac{1}{T}\sum_{t=1}^{T-1}\left(\frac{f_{t+1} - y_{t+1}}{y_t}\right)^2 \cdot \left[\frac{1}{T}\sum_{t=1}^{T-1}\left(\frac{y_{t+1} - y_t}{y_t}\right)^2\right]^{-1}$$

The more accurate the forecasts, the lower the value of Theil's U, which has a minimum of 0.[1] This measure can be interpreted as the ratio of the RMSE of the proposed forecasting model to the RMSE of a naïve model which simply predicts $y_{t+1} = y_t$ for all t. The naïve model yields $U = 1$; values less than 1 indicate an improvement relative to this benchmark and values greater than 1 a deterioration.

In addition, Theil (1966, pp. 33–36) proposed a decomposition of the MSE which can be useful in evaluating a set of forecasts. He showed that the MSE could be broken down into three non-negative components as follows

$$\text{MSE} = \left(\bar{f} - \bar{y}\right)^2 + \left(s_f - r s_y\right)^2 + \left(1 - r^2\right) s_y^2$$

where $\bar{f}$ and $\bar{y}$ are the sample means of the forecasts and the observations, s_f and s_y are the respective standard deviations (using T in the denominator), and r is the sample correlation between y and f. Dividing through by MSE we get

$$\frac{\left(\bar{f} - \bar{y}\right)^2}{\text{MSE}} + \frac{\left(s_f - r s_y\right)^2}{\text{MSE}} + \frac{\left(1 - r^2\right) s_y^2}{\text{MSE}} = 1 \tag{29.2}$$

[1] This statistic is sometimes called U_2, to distinguish it from a related but different U defined in an earlier work by Theil (1961). It seems to be generally accepted that the later version of Theil's U is a superior statistic, so we ignore the earlier version here.

Theil labeled the three terms on the left-hand side of (29.2) the bias proportion (U^M), regression proportion (U^R) and disturbance proportion (U^D), respectively. If y and f represent the in-sample observations of the dependent variable and the fitted values from a linear regression then the first two components, U^M and U^R, will be zero (apart from rounding error), and the entire MSE will be accounted for by the unsystematic part, U^D. In the case of out-of-sample prediction, however (or "prediction" over a sub-sample of the data used in the regression), U^M and U^R are not necessarily close to zero, although this is a desirable property for a forecast to have. U^M differs from zero if and only if the mean of the forecasts differs from the mean of the realizations, and U^R is non-zero if and only if the slope of a simple regression of the realizations on the forecasts differs from 1.

The above-mentioned statistics are printed as part of the output of the `fcast` command. They can also be retrieved in the form of a column vector using the function `fcstats`, which takes two series arguments corresponding to y and f. The vector returned is

$$\left(\text{ME} \quad \text{MSE} \quad \text{MAE} \quad \text{MPE} \quad \text{MAPE} \quad U \quad U^M \quad U^R \quad U^D \right)'$$

(Note that the RMSE is not included since it can easily be obtained given the MSE.) The series given as arguments to `fcstats` must not contain any missing values in the currently defined sample range; use the `smpl` command to adjust the range if needed.

29.5 Forecasts based on VAR models

To be written.

29.6 Forecasting from simultaneous systems

To be written.

Chapter 30

The Kalman Filter

30.1 Preamble

The Kalman filter has been used "behind the scenes" in gretl for quite some time, in computing ARMA estimates. But user access to the Kalman filter is new and it has not yet been tested to any great extent. We have run some tests of relatively simple cases against the benchmark of SsfPack Basic. This is state-space software written by Koopman, Shephard and Doornik and documented in Koopman, Shephard and Doornik (1999). It requires Doornik's ox program. Both ox and SsfPack are available as free downloads for academic use but neither is open-source; see http://www.ssfpack.com. Since Koopman is one of the leading researchers in this area, presumably the results from SsfPack are generally reliable. To date we have been able to replicate the SsfPack results in gretl with a high degree of precision.

We welcome both success reports and bug reports.

30.2 Notation

It seems that in econometrics everyone is happy with $y = X\beta + u$, but we can't, as a community, make up our minds on a standard notation for state-space models. Harvey (1989), Hamilton (1994), Harvey and Proietti (2005) and Pollock (1999) all use different conventions. The notation used here is based on James Hamilton's, with slight variations.

A state-space model can be written as

$$\boldsymbol{\xi}_{t+1} = \mathbf{F}_t \boldsymbol{\xi}_t + \mathbf{v}_t \tag{30.1}$$

$$\mathbf{y}_t = \mathbf{A}_t' \mathbf{x}_t + \mathbf{H}_t' \boldsymbol{\xi}_t + \mathbf{w}_t \tag{30.2}$$

where (30.1) is the state transition equation and (30.2) is the observation or measurement equation. The state vector, $\boldsymbol{\xi}_t$, is $(r \times 1)$ and the vector of observables, $\mathbf{y}_t$, is $(n \times 1)$; $\mathbf{x}_t$ is a $(k \times 1)$ vector of exogenous variables. The $(r \times 1)$ vector $\mathbf{v}_t$ and the $(n \times 1)$ vector $\mathbf{w}_t$ are assumed to be vector white noise:

$$E(\mathbf{v}_t \mathbf{v}_s') = \mathbf{Q}_t \text{ for } t = s, \text{ otherwise } \mathbf{0}$$

$$E(\mathbf{w}_t \mathbf{w}_s') = \mathbf{R}_t \text{ for } t = s, \text{ otherwise } \mathbf{0}$$

The number of time-series observations will be denoted by T. In the special case when $\mathbf{F}_t = \mathbf{F}$, $\mathbf{H}_t = \mathbf{H}$, $\mathbf{A}_t = \mathbf{A}$, $\mathbf{Q}_t = \mathbf{Q}$ and $\mathbf{R}_t = \mathbf{R}$, the model is said to be *time-invariant.*

The Kalman recursions

Using this notation, and assuming for the moment that $\mathbf{v}_t$ and $\mathbf{w}_t$ are mutually independent, the Kalman recursions can be written as follows.

Initialization is via the unconditional mean and variance of $\boldsymbol{\xi}_1$:

$$\hat{\boldsymbol{\xi}}_{1|0} = E(\boldsymbol{\xi}_1)$$

$$\mathbf{P}_{1|0} = E\left\{[\boldsymbol{\xi}_1 - E(\boldsymbol{\xi}_1)]\,[\boldsymbol{\xi}_1 - E(\boldsymbol{\xi}_1)]'\right\}$$

Usually these are given by $\hat{\boldsymbol{\xi}}_{1|0} = \mathbf{0}$ and

$$\text{vec}(\mathbf{P}_{1|0}) = [\mathbf{I}_{r^2} - \mathbf{F} \otimes \mathbf{F}]^{-1} \cdot \text{vec}(\mathbf{Q}) \tag{30.3}$$

but see below for further discussion of the initial variance.

Iteration then proceeds in two steps.[1] First we update the estimate of the state

$$\hat{\boldsymbol{\xi}}_{t+1|t} = \mathbf{F}_t \hat{\boldsymbol{\xi}}_{t|t-1} + \mathbf{K}_t \mathbf{e}_t \tag{30.4}$$

where $\mathbf{e}_t$ is the prediction error for the observable:

$$\mathbf{e}_t = \mathbf{y}_t - \mathbf{A}_t'\mathbf{x}_t - \mathbf{H}_t'\hat{\boldsymbol{\xi}}_{t|t-1}$$

and $\mathbf{K}_t$ is the gain matrix, given by

$$\mathbf{K}_t = \mathbf{F}_t \mathbf{P}_{t|t-1} \mathbf{H}_t \boldsymbol{\Sigma}_t^{-1} \tag{30.5}$$

with

$$\boldsymbol{\Sigma}_t = \mathbf{H}_t' \mathbf{P}_{t|t-1} \mathbf{H}_t + \mathbf{R}_t$$

The second step then updates the estimate of the variance of the state using

$$\mathbf{P}_{t+1|t} = \mathbf{F}_t \mathbf{P}_{t|t-1} \mathbf{F}_t' - \mathbf{K}_t \boldsymbol{\Sigma}_t \mathbf{K}_t' + \mathbf{Q}_t \tag{30.6}$$

Cross-correlated disturbances

The formulation given above assumes mutual independence of the disturbances in the state and observation equations, $\mathbf{v}_t$ and $\mathbf{w}_t$. This assumption holds good in many practical applications, but a more general formulation allows for cross-correlation. In place of (30.1)–(30.2) we may write

$$\begin{aligned} \boldsymbol{\xi}_{t+1} &= \mathbf{F}_t \boldsymbol{\xi}_t + \mathbf{B}_t \boldsymbol{\varepsilon}_t \\ \mathbf{y}_t &= \mathbf{A}_t'\mathbf{x}_t + \mathbf{H}_t'\boldsymbol{\xi}_t + \mathbf{C}_t \boldsymbol{\varepsilon}_t \end{aligned}$$

where $\boldsymbol{\varepsilon}_t$ is a $(p \times 1)$ disturbance vector, all the elements of which have unit variance, $\mathbf{B}_t$ is $(r \times p)$ and $\mathbf{C}_t$ is $(n \times p)$.

The no-correlation case is nested thus: define $\mathbf{v}_t^*$ and $\mathbf{w}_t^*$ as modified versions of $\mathbf{v}_t$ and $\mathbf{w}_t$, scaled such that each element has unit variance, and let

$$\boldsymbol{\varepsilon}_t = \begin{bmatrix} \mathbf{v}_t^* \\ \mathbf{w}_t^* \end{bmatrix}$$

so that $p = r + n$. Then (suppressing time subscripts for simplicity) let

$$\begin{aligned} \mathbf{B} &= \begin{bmatrix} \boldsymbol{\Gamma}_{r\times r} & \vdots & \mathbf{0}_{r\times n} \end{bmatrix} \\ \mathbf{C} &= \begin{bmatrix} \mathbf{0}_{n\times r} & \vdots & \boldsymbol{\Lambda}_{n\times n} \end{bmatrix} \end{aligned}$$

where $\boldsymbol{\Gamma}$ and $\boldsymbol{\Lambda}$ are lower triangular matrices satisfying $\mathbf{Q} = \boldsymbol{\Gamma}\boldsymbol{\Gamma}'$ and $\mathbf{R} = \boldsymbol{\Lambda}\boldsymbol{\Lambda}'$ respectively. The zero sub-matrices in the above expressions for $\mathbf{B}$ and $\mathbf{C}$ produce the case of mutual independence; this corresponds to the condition $\mathbf{BC}' = \mathbf{0}$.

In the general case p is not necessarily equal to $r + n$, and $\mathbf{BC}'$ may be non-zero. This means that the Kalman gain equation (30.5) must be modified as

$$\mathbf{K}_t = (\mathbf{F}_t \mathbf{P}_{t|t-1} \mathbf{H}_t + \mathbf{B}_t \mathbf{C}_t') \boldsymbol{\Sigma}_t^{-1} \tag{30.7}$$

Otherwise, the equations given earlier hold good, if we write $\mathbf{BB}'$ in place of $\mathbf{Q}$ and $\mathbf{CC}'$ in place of $\mathbf{R}$.

In the account of gretl's Kalman facility below we take the uncorrelated case as the baseline, but add remarks on how to handle the correlated case where applicable.

---

[1] For a justification of the following formulae see the classic book by Anderson and Moore (1979) or, for a more modern treatment, Pollock (1999) or Hamilton (1994). A transcription of R. E. Kalman's original paper (Kalman, 1960) is available at `http://www.cs.unc.edu/~welch/kalman/kalmanPaper.html`.

30.3 Intended usage

The Kalman filter can be used in three ways: two of these are the classic forward and backward pass, or filtering and smoothing respectively; the third use is simulation. In the filtering/smoothing case you have the data $\mathbf{y}_t$ and you want to reconstruct the states $\boldsymbol{\xi}_t$ (and the forecast errors as a by-product), but we may also have a computational apparatus that does the reverse: given artificially-generated series $\mathbf{w}_t$ and $\mathbf{v}_t$, generate the states $\boldsymbol{\xi}_t$ (and the observables $\mathbf{y}_t$ as a by-product).

The usefulness of the classical filter is well known; the usefulness of the Kalman filter as a simulation tool may be huge too. Think for instance of Monte Carlo experiments, simulation-based inference—see Gourieroux and Monfort (1996)—or Bayesian methods, especially in the context of the estimation of DSGE models.

30.4 Overview of syntax

Using the Kalman filter in gretl is a two-step process. First you set up your filter, using a block of commands starting with `kalman` and ending with `end kalman`—much like the `gmm` command. Then you invoke the functions `kfilter`, `ksmooth` or `ksimul` to do the actual work. The next two sections expand on these points.

30.5 Defining the filter

Each line within the `kalman` ... `end kalman` block takes the form

keyword value

where *keyword* represents a matrix, as shown below. (An additional matrix which may be useful in some cases is introduced later under the heading "Constant term in the state transition".)

Keyword	Symbol	Dimensions
`obsy`	$\mathbf{y}$	$T \times n$
`obsymat`	$\mathbf{H}$	$r \times n$
`obsx`	$\mathbf{x}$	$T \times k$
`obsxmat`	$\mathbf{A}$	$k \times n$
`obsvar`	$\mathbf{R}$	$n \times n$
`statemat`	$\mathbf{F}$	$r \times r$
`statevar`	$\mathbf{Q}$	$r \times r$
`inistate`	$\hat{\boldsymbol{\xi}}_{1\|0}$	$r \times 1$
`inivar`	$\mathbf{P}_{1\|0}$	$r \times r$

For the data matrices $\mathbf{y}$ and $\mathbf{x}$ the corresponding *value* may be the name of a predefined matrix, the name of a data series, or the name of a list of series.[2]

For the other inputs, *value* may be the name of a predefined matrix or, if the input in question happens to be (1×1), the name of a scalar variable or a numerical constant. If the *value* of a coefficient matrix is given as the name of a matrix or scalar variable, the input is not "hard-wired" into the Kalman structure, rather a record is made of the *name* of the variable and on each run of a Kalman function (as described below) its value is re-read. It is therefore possible to write one `kalman` block and then do several filtering or smoothing passes using different sets of coefficients.[3] An example of this technique is provided later, in the example scripts 30.1 and 30.2. This facility to alter the values of the coefficients between runs of the filter is to be distinguished from the case of *time-varying* matrices, which is discussed below.

[2]Note that the data matrices `obsy` and `obsx` have T rows. That is, the column vectors $\mathbf{y}_t$ and $\mathbf{x}_t$ in (30.1) and (30.2) are in fact the transposes of the t-dated rows of the full matrices.

[3]Note, however, that the dimensions of the various input matrices are defined via the initial `kalman` set-up and it is an error if any of the matrices are changed in size.

Not all of the above-mentioned inputs need be specified in every case; some are optional. (In addition, you can specify the matrices in any order.) The mandatory elements are **y**, **H**, **F** and **Q**, so the minimal kalman block looks like this:

```
kalman
  obsy y
  obsymat H
  statemat F
  statevar Q
end kalman
```

The optional matrices are listed below, along with the implication of omitting the given matrix.

Keyword	If omitted...
obsx	no exogenous variables in observation equation
obsxmat	no exogenous variables in observation equation
obsvar	no disturbance term in observation equation
inistate	$\hat{\boldsymbol{\xi}}_{1\|0}$ is set to a zero vector
inivar	$\mathbf{P}_{1\|0}$ is set automatically

It might appear that the obsx (**x**) and obsxmat (**A**) matrices must go together—either both are given or neither is given. But an exception is granted for convenience. If the observation equation includes a constant but no additional exogenous variables, you can give a $(1 \times n)$ value for **A** without having to specify obsx. More generally, if the row dimension of **A** is 1 greater than the column dimension of **x**, it is assumed that the first element of **A** is associated with an implicit column of 1s.

Regarding the automatic initialization of $\mathbf{P}_{1|0}$ (in case no inivar input is given): by default this is done as in equation (30.3). However, this method is applicable only if all the eigenvalues of **F** lie inside the unit circle. If this condition is not satisfied we instead apply a diffuse prior, setting $\mathbf{P}_{1|0} = \kappa \mathbf{I}_r$ with $\kappa = 10^7$. If you wish to impose this diffuse prior from the outset, append the option flag --diffuse to the end kalman statement.[4]

Time-varying matrices

Any or all of the matrices obsymat, obsxmat, obsvar, statemat and statevar may be time-varying. In that case the *value* corresponding to the matrix keyword should be given in a special form: the name of an existing matrix plus a function call which modifies that matrix, separated by a semicolon. Note that in this case you must use a matrix variable, even if the matrix in question happens to be 1×1.

For example, suppose the matrix **H** is time-varying. Then we might write

```
obsymat H ; modify_H(&H, theta)
```

where modify_H is a user-defined function which modifies matrix H (and theta is a suitable additional argument to that function, if required).

The above is just an illustration: the matrix argument does not have to come first, and the function can have as many arguments as you like. The essential point is that the function must modify the specified matrix, which requires that it be given as an argument in "pointer" form (preceded by &). The function need not return any value directly; if it does, that value is ignored.

Such matrix-modifying functions will be called at each time-step of the filter operation, prior to performing any calculations. They have access to the current time-step of the Kalman filter via the internal variable $kalman_t, which has value 1 on the first step, 2 on the second, and so on, up to step T. They also have access to the previous n-vector of forecast errors, $\mathbf{e}_{t-1}$, under the name $kalman_uhat. When $t = 1$ this will be a zero vector.

[4] Initialization of the Kalman filter outside of the case where equation (30.3) applies has been the subject of much discussion in the literature—see for example de Jong (1991), Koopman (1997). At present gretl does not implement any of the more elaborate proposals that have been made.

Correlated disturbances

Defining a filter in which the disturbances $\mathbf{v}_t$ and $\mathbf{w}_t$ are correlated involves one modification to the account given above. If you append the `--cross` option flag to the `end kalman` statement, then the matrices corresponding to the keywords `statevar` and `obsvar` are interpreted not as $\mathbf{Q}$ and $\mathbf{R}$ but rather as $\mathbf{B}$ and $\mathbf{C}$ as discussed in section 30.2. Gretl then computes $\mathbf{Q} = \mathbf{BB}'$ and $\mathbf{R} = \mathbf{CC}'$ as well as the cross-product $\mathbf{BC}'$ and utilizes the modified expression for the gain as given in equation (30.7). As mentioned above, $\mathbf{B}$ should be $(r \times p)$ and $\mathbf{C}$ should be $(n \times p)$, where p is the number of elements in the combined disturbance vector $\boldsymbol{\varepsilon}_t$.

Constant term in the state transition

In some applications it is useful to be able to represent a constant term in the state transition equation explicitly; that is, equation (30.1) becomes

$$\boldsymbol{\xi}_{t+1} = \boldsymbol{\mu} + \mathbf{F}_t\boldsymbol{\xi}_t + \mathbf{v}_t \tag{30.8}$$

This is never strictly necessary; the system (30.1) and (30.2) is general enough to accommodate such a term, by absorbing it as an extra (unvarying) element in the state vector. But this comes at the cost of expanding all the matrices that touch the state ($\boldsymbol{\xi}$, $\mathbf{F}$, $\mathbf{v}$, $\mathbf{Q}$, $\mathbf{H}$), making the model relatively awkward to formulate and forecasts relatively expensive to compute.

As a simple illustration, consider a univariate model in which the state, s_t, is just a random walk with drift μ and the observed variable, y_t, is the state plus white noise:

$$s_{t+1} = \mu + s_t + v_t \tag{30.9}$$

$$y_t = s_t + w_t \tag{30.10}$$

Putting this into the standard form of (30.1) and (30.2) we get:

$$\begin{bmatrix} s_{t+1} \\ \mu \end{bmatrix} = \begin{bmatrix} 1 & 1 \\ 0 & 1 \end{bmatrix} \begin{bmatrix} s_t \\ \mu \end{bmatrix} + \begin{bmatrix} v_t \\ 0 \end{bmatrix}, \qquad \mathbf{Q} = \begin{bmatrix} \sigma_v^2 & 0 \\ 0 & 0 \end{bmatrix}$$

$$y_t = \begin{bmatrix} 1 & 0 \end{bmatrix} \begin{bmatrix} s_t \\ \mu \end{bmatrix} + w_t$$

In such a simple case the notational and computational burden is not very great; nonetheless it is clearly more "natural" to express this system in the form of (30.9) and (30.10) and in a multivariate model the gain in parsimony could be substantial.

For this reason we support the use of an additional named matrix in the `kalman` setup, namely `stconst`. This corresponds to $\boldsymbol{\mu}$ in equation (30.8); it should be an $r \times 1$ vector (or if $r = 1$ may be given as the name of a scalar variable). The use of `stconst` in setting up a filter corresponding to (30.9) and (30.10) is shown below.

```
matrix H = {1}
matrix R = {1}
matrix F = {1}
matrix Q = {1}
matrix mu = {0.05}

kalman
  obsy y
  obsymat H
  obsvar R
  statemat F
  statevar Q
  stconst mu
end kalman
```

Handling of missing values

It is acceptable for the data matrices, obsy and obsx, to contain missing values. In this case the filtering operation will work around the missing values, and the ksmooth function can be used to obtain estimates of these values. However, there are two points to note.

First, gretl's default behavior is to skip missing observations when constructing matrices from data series. To change this, use the set command thus:

```
set skip_missing off
```

Second, the handling of missing values is not yet quite right for the case where the observable vector $\mathbf{y}_t$ contains more than one element. At present, if any of the elements of $\mathbf{y}_t$ are missing the entire observation is ignored. Clearly it should be possible to make use of any non-missing elements, and this is not very difficult in principle, it's just awkward and is not implemented yet.

Persistence and identity of the filter

At present there is no facility to create a "named filter". Only one filter can exist at any point in time, namely the one created by the last kalman block.[5] If a filter is already defined, and you give a new kalman block, the old filter is over-written. Otherwise the existing filter persists (and remains available for the kfilter, ksmooth and ksimul functions) until either (a) the gretl session is terminated or (b) the command delete kalman is given.

30.6 The kfilter function

Once a filter is established, as discussed in the previous section, kfilter can be used to run a forward, forecasting pass. This function returns a scalar code: 0 for successful completion, or 1 if numerical problems were encountered. On successful completion, two scalar accessor variables become available: $kalman_lnl, which gives the overall log-likelihood under the joint normality assumption,

$$\ell = -\frac{1}{2}\left[nT\log(2\pi) + \sum_{t=1}^{T}\log|\mathbf{\Sigma}_t| + \sum_{t=1}^{T}\mathbf{e}_t'\mathbf{\Sigma}_t^{-1}\mathbf{e}_t\right]$$

and $kalman_s2, which gives the estimated variance,

$$\hat{\sigma}^2 = \frac{1}{nT}\sum_{t=1}^{T}\mathbf{e}_t'\mathbf{\Sigma}_t^{-1}\mathbf{e}_t$$

(but see below for modifications to these formulae for the case of a diffuse prior). In addition the accessor $kalman_llt gives a $(T \times 1)$ vector, element t of which is

$$\ell_t = -\frac{1}{2}\left[n\log(2\pi) + \log|\mathbf{\Sigma}_t| + \mathbf{e}_t'\mathbf{\Sigma}_t^{-1}\mathbf{e}_t\right]$$

The kfilter function does not require any arguments, but up to five matrix quantities may be retrieved via optional pointer arguments. Each of these matrices has T rows, one for each time-step; the contents of the rows are shown in the following listing.

1. Forecast errors for the observable variables: $\mathbf{e}_t'$, n columns.
2. Variance matrix for the forecast errors: $\text{vech}(\mathbf{\Sigma}_t)'$, $n(n+1)/2$ columns.
3. Estimate of the state vector: $\hat{\boldsymbol{\xi}}_{t|t-1}'$, r columns.

[5]This is not quite true: more precisely, there can be no more than one Kalman filter *at each level of function execution.* That is, if a gretl script creates a Kalman filter, a user-defined function called from that script may also create a filter, without interfering with the original one.

4. MSE of estimate of the state vector: $\text{vech}(\mathbf{P}_{t|t-1})'$, $r(r+1)/2$ columns.

5. Kalman gain: $\text{vec}(\mathbf{K}_t)'$, rn columns.

Unwanted trailing arguments can be omitted, otherwise unwanted arguments can be skipped by using the keyword `null`. For example, the following call retrieves the forecast errors in the matrix E and the estimate of the state vector in S:

```
matrix E S
kfilter(&E, null, &S)
```

Matrices given as pointer arguments do not have to be correctly dimensioned in advance; they will be resized to receive the specified content.

Further note: in general, the arguments to `kfilter` should all be matrix-pointers, but under two conditions you can give a pointer to a series variable instead. The conditions are: (i) the matrix in question has just one column in context (for example, the first two matrices will have a single column if the length of the observables vector, n, equals 1) and (ii) the time-series length of the filter is equal to the current gretl sample size.

Likelihood under the diffuse prior

There seems to be general agreement in the literature that the log-likelihood calculation should be modified in the case of a diffuse prior for $\mathbf{P}_{1|0}$. However, it is not clear to us that there is a well-defined "correct" method for this. At present we emulate SsfPack (see Koopman *et al.* (1999) and section 30.1). In case $\mathbf{P}_{1|0} = \kappa \mathbf{I}_r$, we set $d = r$ and calculate

$$\ell = -\frac{1}{2}\left[(nT-d)\log(2\pi) + \sum_{t=1}^{T}\log|\boldsymbol{\Sigma}_t| + \sum_{t=1}^{T}\mathbf{e}_t'\boldsymbol{\Sigma}_t^{-1}\mathbf{e}_t - d\log(\kappa)\right]$$

and

$$\hat{\sigma}^2 = \frac{1}{nT-d}\sum_{t=1}^{T}\mathbf{e}_t'\boldsymbol{\Sigma}_t^{-1}\mathbf{e}_t$$

30.7 The ksmooth function

This function returns the $(T \times r)$ matrix of smoothed estimates of the state vector—that is, estimates based on all T observations: row t of this matrix holds $\hat{\boldsymbol{\xi}}_{t|T}'$. This function has no required arguments but it offers one optional matrix-pointer argument, which retrieves the variance of the smoothed state estimate, $\mathbf{P}_{t|T}$. The latter matrix is $(T \times r(r+1)/2)$; each row is in transposed vech form. Examples:

```
matrix S = ksmooth()   # smoothed state only
matrix P
S = ksmooth(&P)        # the variance is wanted
```

These values are computed via a backward pass of the filter, from $t = T$ to $t = 1$, as follows:

$$\begin{aligned}
\mathbf{L}_t &= \mathbf{F}_t - \mathbf{K}_t\mathbf{H}_t' \\
\mathbf{u}_{t-1} &= \mathbf{H}_t\boldsymbol{\Sigma}_t^{-1}\mathbf{e}_t + \mathbf{L}_t'\mathbf{u}_t \\
\mathbf{U}_{t-1} &= \mathbf{H}_t\boldsymbol{\Sigma}_t^{-1}\mathbf{H}_t' + \mathbf{L}_t'\mathbf{U}_t\mathbf{L}_t \\
\hat{\boldsymbol{\xi}}_{t|T} &= \hat{\boldsymbol{\xi}}_{t|t-1} + \mathbf{P}_{t|t-1}\mathbf{u}_{t-1} \\
\mathbf{P}_{t|T} &= \mathbf{P}_{t|t-1} - \mathbf{P}_{t|t-1}\mathbf{U}_{t-1}\mathbf{P}_{t|t-1}
\end{aligned}$$

with initial values $\mathbf{u}_T = 0$ and $\mathbf{U}_T = 0$.[6]

This iteration is preceded by a special forward pass in which the matrices $\mathbf{K}_t$, $\boldsymbol{\Sigma}_t^{-1}$, $\hat{\boldsymbol{\xi}}_{t|t-1}$ and $\mathbf{P}_{t|t-1}$ are stored for all t. If $\mathbf{F}$ is time-varying, its values for all t are stored on the forward pass, and similarly for $\mathbf{H}$.

[6] See I. Karibzhanov's exposition at `http://www.econ.umn.edu/~karib003/help/kalcvs.htm`.

30.8 The ksimul function

This simulation function takes up to three arguments. The first, mandatory, argument is a $(T \times r)$ matrix containing artificial disturbances for the state transition equation: row t of this matrix represents $\mathbf{v}_t'$. If the current filter has a non-null $\mathbf{R}$ (obsvar) matrix, then the second argument should be a $(T \times n)$ matrix containing artificial disturbances for the observation equation, on the same pattern. Otherwise the second argument should be given as null. If $r = 1$ you may give a series for the first argument, and if $n = 1$ a series is acceptable for the second argument.

Provided that the current filter does not include exogenous variables in the observation equation (obsx), the T for simulation need not equal that defined by the original obsy data matrix: in effect T is temporarily redefined by the row dimension of the first argument to ksimul. Once the simulation is completed, the T value associated with the original data is restored.

The value returned by ksimul is a $(T \times n)$ matrix holding simulated values for the observables at each time step. A third optional matrix-pointer argument allows you to retrieve a $(T \times r)$ matrix holding the simulated state vector. Examples:

```
matrix Y = ksimul(V)         # obsvar is null
Y = ksimul(V, W)             # obsvar is non-null
matrix S
Y = ksimul(V, null, &S) # the simulated state is wanted
```

The initial value $\boldsymbol{\xi}_1$ is calculated thus: we find the matrix $\mathbf{T}$ such that $\mathbf{TT}' = \mathbf{P}_{1|0}$ (as given by the inivar element in the kalman block), multiply it into $\mathbf{v}_1$, and add the result to $\boldsymbol{\xi}_{1|0}$ (as given by inistate).

If the disturbances are correlated across the two equations the arguments to ksimul must be revised: the first argument should be a $(T \times p)$ matrix, each row of which represents $\boldsymbol{\varepsilon}_t'$ (see section 30.2), and the second argument should be given as null.

30.9 Example 1: ARMA estimation

As is well known, the Kalman filter provides a very efficient way to compute the likelihood of ARMA models; as an example, take an ARMA(1,1) model

$$y_t = \phi y_{t-1} + \varepsilon_t + \theta \varepsilon_{t-1}$$

One of the ways the above equation can be cast in state-space form is by defining a latent process $\xi_t = (1 - \phi L)^{-1} \varepsilon_t$. The observation equation corresponding to (30.2) is then

$$y_t = \xi_t + \theta \xi_{t-1} \tag{30.11}$$

and the state transition equation corresponding to (30.1) is

$$\begin{bmatrix} \xi_t \\ \xi_{t-1} \end{bmatrix} = \begin{bmatrix} \phi & 0 \\ 1 & 0 \end{bmatrix} \begin{bmatrix} \xi_{t-1} \\ \xi_{t-2} \end{bmatrix} + \begin{bmatrix} \varepsilon_t \\ 0 \end{bmatrix}$$

The gretl syntax for a corresponding kalman block would be

```
matrix H = {1; theta}
matrix F = {phi, 0; 1, 0}
matrix Q = {s^2, 0; 0, 0}

kalman
    obsy y
    obsymat H
    statemat F
    statevar Q
end kalman
```

Note that the observation equation (30.11) does not include an "error term"; this is equivalent to saying that $V(\mathbf{w}_t) = 0$ and, as a consequence, the kalman block does not include an obsvar keyword.

Once the filter is set up, all it takes to compute the log-likelihood for given values of ϕ, θ and σ^2 is to execute the kfilter() function and use the $kalman_lnl accessor (which returns the total log-likelihood) or, more appropriately if the likelihood has to be maximized through mle, the $kalman_llt accessor, which returns the series of individual contribution to the log-likelihood for each observation. An example is shown in script 30.1.

Example 30.1: ARMA estimation

```
function void arma11_via_kalman(series y)
    /* parameter initalization */
    phi = 0
    theta = 0
    sigma = 1

    /* Kalman filter setup */
    matrix H = {1; theta}
    matrix F = {phi, 0; 1, 0}
    matrix Q = {sigma^2, 0; 0, 0}

    kalman
        obsy y
        obsymat H
        statemat F
        statevar Q
    end kalman

    /* maximum likelihood estimation */
    mle logl = ERR ? NA : $kalman_llt
        H[2] = theta
        F[1,1] = phi
        Q[1,1] = sigma^2
        ERR = kfilter()
        params phi theta sigma
    end mle -h
end function

# ----------------------- main --------------------------

open arma.gdt         # open the "arma" example dataset
arma11_via_kalman(y) # estimate an arma(1,1) model
arma 1 1 ; y --nc     # check via native command
```

30.10 Example 2: local level model

Suppose we have a series $y_t = \mu_t + \varepsilon_t$, where μ_t is a random walk with normal increments of variance σ_1^2 and ε_t is a normal white noise with variance σ_2^2, independent of μ_t. This is known as the "local level" model in Harvey's (1989) terminology, and it can be cast in state-space form as equations (30.1)-(30.2) with $\mathbf{F} = 1$, $\mathbf{v}_t \sim N(0, \sigma_1^2)$, $\mathbf{H} = 1$ and $\mathbf{w}_t \sim N(0, \sigma_2^2)$. The translation to a kalman block is

```
kalman
   obsy y
   obsymat 1
   statemat 1
```

```
    statevar s2
    obsvar s1
end kalman --diffuse
```

The two unknown parameters σ_1^2 and σ_2^2 can be estimated via maximum likelihood. Script 30.2 provides an example of simulation and estimation of such a model. For the sake of brevity, simulation is carried out via ordinary gretl commands, rather than the state-space apparatus described above.

The example contains two functions: the first one carries out the estimation of the unknown parameters σ_1^2 and σ_2^2 via maximum likelihood; the second one uses these estimates to compute a smoothed estimate of the unobservable series μ_t under the name muhat. A plot of μ_t and its estimate is presented in Figure 30.1.

By appending the following code snippet to the example in Table 30.2, one may check the results against the R command StructTS.

```
foreign language=R --send-data
  y <- gretldata[,"y"]
  a <- StructTS(y, type="level")
  a
  StateFromR <- as.ts(tsSmooth(a))
  gretl.export(StateFromR)
end foreign

append @dotdir/StateFromR.csv

ols muhat 0 StateFromR --simple
```

Example 30.2: Local level model

```
function matrix local_level (series y)
    /* starting values */
    scalar s1 = 1
    scalar s2 = 1

    /* Kalman filter set-up */
    kalman
       obsy y
       obsymat 1
       statemat 1
       statevar s2
       obsvar s1
    end kalman --diffuse

    /* ML estimation */
    mle ll = ERR ? NA : $kalman_llt
        ERR = kfilter()
        params s1 s2
    end mle

    return s1 ~ s2
end function

function series loclev_sm (series y, scalar s1, scalar s2)
    /* return the smoothed estimate of \mu_t */
    kalman
       obsy y
       obsymat 1
       statemat 1
       statevar s2
       obsvar s1
    end kalman --diffuse
    series ret = ksmooth()
    return ret
end function

/* -------------------- main script -------------------- */

nulldata 200
set seed 202020
setobs 1 1 --special
true_s1 = 0.25
true_s2 = 0.5
v = normal() * sqrt(true_s1)
w = normal() * sqrt(true_s2)
mu = 2 + cum(w)
y = mu + v

matrix Vars = local_level(y)           # estimate the variances
muhat = loclev_sm(y, Vars[1], Vars[2]) # compute the smoothed state
```

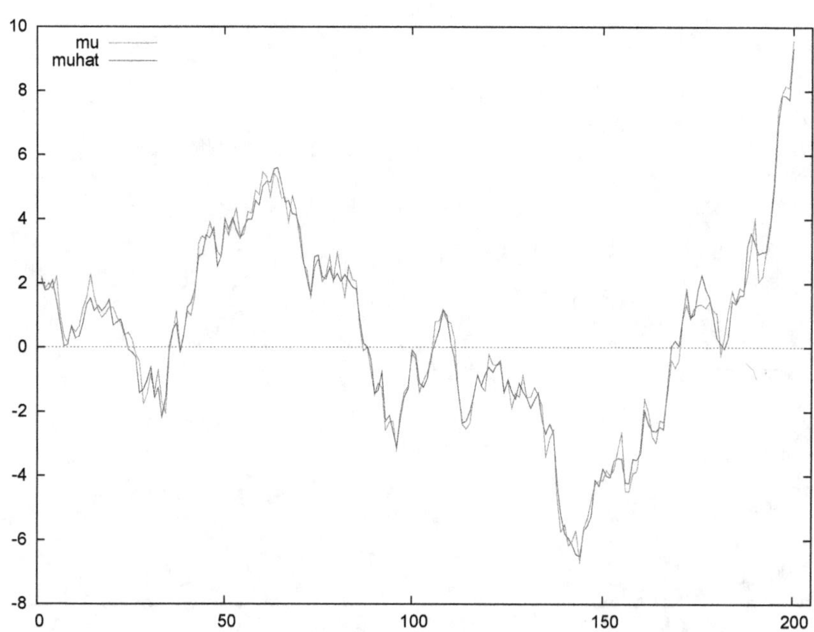

Figure 30.1: Local level model: μ_t and its smoothed estimate

Chapter 31

Numerical methods

Several functions are available to aid in the construction of special-purpose estimators: one group of functions are used to maximize user-supplied functions via numerical methods: BFGS, Newton-Raphson and Simulated Annealing. Another relevant function is `fdjac`, which produces a forward-difference approximation to the Jacobian.

31.1 BFGS

The `BFGSmax` function has two required arguments: a vector holding the initial values of a set of parameters, and a call to a function that calculates the (scalar) criterion to be maximized, given the current parameter values and any other relevant data. If the object is in fact minimization, this function should return the negative of the criterion. On successful completion, `BFGSmax` returns the maximized value of the criterion and the vector given via the first argument holds the parameter values which produce the maximum. It is assumed here that the objective function is a user-defined function (see Chapter 13) with the following general set-up:

```
function scalar ObjFunc (matrix *theta, matrix *X)
  scalar val = ...  # do some computation
  return val
end function
```

The first argument contains the parameter vector and the second may be used to hold "extra" values that are necessary to compute the objective function, but are not the variables of the optimization problem. For example, if the objective function were a loglikelihood, the first argument would contain the parameters and the second one the data. Or, for more economic-theory inclined readers, if the objective function were the utility of a consumer, the first argument might contain the quantities of goods and the second one their prices and disposable income.

Example 31.1: Finding the minimum of the Rosenbrock function

```
function scalar Rosenbrock(matrix *param)
  scalar x = param[1]
  scalar y = param[2]
  return -(1-x)^2 - 100 * (y - x^2)^2
end function

matrix theta = { 0, 0 }

set max_verbose 1
M = BFGSmax(&theta, Rosenbrock(&theta))

print theta
```

The operation of BFGS can be adjusted using the `set` variables `bfgs_maxiter` and `bfgs_toler` (see Chapter 21). In addition you can provoke verbose output from the maximizer by assigning a positive value to `max_verbose`, again via the `set` command.

The Rosenbrock function is often used as a test problem for optimization algorithms. It is also known as "Rosenbrock's Valley" or "Rosenbrock's Banana Function", on account of the fact that its contour lines are banana-shaped. It is defined by:

$$f(x,y) = (1-x)^2 + 100(y-x^2)^2$$

The function has a global minimum at $(x,y) = (1,1)$ where $f(x,y) = 0$. Example 31.1 shows a gretl script that discovers the minimum using `BFGSmax` (giving a verbose account of progress). Note that, in this particular case, the function to be maximized only depends on the parameters, so the second parameter is omitted from the definition of the function `Rosenbrock`.

Limited-memory variant

See Byrd *et al.* (1995) (... FIXME: expand a little here ...)

Supplying analytical derivatives for BFGS

An optional third argument to the `BFGSmax` function enables the user to supply analytical derivatives of the criterion function with respect to the parameters (without which a numerical approximation to the gradient is computed). This argument is similar to the second one in that it specifies a function call. In this case the function that is called must have the following signature.

Its first argument should be a pre-defined matrix correctly dimensioned to hold the gradient; that is, if the parameter vector contains k elements, the gradient matrix must also be a k-vector. This matrix argument must be given in "pointer" form so that its content can be modified by the function. (Note that unlike the parameter vector, where the choice of initial values can be important, the initial values given to the gradient are immaterial and do not affect the results.)

In addition the gradient function must have as one of its argument the parameter vector. This may be given in pointer form (which enhances efficiency) but that is not required. Additional arguments may be specified if necessary.

Given the current parameter values, the function call must fill out the gradient vector appropriately. It is not required that the gradient function returns any value directly; if it does, that value is ignored.

Example 31.2 illustrates, showing how the Rosenbrock script can be modified to use analytical derivatives. (Note that since this is a minimization problem the values written into `g[1]` and `g[2]` in the function `Rosen_grad` are in fact the derivatives of the negative of the Rosenbrock function.)

31.2 Newton-Raphson

BFGS, discussed above, is an excellent all-purpose maximizer, and about as robust as possible given the limitations of digital computer arithmetic. The Newton-Raphson maximizer is not as robust, but may converge much faster than BFGS for problems where the maximand is reasonably well behaved—in particular, where it is anything like quadratic (see below). The case for using Newton-Raphson is enhanced if it is possible to supply a function to calculate the Hessian analytically.

The gretl function `NRmax`, which implements the Newton-Raphson method, has a maximum of four arguments. The first two (required) arguments are exactly as for BFGS: an initial parameter vector, and a function call which returns the maximand given the parameters. The (optional) third argument is again as in BFGS: a function call that calculates the gradient. Specific to `NRmax` is an optional fourth argument, namely a function call to calculate the (negative) Hessian. The first argument of this function must be a pre-defined matrix of the right dimension to hold the Hessian—that is, a $k \times k$ matrix, where k is the length of the parameter vector—given in "pointer" form. The second argument should be the parameter vector (optionally in pointer form). Other data may be passed as additional arguments as needed. Similarly to the case with the gradient, if the fourth argument to `NRmax` is omitted then a numerical approximation to the Hessian is constructed.

Example 31.2: Rosenbrock function with analytical gradient

```
function scalar Rosenbrock (matrix *param)
  scalar x = param[1]
  scalar y = param[2]
  return -(1-x)^2 - 100 * (y - x^2)^2
end function

function void Rosen_grad (matrix *g, matrix *param)
  scalar x = param[1]
  scalar y = param[2]
  g[1] = 2*(1-x) + 2*x*(200*(y-x^2))
  g[2] = -200*(y - x^2)
end function

matrix theta = { 0, 0 }
matrix grad = { 0, 0 }

set max_verbose 1
M = BFGSmax(&theta, Rosenbrock(&theta), Rosen_grad(&grad, &theta))

print theta
print grad
```

What is ultimately required in Newton-Raphson is the negative inverse of the Hessian. Note that if you give the optional fourth argument, your function should compute the negative Hessian, but should not invert it; `NRmax` takes care of inversion, with special handling for the case where the matrix is not negative definite, which can happen far from the maximum.

Script 31.3 extends the Rosenbrock example, using `NRmax` with a function `Rosen_hess` to compute the Hessian. The functions `Rosenbrock` and `Rosen_grad` are just the same as in Example 31.2 and are omitted for brevity.

Example 31.3: Rosenbrock function via Newton-Raphson

```
function void Rosen_hess (matrix *H, matrix *param)
  scalar x = param[1]
  scalar y = param[2]
  H[1,1] = 2 - 400*y + 1200*x^2
  H[1,2] = -400*x
  H[2,1] = -400*x
  H[2,2] = 200
end function

matrix theta = { 0, 0 }
matrix grad = { 0, 0 }
matrix H = zeros(2, 2)

set max_verbose 1
M = NRmax(&theta, Rosenbrock(&theta), Rosen_grad(&grad, &theta),
          Rosen_hess(&H, &theta))

print theta
print grad
```

The idea behind Newton-Raphson is to exploit a quadratic approximation to the maximand, under the assumption that it is concave. If this is true, the method is very effective. However, if the algorithm happens to evaluate the function at a point where the Hessian is not negative definite, things may go wrong. Script 31.4 exemplifies this by using a normal density, which is concave in the interval $(-1, 1)$ and convex elsewhere. If the algorithm is started from within the interval everything goes well and NR is (slightly) more effective than BFGS. If, however, the Hessian is positive at the starting point BFGS converges with only little more difficulty, while Newton-Raphson fails.

Example 31.4: Maximization of a Gaussian density

```
function scalar ND(matrix x)
    scalar z = x[1]
    return exp(-0.5*z*z)
end function

set max_verbose 1

x = {0.75}
A = BFGSmax(&x, ND(x))

x = {0.75}
A = NRmax(&x, ND(x))

x = {1.5}
A = BFGSmax(&x, ND(x))

x = {1.5}
A = NRmax(&x, ND(x))
```

31.3 Simulated Annealing

Simulated annealing—as implemented by the gretl function `simann`—is not a full-blown maximization method in its own right, but can be a useful auxiliary tool in problems where convergence depends sensitively on the initial values of the parameters. The idea is that you supply initial values and the simulated annealing mechanism tries to improve on them via controlled randomization.

The `simann` function takes up to three arguments. The first two (required) are the same as for `BFGSmax` and `NRmax`: an initial parameter vector and a function that computes the maximand. The optional third argument is a positive integer giving the maximum number of iterations, n, which defaults to 1024.

Starting from the specified point in the parameter space, for each of n iterations we select at random a new point within a certain radius of the previous one and determine the value of the criterion at the new point. If the criterion is higher we jump to the new point; otherwise, we jump with probability P (and remain at the previous point with probability $1 - P$). As the iterations proceed, the system gradually "cools"—that is, the radius of the random perturbation is reduced, as is the probability of making a jump when the criterion fails to increase.

In the course of this procedure $n + 1$ points in the parameter space are evaluated: call them $\theta_i, i = 0, \ldots, n$, where θ_0 is the initial value given by the user. Let θ^* denote the "best" point among $\theta_1, \ldots, \theta_n$ (highest criterion value). The value written into the parameter vector on completion is then θ^* if θ^* is better than θ_0, otherwise θ_n. In other words, failing an actual improvement in the criterion, `simann` randomizes the starting point, which may be helpful in tricky optimization problems.

Example 31.5 shows `simann` at work as a helper for `BFGSmax` in finding the maximum of a bimodal function. Unaided, BFGSmax requires 60 function evaluations and 55 evaluations of the gradient, while after simulated annealing the maximum is found with 7 function evaluations and 6 evaluations of the gradient.[1]

Example 31.5: BFGS with initialization via Simulated Annealing

```
function scalar bimodal (matrix x, matrix A)
    scalar ret = exp(-qform((x-1)', A))
    ret += 2*exp(-qform((x+4)', A))
    return ret
end function

set seed 12334
set max_verbose on

scalar k = 2
matrix A = 0.1 * I(k)
matrix x0 = {3; -5}

x = x0
u = BFGSmax(&x, bimodal(x, A))
print x

x = x0
u = simann(&x, bimodal(x, A), 1000)
print x
u = BFGSmax(&x, bimodal(x, A))
print x
```

31.4 Computing a Jacobian

Gretl offers the possibility of differentiating numerically a user-defined function via the fdjac function.

This function takes two arguments: an $n \times 1$ matrix holding initial parameter values and a function call that calculates and returns an $m \times 1$ matrix, given the current parameter values and any other relevant data. On successful completion it returns an $m \times n$ matrix holding the Jacobian. For example,

```
matrix Jac = fdjac(theta, SumOC(&theta, &X))
```

where we assume that SumOC is a user-defined function with the following structure:

```
function matrix SumOC (matrix *theta, matrix *X)
  matrix V = ...   # do some computation
  return V
end function
```

This may come in handy in several cases: for example, if you use BFGSmax to estimate a model, you may wish to calculate a numerical approximation to the relevant Jacobian to construct a covariance matrix for your estimates.

Another example is the delta method: if you have a consistent estimator of a vector of parameters $\hat{\theta}$, and a consistent estimate of its covariance matrix Σ, you may need to compute estimates for a nonlinear continuous transformation $\psi = g(\theta)$. In this case, a standard result

[1]Your mileage may vary: these figures are somewhat compiler- and machine-dependent.

in asymptotic theory is that

$$\left\{ \begin{array}{c} \hat{\theta} \xrightarrow{\text{p}} \theta \\ \sqrt{T}\left(\hat{\theta} - \theta\right) \xrightarrow{\text{d}} N(0, \Sigma) \end{array} \right\} \Longrightarrow \left\{ \begin{array}{c} \hat{\psi} = g(\hat{\theta}) \xrightarrow{\text{p}} \psi = g(\theta) \\ \sqrt{T}\left(\hat{\psi} - \psi\right) \xrightarrow{\text{d}} N(0, J\Sigma J') \end{array} \right\}$$

where T is the sample size and J is the Jacobian $\left.\frac{\partial g(x)}{\partial x}\right|_{x=\theta}$.

Example 31.6: Delta Method

```
function matrix MPC(matrix *param, matrix *Y)
  beta = param[2]
  gamma = param[3]
  y = Y[1]
  return beta*gamma*y^(gamma-1)
end function

# William Greene, Econometric Analysis, 5e, Chapter 9
set echo off
set messages off
open greene5_1.gdt

# Use OLS to initialize the parameters
ols realcons 0 realdpi --quiet
genr a = $coeff(0)
genr b = $coeff(realdpi)
genr g = 1.0

# Run NLS with analytical derivatives
nls realcons = a + b * (realdpi^g)
  deriv a = 1
  deriv b = realdpi^g
  deriv g = b * realdpi^g * log(realdpi)
end nls

matrix Y = realdpi[2000:4]
matrix theta = $coeff
matrix V = $vcv

mpc = MPC(&theta, &Y)
matrix Jac = fdjac(theta, MPC(&theta, &Y))
Sigma = qform(Jac, V)

printf "\nmpc = %g, std.err = %g\n", mpc, sqrt(Sigma)
scalar teststat = (mpc-1)/sqrt(Sigma)
printf "\nTest for MPC = 1: %g (p-value = %g)\n", \
        teststat, pvalue(n,abs(teststat))
```

Script 31.6 exemplifies such a case: the example is taken from Greene (2003), section 9.3.1. The slight differences between the results reported in the original source and what gretl returns are due to the fact that the Jacobian is computed numerically, rather than analytically as in the book.

On the subject of numerical versus analytical derivatives, one may wonder what difference it makes to use one method or another. Simply put, the answer is: analytical derivatives may be painful to derive and to translate into code, but in most cases they are much faster than using `fdjac`; as a consequence, if you need to use derivatives as part of an algorithm that requires iteration (such as numerical optimization, or a Monte Carlo experiment), you'll definitely want to use analytical derivatives.

Analytical derivatives are also, in most cases, more precise than numerical ones, but this advantage may or may not be negligible in practice depending on the practical details: the two fundamental aspects to take in consideration are nonlinearity and machine precision.

As an example, consider the derivative of a highly nonlinear function such as the matrix inverse. In order to keep the example simple, let's focus on 2×2 matrices and define the function

```
function matrix vecinv(matrix x)
    A = mshape(x,2,2)
    return vec(inv(A))'
end function
```

which, given $\text{vec}(A)$, returns $\text{vec}(A^{-1})$. As is well known (see for example Magnus and Neudecker (1988)),

$$\frac{\partial \text{vec}(A^{-1})}{\partial \text{vec}(A)} = -(A^{-1})' \otimes (A^{-1}),$$

which is rather easy to code in hansl as

```
function matrix grad(matrix x)
    iA = inv(mshape(x,2,2))
    return -iA' ** iA
end function
```

Using the `fdjac` function to obtain the same result is even easier: you just invoke it like

```
fdjac(a, "vecinv(a)")
```

In order to see what the difference is, in terms of precision, between analytical and numerical Jacobians, let's start from $A = \begin{bmatrix} 2 & 1 \\ 1 & 1 \end{bmatrix}$. The following code

```
a = {2; 1; 1; 1}
ia = vecinv(a)
ag = grad(a)
ng = fdjac(a, "vecinv(a)")
dg = ag - ng
print ag ng dg
```

gives

```
ag (4 x 4)

  -1    1    1   -1
   1   -2   -1    2
   1   -1   -2    2
  -1    2    2   -4

ng (4 x 4)

  -1    1    1   -1
   1   -2   -1    2
   1   -1   -2    2
  -1    2    2   -4

dg (4 x 4)

  -3.3530e-08  -3.7251e-08  -3.7251e-08  -3.7255e-08
   2.6079e-08   5.2150e-08   3.7251e-08   6.7060e-08
   2.6079e-08   3.7251e-08   5.2150e-08   6.7060e-08
  -2.2354e-08  -5.9600e-08  -5.9600e-08  -1.4902e-07
```

in which the analytically-computed derivative and its numerical approximation are essentially the same. If, however, you set $A = \begin{bmatrix} 1.0001 & 1 \\ 1 & 1 \end{bmatrix}$ you end up evaluating the function at a point in which the function itself is considerably more nonlinear since the matrix is much closer to being singular. As a consequence, the numerical approximation becomes much less satisfactory:

```
ag (4 x 4)

 -1.0000e+08   1.0000e+08   1.0000e+08  -1.0000e+08
  1.0000e+08  -1.0001e+08  -1.0000e+08   1.0001e+08
  1.0000e+08  -1.0000e+08  -1.0001e+08   1.0001e+08
 -1.0000e+08   1.0001e+08   1.0001e+08  -1.0002e+08

ng (4 x 4)

 -9.9985e+07   1.0001e+08   1.0001e+08  -9.9985e+07
  9.9985e+07  -1.0002e+08  -1.0001e+08   9.9995e+07
  9.9985e+07  -1.0001e+08  -1.0002e+08   9.9995e+07
 -9.9985e+07   1.0002e+08   1.0002e+08  -1.0001e+08

dg (4 x 4)

     -14899.       -14901.       -14901.       -14900.
      14899.        14903.        14901.        14902.
      14899.        14901.        14903.        14902.
     -14899.       -14903.       -14903.       -14903.
```

Moreover, machine precision may have its impact: if you take $A = 0.00001 \times \begin{bmatrix} 2 & 1 \\ 1 & 1 \end{bmatrix}$, the matrix itself is not singular at all, but the order of magnitude of its elements is close enough to machine precision to provoke problems:

```
ag (4 x 4)

 -1.0000e+10   1.0000e+10   1.0000e+10  -1.0000e+10
  1.0000e+10  -2.0000e+10  -1.0000e+10   2.0000e+10
  1.0000e+10  -1.0000e+10  -2.0000e+10   2.0000e+10
 -1.0000e+10   2.0000e+10   2.0000e+10  -4.0000e+10

ng (4 x 4)

 -1.0000e+10   1.0000e+10   1.0000e+10  -1.0000e+10
  1.0000e+10  -2.0000e+10  -1.0000e+10   2.0000e+10
  1.0000e+10  -1.0000e+10  -2.0000e+10   2.0000e+10
 -1.0000e+10   2.0000e+10   2.0000e+10  -4.0000e+10

dg (4 x 4)

     -488.30       -390.60       -390.60       -195.33
      634.79        781.21        390.60        585.98
      634.79        488.26        683.55        585.98
     -781.27       -976.52       -781.21       -1367.3
```

Chapter 32

Discrete and censored dependent variables

This chapter deals with models for dependent variables that are discrete or censored or otherwise limited (as in event counts or durations, which must be positive) and that therefore call for estimation methods other than the classical linear model. We discuss several estimators (mostly based on the Maximum Likelihood principle), adding some details and examples to complement the material on these methods in the *Gretl Command Reference.*

32.1 Logit and probit models

It often happens that one wants to specify and estimate a model in which the dependent variable is not continuous, but discrete. A typical example is a model in which the dependent variable is the occupational status of an individual (1 = employed, 0 = unemployed). A convenient way of formalizing this situation is to consider the variable y_i as a Bernoulli random variable and analyze its distribution conditional on the explanatory variables x_i. That is,

$$y_i = \begin{cases} 1 & P_i \\ 0 & 1 - P_i \end{cases} \tag{32.1}$$

where $P_i = P(y_i = 1|x_i)$ is a given function of the explanatory variables x_i.

In most cases, the function P_i is a cumulative distribution function F, applied to a linear combination of the x_is. In the probit model, the normal cdf is used, while the logit model employs the logistic function $\Lambda()$. Therefore, we have

$$\text{probit} \qquad P_i = F(z_i) = \Phi(z_i) \tag{32.2}$$

$$\text{logit} \qquad P_i = F(z_i) = \Lambda(z_i) = \frac{1}{1 + e^{-z_i}} \tag{32.3}$$

$$z_i = \sum_{j=1}^{k} x_{ij}\beta_j \tag{32.4}$$

where z_i is commonly known as the *index* function. Note that in this case the coefficients β_j cannot be interpreted as the partial derivatives of $E(y_i|x_i)$ with respect to x_{ij}. However, for a given value of x_i it is possible to compute the vector of "slopes", that is

$$\text{slope}_j(\bar{x}) = \left. \frac{\partial F(z)}{\partial x_j} \right|_{z=\bar{z}}$$

Gretl automatically computes the slopes, setting each explanatory variable at its sample mean.

Another, equivalent way of thinking about this model is in terms of an unobserved variable y_i^* which can be described thus:

$$y_i^* = \sum_{j=1}^{k} x_{ij}\beta_j + \varepsilon_i = z_i + \varepsilon_i \tag{32.5}$$

We observe $y_i = 1$ whenever $y_i^* > 0$ and $y_i = 0$ otherwise. If ε_i is assumed to be normal, then we have the probit model. The logit model arises if we assume that the density function of ε_i is

$$\lambda(\varepsilon_i) = \frac{\partial \Lambda(\varepsilon_i)}{\partial \varepsilon_i} = \frac{e^{-\varepsilon_i}}{(1 + e^{-\varepsilon_i})^2}$$

Both the probit and logit model are estimated in gretl via maximum likelihood, where the log-likelihood can be written as

$$L(\beta) = \sum_{y_i=0} \ln[1 - F(z_i)] + \sum_{y_i=1} \ln F(z_i), \tag{32.6}$$

which is always negative, since $0 < F(\cdot) < 1$. Since the score equations do not have a closed form solution, numerical optimization is used. However, in most cases this is totally transparent to the user, since usually only a few iterations are needed to ensure convergence. The `--verbose` switch can be used to track the maximization algorithm.

Example 32.1: Estimation of simple logit and probit models

```
open greene19_1

logit GRADE const GPA TUCE PSI
probit GRADE const GPA TUCE PSI
```

As an example, we reproduce the results given in chapter 21 of Greene (2000), where the effectiveness of a program for teaching economics is evaluated by the improvements of students' grades. Running the code in example 32.1 gives the output reported in Table 32.1; note that, for the probit model, a conditional moment test on skewness and kurtosis is printed out automatically as a test for normality.

In this context, the `$uhat` accessor function takes a special meaning: it returns generalized residuals as defined in Gourieroux, Monfort, Renault and Trognon (1987), which can be interpreted as unbiased estimators of the latent disturbances ε_i. These are defined as

$$u_i = \begin{cases} y_i - \hat{P}_i & \text{for the logit model} \\ y_i \cdot \frac{\phi(\hat{z}_i)}{\Phi(\hat{z}_i)} - (1 - y_i) \cdot \frac{\phi(\hat{z}_i)}{1-\Phi(\hat{z}_i)} & \text{for the probit model} \end{cases} \tag{32.7}$$

Among other uses, generalized residuals are often used for diagnostic purposes. For example, it is very easy to set up an omitted variables test equivalent to the familiar LM test in the context of a linear regression; example 32.2 shows how to perform a variable addition test.

Example 32.2: Variable addition test in a probit model

```
open greene19_1

probit GRADE const GPA PSI
series u = $uhat
ols u const GPA PSI TUCE -q
printf "Variable addition test for TUCE:\n"
printf "Rsq * T = %g (p. val. = %g)\n", $trsq, pvalue(X,1,$trsq)
```

The perfect prediction problem

One curious characteristic of logit and probit models is that (quite paradoxically) estimation is not feasible if a model fits the data perfectly; this is called the *perfect prediction problem.* The reason why this problem arises is easy to see by considering equation (32.6): if for some vector β and scalar k it's the case that $z_i < k$ whenever $y_i = 0$ and $z_i > k$ whenever $y_i = 1$, the same thing is true for any multiple of β. Hence, $L(\beta)$ can be made arbitrarily close to 0 simply

```
Model 1: Logit estimates using the 32 observations 1-32
Dependent variable: GRADE

      VARIABLE        COEFFICIENT         STDERROR      T STAT       SLOPE
                                                                   (at mean)
  const               -13.0213            4.93132      -2.641
  GPA                   2.82611           1.26294       2.238      0.533859
  TUCE                  0.0951577         0.141554      0.672      0.0179755
  PSI                   2.37869           1.06456       2.234      0.449339

  Mean of GRADE = 0.344
  Number of cases 'correctly predicted' = 26 (81.2%)
  f(beta'x) at mean of independent vars = 0.189
  McFadden's pseudo-R-squared = 0.374038
  Log-likelihood = -12.8896
  Likelihood ratio test: Chi-square(3) = 15.4042 (p-value 0.001502)
  Akaike information criterion (AIC) = 33.7793
  Schwarz Bayesian criterion (BIC) = 39.6422
  Hannan-Quinn criterion (HQC) = 35.7227

           Predicted
             0    1
  Actual 0  18    3
         1   3    8

Model 2: Probit estimates using the 32 observations 1-32
Dependent variable: GRADE

      VARIABLE        COEFFICIENT         STDERROR      T STAT       SLOPE
                                                                   (at mean)
  const                -7.45232           2.54247      -2.931
  GPA                   1.62581           0.693883      2.343      0.533347
  TUCE                  0.0517288         0.0838903     0.617      0.0169697
  PSI                   1.42633           0.595038      2.397      0.467908

  Mean of GRADE = 0.344
  Number of cases 'correctly predicted' = 26 (81.2%)
  f(beta'x) at mean of independent vars = 0.328
  McFadden's pseudo-R-squared = 0.377478
  Log-likelihood = -12.8188
  Likelihood ratio test: Chi-square(3) = 15.5459 (p-value 0.001405)
  Akaike information criterion (AIC) = 33.6376
  Schwarz Bayesian criterion (BIC) = 39.5006
  Hannan-Quinn criterion (HQC) = 35.581

           Predicted
             0    1
  Actual 0  18    3
         1   3    8

Test for normality of residual -
  Null hypothesis: error is normally distributed
  Test statistic: Chi-square(2) = 3.61059
  with p-value = 0.164426
```

Table 32.1: Example logit and probit

by choosing enormous values for β. As a consequence, the log-likelihood has no maximum, despite being bounded.

Gretl has a mechanism for preventing the algorithm from iterating endlessly in search of a non-existent maximum. One sub-case of interest is when the perfect prediction problem arises because of a single binary explanatory variable. In this case, the offending variable is dropped from the model and estimation proceeds with the reduced specification. Nevertheless, it may happen that no single "perfect classifier" exists among the regressors, in which case estimation is simply impossible and the algorithm stops with an error. This behavior is triggered during the iteration process if

$$\max_{i:y_i=0} z_i < \min_{i:y_i=1} z_i$$

If this happens, unless your model is trivially mis-specified (like predicting if a country is an oil exporter on the basis of oil revenues), it is normally a small-sample problem: you probably just don't have enough data to estimate your model. You may want to drop some of your explanatory variables.

This problem is well analyzed in Stokes (2004); the results therein are replicated in the example script `murder_rates.inp`.

32.2 Ordered response models

These models constitute a simple variation on ordinary logit/probit models, and are usually applied when the dependent variable is a discrete and ordered measurement—not simply binary, but on an ordinal rather than an interval scale. For example, this sort of model may be applied when the dependent variable is a qualitative assessment such as "Good", "Average" and "Bad".

In the general case, consider an ordered response variable, y, that can take on any of the $J + 1$ values $0, 1, 2, \ldots, J$. We suppose, as before, that underlying the observed response is a latent variable,

$$y^* = X\beta + \varepsilon = z + \varepsilon$$

Now define "cut points", $\alpha_1 < \alpha_2 < \cdots < \alpha_J$, such that

$$\begin{aligned} y &= 0 \quad \text{if } y^* \le \alpha_1 \\ y &= 1 \quad \text{if } \alpha_1 < y^* \le \alpha_2 \\ &\vdots \\ y &= J \quad \text{if } y^* > \alpha_J \end{aligned}$$

For example, if the response takes on three values there will be two such cut points, α_1 and α_2.

The probability that individual i exhibits response j, conditional on the characteristics x_i, is then given by

$$P(y_i = j \,|\, x_i) = \begin{cases} P(y^* \le \alpha_1 \,|\, x_i) = F(\alpha_1 - z_i) & \text{for } j = 0 \\ P(\alpha_j < y^* \le \alpha_{j+1} \,|\, x_i) = F(\alpha_{j+1} - z_i) - F(\alpha_j - z_i) & \text{for } 0 < j < J \\ P(y^* > \alpha_J \,|\, x_i) = 1 - F(\alpha_J - z_i) & \text{for } j = J \end{cases} \tag{32.8}$$

The unknown parameters α_j are estimated jointly with the βs via maximum likelihood. The $\hat{\alpha}_j$ estimates are reported by gretl as `cut1`, `cut2` and so on. For the probit variant, a conditional moment test for normality constructed in the spirit of Chesher and Irish (1987) is also included.

Note that the α_j parameters can be shifted arbitrarily by adding a constant to z_i, so the model is under-identified if there is some linear combination of the explanatory variables which is constant. The most obvious case in which this occurs is when the model contains a constant term; for this reason, gretl drops automatically the intercept if present. However, it may happen that the user inadventently specifies a list of regressors that may be combined in such a way to produce a constant (for example, by using a full set of dummy variables for a discrete factor). If this happens, gretl will also drop any offending regressors.

In order to apply these models in gretl, the dependent variable must either take on only non-negative integer values, or be explicitly marked as discrete. (In case the variable has non-integer values, it will be recoded internally.) Note that gretl does not provide a separate command for ordered models: the `logit` and `probit` commands automatically estimate the ordered version if the dependent variable is acceptable, but not binary.

Example 32.3 reproduces the results presented in section 15.10 of Wooldridge (2002a). The question of interest in this analysis is what difference it makes, to the allocation of assets in pension funds, whether individual plan participants have a choice in the matter. The response variable is an ordinal measure of the weight of stocks in the pension portfolio. Having reported the results of estimation of the ordered model, Wooldridge illustrates the effect of the `choice` variable by reference to an "average" participant. The example script shows how one can compute this effect in gretl.

After estimating ordered models, the `$uhat` accessor yields generalized residuals as in binary models; additionally, the `$yhat` accessor function returns $\hat{z}_i$, so it is possible to compute an unbiased estimator of the latent variable y_i^* simply by adding the two together.

32.3 Multinomial logit

When the dependent variable is not binary and does not have a natural ordering, *multinomial* models are used. Multinomial logit is supported in gretl via the `--multinomial` option to the `logit` command. Simple models can also be handled via the `mle` command (see chapter 21). We give here an example of such a model. Let the dependent variable, y_i, take on integer values $0, 1, \ldots p$. The probability that $y_i = k$ is given by

$$P(y_i = k|x_i) = \frac{\exp(x_i\beta_k)}{\sum_{j=0}^{p}\exp(x_i\beta_j)}$$

For the purpose of identification one of the outcomes must be taken as the "baseline"; it is usually assumed that $\beta_0 = 0$, in which case

$$P(y_i = k|x_i) = \frac{\exp(x_i\beta_k)}{1 + \sum_{j=1}^{p}\exp(x_i\beta_j)}$$

and

$$P(y_i = 0|x_i) = \frac{1}{1 + \sum_{j=1}^{p}\exp(x_i\beta_j)}.$$

Example 32.4 reproduces Table 15.2 in Wooldridge (2002a), based on data on career choice from Keane and Wolpin (1997). The dependent variable is the occupational status of an individual (0 = in school; 1 = not in school and not working; 2 = working), and the explanatory variables are education and work experience (linear and square) plus a "black" binary variable. The full data set is a panel; here the analysis is confined to a cross-section for 1987.

32.4 Bivariate probit

The bivariate probit model is simply a two-equation system in which each equation is a probit model, but the two disturbance terms may not be independent. In formulae,

$$y_{1,i}^* = \sum_{j=1}^{k_1} x_{ij}\beta_j + \varepsilon_{1,i} \qquad y_{1,i} = 1 \Longleftrightarrow y_{1,i}^* > 0 \tag{32.9}$$

$$y_{2,i}^* = \sum_{j=1}^{k_2} z_{ij}\gamma_j + \varepsilon_{2,i} \qquad y_{2,i} = 1 \Longleftrightarrow y_{2,i}^* > 0 \tag{32.10}$$

$$\begin{bmatrix} \varepsilon_{2,i} \\ \varepsilon_{2,i} \end{bmatrix} \sim N\left[0, \begin{pmatrix} 1 & \rho \\ \rho & 1 \end{pmatrix}\right] \tag{32.11}$$

- The explanatory variables for the first equation x and for the second equation z may overlap

Example 32.3: Ordered probit model

```
/*
  Replicate the results in Wooldridge, Econometric Analysis of Cross
  Section and Panel Data, section 15.10, using pension-plan data from
  Papke (AER, 1998).

  The dependent variable, pctstck (percent stocks), codes the asset
  allocation responses of "mostly bonds", "mixed" and "mostly stocks"
  as {0, 50, 100}.

  The independent variable of interest is "choice", a dummy indicating
  whether individuals are able to choose their own asset allocations.
*/

open pension.gdt

# demographic characteristics of participant
list DEMOG = age educ female black married
# dummies coding for income level
list INCOME = finc25 finc35 finc50 finc75 finc100 finc101

# Papke's OLS approach
ols pctstck const choice DEMOG INCOME wealth89 prftshr
# save the OLS choice coefficient
choice_ols = $coeff(choice)

# estimate ordered probit
probit pctstck choice DEMOG INCOME wealth89 prftshr

k = $ncoeff
matrix b = $coeff[1:k-2]
a1 = $coeff[k-1]
a2 = $coeff[k]

/*
   Wooldridge illustrates the 'choice' effect in the ordered probit
   by reference to a single, non-black male aged 60, with 13.5 years
   of education, income in the range $50K - $75K and wealth of $200K,
   participating in a plan with profit sharing.
*/
matrix X = {60, 13.5, 0, 0, 0, 0, 0, 0, 1, 0, 0, 200, 1}

# with 'choice' = 0
scalar Xb = (0 ~ X) * b
P0 = cdf(N, a1 - Xb)
P50 = cdf(N, a2 - Xb) - P0
P100 = 1 - cdf(N, a2 - Xb)
E0 = 50 * P50 + 100 * P100

# with 'choice' = 1
Xb = (1 ~ X) * b
P0 = cdf(N, a1 - Xb)
P50 = cdf(N, a2 - Xb) - P0
P100 = 1 - cdf(N, a2 - Xb)
E1 = 50 * P50 + 100 * P100

printf "\nWith choice, E(y) = %.2f, without E(y) = %.2f\n", E1, E0
printf "Estimated choice effect via ML = %.2f (OLS = %.2f)\n", E1 - E0,
  choice_ols
```

Example 32.4: Multinomial logit

Input:

```
open keane.gdt
smpl (year=87) --restrict
logit status 0 educ exper expersq black --multinomial
```

Output (selected portions):

```
Model 1: Multinomial Logit, using observations 1-1738 (n = 1717)
Missing or incomplete observations dropped: 21
Dependent variable: status
Standard errors based on Hessian

             coefficient   std. error      z      p-value
  --------------------------------------------------------
  status = 2
  const       10.2779       1.13334       9.069   1.20e-19 ***
  educ        -0.673631     0.0698999    -9.637   5.57e-22 ***
  exper       -0.106215     0.173282     -0.6130  0.5399
  expersq     -0.0125152    0.0252291    -0.4961  0.6199
  black        0.813017     0.302723      2.686   0.0072   ***
  status = 3
  const        5.54380      1.08641       5.103   3.35e-07 ***
  educ        -0.314657     0.0651096    -4.833   1.35e-06 ***
  exper        0.848737     0.156986      5.406   6.43e-08 ***
  expersq     -0.0773003    0.0229217    -3.372   0.0007   ***
  black        0.311361     0.281534      1.106   0.2687

Mean dependent var   2.691322   S.D. dependent var   0.573502
Log-likelihood      -907.8572   Akaike criterion     1835.714
Schwarz criterion    1890.198   Hannan-Quinn         1855.874

Number of cases 'correctly predicted' = 1366 (79.6%)
Likelihood ratio test: Chi-square(8) = 583.722 [0.0000]
```

- example contained in the `biprobit.inp` sample script.
- `$uhat` and `$yhat` are matrices

FIXME: expand.

32.5 Panel estimators

When your dataset is a panel, the traditional choice for binary dependent variable models was, for many years, to use logit with fixed effects and probit with random effects (see 18.1 for a brief discussion of this dichotomy in the context of linear models). Nowadays, the choice is somewhat wider, but the two traditional models are by and large what practitioners use as routine tools.

Gretl provides FE logit as a function package[1] and RE probit natively. Provided your dataset has a panel structure, the latter option can be obtained by adding the `--random` option to the `probit` command:

```
probit depvar const indvar1 indvar2 --random
```

as exemplified in the `reprobit.inp` sample script. The numerical technique used for this particular estimator is *Gauss-Hermite quadrature*, which we'll now briefly describe. Generalizing equation (32.5) to a panel context, we get

$$y^*_{i,t} = \sum_{j=1}^{k} x_{ijt}\beta_j + \alpha_i + \varepsilon_{i,t} = z_{i,t} + \omega_{i,t} \tag{32.12}$$

in which we assume that the individual effect, α_i, and the disturbance term, $\varepsilon_{i,t}$, are mutually independent zero-mean Gaussian random variables. The composite error term, $\omega_{i,t} = \alpha_i + \varepsilon_{i,t}$, is therefore a normal r. v. with mean zero and variance $1 + \sigma^2_\alpha$. Because of the individual effect, α_i, observations for the same unit are not independent; the likelihood therefore has to be evaluated on a per-unit basis, as

$$\ell_i = \log P\left[y_{i,1}, y_{i,2}, \ldots, y_{i,T}\right].$$

and there's no way to write the above as a product of individual terms.

However, the above probability *could* be written as a product if we were to treat α_i as a constant; in that case we would have

$$\ell_i|\alpha_i = \sum_{t=1}^{T} \Phi\left[(2y_{i,t} - 1)\frac{x_{ijt}\beta_j + \alpha_i}{\sqrt{1+\sigma^2_\alpha}}\right]$$

so that we can compute ℓ_i by integrating α_i out as

$$\ell_i = E\left(\ell_i|\alpha_i\right) = \int_{-\infty}^{\infty} (\ell_i|\alpha_i)\frac{\varphi(\alpha_i)}{\sqrt{1+\sigma^2_\alpha}}\mathrm{d}\alpha_i$$

The technique known as Gauss-Hermite quadrature is simply a way of approximating the above integral via a sum of carefully chosen terms:[2]

$$\ell_i \simeq \sum_{k=1}^{m} (\ell_i|\alpha_i = n_k)w_k$$

where the numbers n_k and w_k are known as *quadrature points* and *weights*, respectively. Of course, accuracy improves with higher values of m, but so does CPU usage. Note that this

[1] Although this may change in the near future.

[2] Some have suggested using a more refined method called *adaptive* Gauss-Hermite quadrature; this is not implemented in gretl.

technique can also be used in more general cases by using the quadtable() function and the mle command via the apparatus described in chapter 21. Here, however, the calculations were hard-coded in C for maximal speed and efficiency.

Experience shows that a reasonable compromise can be achieved in most cases by choosing m in the order of 20 or so; gretl uses 32 as a default value, but this can be changed via the --quadpoints option, as in

```
probit y const x1 x2 x3 --random --quadpoints=48
```

32.6 The Tobit model

The Tobit model is used when the dependent variable of a model is *censored*. Assume a latent variable y_i^* can be described as

$$y_i^* = \sum_{j=1}^{k} x_{ij}\beta_j + \varepsilon_i,$$

where $\varepsilon_i \sim N(0, \sigma^2)$. If y_i^* were observable, the model's parameters could be estimated via ordinary least squares. On the contrary, suppose that we observe y_i, defined as

$$y_i = \begin{cases} a & \text{for} \quad y_i^* \le a \\ y_i^* & \text{for} \quad a < y_i^* < b \\ b & \text{for} \quad y_i^* \ge b \end{cases} \tag{32.13}$$

In most cases found in the applied literature, $a = 0$ and $b = \infty$, so in practice negative values of y_i^* are not observed and are replaced by zeros.

In this case, regressing y_i on the x_is does not yield consistent estimates of the parameters β, because the conditional mean $E(y_i|x_i)$ is not equal to $\sum_{j=1}^{k} x_{ij}\beta_j$. It can be shown that restricting the sample to non-zero observations would not yield consistent estimates either. The solution is to estimate the parameters via maximum likelihood. The syntax is simply

```
tobit depvar indvars
```

As usual, progress of the maximization algorithm can be tracked via the --verbose switch, while $uhat returns the generalized residuals. Note that in this case the generalized residual is defined as $\hat{u}_i = E(\varepsilon_i|y_i = 0)$ for censored observations, so the familiar equality $\hat{u}_i = y_i - \hat{y}_i$ only holds for uncensored observations, that is, when $y_i > 0$.

An important difference between the Tobit estimator and OLS is that the consequences of non-normality of the disturbance term are much more severe: non-normality implies inconsistency for the Tobit estimator. For this reason, the output for the Tobit model includes the Chesher and Irish (1987) normality test by default.

The general case in which a is nonzero and/or b is finite can be handled by using the options --llimit and --rlimit. So, for example,

```
tobit depvar indvars --llimit=10
```

would tell gretl that the left bound a is set to 10.

32.7 Interval regression

The interval regression model arises when the dependent variable is unobserved for some (possibly all) observations; what we observe instead is an interval in which the dependent variable lies. In other words, the data generating process is assumed to be

$$y_i^* = x_i\beta + \epsilon_i$$

but we only know that $m_i \le y_i^* \le M_i$, where the interval may be left- or right-unbounded (but not both). If $m_i = M_i$, we effectively observe y_i^* and no information loss occurs. In practice, each observation belongs to one of four categories:

1. left-unbounded, when $m_i = -\infty$,
2. right-unbounded, when $M_i = \infty$,
3. bounded, when $-\infty < m_i < M_i < \infty$ and
4. point observations when $m_i = M_i$.

It is interesting to note that this model bears similarities to other models in several special cases:

- When all observations are point observations the model trivially reduces to the ordinary linear regression model.
- The interval model could be thought of an ordered probit model (see 32.2) in which the cut points (the α_j coefficients in eq. 32.8) are observed and don't need to be estimated.
- The Tobit model (see 32.6) is a special case of the interval model in which m_i and M_i do not depend on i, that is, the censoring limits are the same for all observations. As a matter of fact, gretl's `tobit` commands is handled internally as a special case of the interval model.

The gretl command `intreg` estimates interval models by maximum likelihood, assuming normality of the disturbance term ϵ_i. Its syntax is

```
intreg minvar maxvar X
```

where `minvar` contains the m_i series, with NAs for left-unbounded observations, and `maxvar` contains M_i, with NAs for right-unbounded observations. By default, standard errors are computed using the negative inverse of the Hessian. If the `--robust` flag is given, then QML or Huber-White standard errors are calculated instead. In this case the estimated covariance matrix is a "sandwich" of the inverse of the estimated Hessian and the outer product of the gradient.

If the model specification contains regressors other than just a constant, the output includes a chi-square statistic for testing the joint null hypothesis that none of these regressors has any effect on the outcome. This is a Wald statistic based on the estimated covariance matrix. If you wish to construct a likelihood ratio test, this is easily done by estimating both the full model and the null model (containing only the constant), saving the log-likelihood in both cases via the `$lnl` accessor, and then referring twice the difference between the two log-likelihoods to the chi-square distribution with k degrees of freedom, where k is the number of additional regressors (see the `pvalue` command in the *Gretl Command Reference*). Also included is a conditional moment normality test, similar to those provided for the probit, ordered probit and Tobit models (see above). An example is contained in the sample script `wtp.inp`, provided with the gretl distribution.

As with the probit and Tobit models, after a model has been estimated the `$uhat` accessor returns the generalized residual, which is an estimate of ϵ_i: more precisely, it equals $y_i - x_i\hat{\beta}$ for point observations and $E(\epsilon_i|m_i, M_i, x_i)$ otherwise. Note that it is possible to compute an unbiased predictor of y_i^* by summing this estimate to $x_i\hat{\beta}$. Script 32.5 shows an example. As a further similarity with Tobit, the interval regression model may deliver inconsistent estimates if the disturbances are non-normal; hence, the Chesher and Irish (1987) test for normality is included by default here too.

32.8 Sample selection model

In the sample selection model (also known as "Tobit II" model), there are two latent variables:

$$y_i^* = \sum_{j=1}^{k} x_{ij}\beta_j + \varepsilon_i \tag{32.14}$$

$$s_i^* = \sum_{j=1}^{p} z_{ij}\gamma_j + \eta_i \tag{32.15}$$

Example 32.5: Interval model on artificial data

Input:

```
nulldata 100
# generate artificial data
set seed 201449
x = normal()
epsilon = 0.2*normal()
ystar = 1 + x + epsilon
lo_bound = floor(ystar)
hi_bound = ceil(ystar)

# run the interval model
intreg lo_bound hi_bound const x

# estimate ystar
gen_resid = $uhat
yhat = $yhat + gen_resid
corr ystar yhat
```

Output (selected portions):

```
Model 1: Interval estimates using the 100 observations 1-100
Lower limit: lo_bound, Upper limit: hi_bound

             coefficient   std. error    t-ratio     p-value
  ----------------------------------------------------------
  const       0.993762     0.0338325      29.37     1.22e-189 ***
  x           0.986662     0.0319959      30.84     8.34e-209 ***

Chi-square(1)          950.9270   p-value               8.3e-209
Log-likelihood        -44.21258   Akaike criterion      94.42517
Schwarz criterion      102.2407   Hannan-Quinn          97.58824

sigma = 0.223273
Left-unbounded observations: 0
Right-unbounded observations: 0
Bounded observations: 100
Point observations: 0

...

corr(ystar, yhat) = 0.98960092
Under the null hypothesis of no correlation:
 t(98) = 68.1071, with two-tailed p-value 0.0000
```

and the observation rule is given by

$$y_i = \begin{cases} y_i^* & \text{for} \quad s_i^* > 0 \\ \blacklozenge & \text{for} \quad s_i^* \leq 0 \end{cases} \tag{32.16}$$

In this context, the ◆ symbol indicates that for some observations we simply do not have data on y: y_i may be 0, or missing, or anything else. A dummy variable d_i is normally used to set censored observations apart.

One of the most popular applications of this model in econometrics is a wage equation coupled with a labor force participation equation: we only observe the wage for the employed. If y_i^* and s_i^* were (conditionally) independent, there would be no reason not to use OLS for estimating equation (32.14); otherwise, OLS does not yield consistent estimates of the parameters β_j.

Since conditional independence between y_i^* and s_i^* is equivalent to conditional independence between ε_i and η_i, one may model the co-dependence between ε_i and η_i as

$$\varepsilon_i = \lambda\eta_i + v_i;$$

substituting the above expression in (32.14), you obtain the model that is actually estimated:

$$y_i = \sum_{j=1}^{k} x_{ij}\beta_j + \lambda\hat{\eta}_i + v_i,$$

so the hypothesis that censoring does not matter is equivalent to the hypothesis $H_0 : \lambda = 0$, which can be easily tested.

The parameters can be estimated via maximum likelihood under the assumption of joint normality of ε_i and η_i; however, a widely used alternative method yields the so-called *Heckit* estimator, named after Heckman (1979). The procedure can be briefly outlined as follows: first, a probit model is fit on equation (32.15); next, the generalized residuals are inserted in equation (32.14) to correct for the effect of sample selection.

Gretl provides the `heckit` command to carry out estimation; its syntax is

```
heckit y X ; d Z
```

where `y` is the dependent variable, `X` is a list of regressors, `d` is a dummy variable holding 1 for uncensored observations and `Z` is a list of explanatory variables for the censoring equation.

Since in most cases maximum likelihood is the method of choice, by default gretl computes ML estimates. The 2-step Heckit estimates can be obtained by using the `--two-step` option. After estimation, the `$uhat` accessor contains the generalized residuals. As in the ordinary Tobit model, the residuals equal the difference between actual and fitted y_i only for uncensored observations (those for which $d_i = 1$).

Example 32.6: Heckit model

```
open mroz87.gdt

genr EXP2 = AX^2
genr WA2 = WA^2
genr KIDS = (KL6+K618)>0

# Greene's specification

list X = const AX EXP2 WE CIT
list Z = const WA WA2 FAMINC KIDS WE

heckit WW X ; LFP Z --two-step
heckit WW X ; LFP Z

# Wooldridge's specification

series NWINC = FAMINC - WW*WHRS
series lww = log(WW)
list X = const WE AX EXP2
list Z = X NWINC WA KL6 K618

heckit lww X ; LFP Z --two-step
```

Example 32.6 shows two estimates from the dataset used in Mroz (1987): the first one replicates Table 22.7 in Greene (2003),[3] while the second one replicates table 17.1 in Wooldridge (2002a).

[3] Note that the estimates given by gretl do not coincide with those found in the printed volume. They do, however, match those found on the errata web page for Greene's book: `http://pages.stern.nyu.edu/~wgreene/Text/Errata/ERRATA5.htm`.

32.9 Count data

Here the dependent variable is assumed to be a non-negative integer, so a probabilistic description of $y_i|x_i$ must hinge on some discrete distribution. The most common model is the Poisson model, in which

$$P(y_i = Y|x_i) = e^{\mu_i}\frac{\mu_i^Y}{Y!}$$

$$\mu_i = \exp\left(\sum_j x_{ij}\beta_j\right)$$

In some cases, an "offset" variable is needed. The number of occurrences of y_i in a given time is assumed to be strictly proportional to the offset variable n_i. In the epidemiology literature, the offset is known as "population at risk". In this case, the model becomes

$$\mu_i = n_i \exp\left(\sum_j x_{ij}\beta_j\right)$$

Another way to look at the offset variable is to consider its natural log as just another explanatory variable whose coefficient is constrained to be one.

Estimation is carried out by maximum likelihood and follows the syntax

```
poisson depvar indep
```

If an offset variable is needed, it has to be specified at the end of the command, separated from the list of explanatory variables by a semicolon, as in

```
poisson depvar indep ; offset
```

It should be noted that the `poisson` command does not use, internally, the same optimization engines as most other gretl command, such as `arma` or `tobit`. As a consequence, some details may differ: the `--verbose` option will yield different output and settings such as `bfgs_toler` will not work.

Overdispersion

In the Poisson model, $E(y_i|x_i) = V(y_i|x_i) = \mu_i$, that is, the conditional mean equals the conditional variance by construction. In many cases, this feature is at odds with the data, as the conditional variance is often larger than the mean; this phenomenon is called "overdispersion". The output from the `poisson` command includes a conditional moment test for overdispersion (as per Davidson and MacKinnon (2004), section 11.5), which is printed automatically after estimation.

Overdispersion can be attributed to unmodeled heterogeneity between individuals. Two data points with the same observable characteristics $x_i = x_j$ may differ because of some unobserved scale factor $s_i \neq s_j$ so that

$$E(y_i|x_i, s_i) = \mu_i s_i \neq \mu_j s_j = E(y_j|x_j, s_j)$$

even though $\mu_i = \mu_j$. In other words, y_i is a Poisson random variable conditional on both x_i and s_i, but since s_i is unobservable, the only thing we can we can use, $P(y_i|x_i)$, will not follow the Poisson distribution.

It is often assumed that s_i can be represented as a gamma random variable with mean 1 and variance α: the parameter α is estimated together with the vector β, and measures the degree of heterogeneity between individuals.

In this case, the conditional probability for y_i given x_i can be shown to be

$$P(y_i|x_i) = \frac{\Gamma(y_i + \alpha^{-1})}{\Gamma(\alpha^{-1})\Gamma(y_i + 1)}\left[\frac{\mu_i}{\mu_i + \alpha^{-1}}\right]^{y_i}\left[\frac{\alpha^{-1}}{\mu_i + \alpha^{-1}}\right]^{\alpha^{-1}} \tag{32.17}$$

which is known as the "Negative Binomial Model". The conditional mean is still $E(y_i|x_i) = \mu_i$, but the variance equals $V(y_i|x_i) = \mu_i(1 + \mu_i\alpha)$. The gretl command for this model is `negbin depvar indep`.

- There is also a less used variant of the negative binomial model, in which the conditional variance is a scalar multiple of the conditional mean, that is $V(y_i|x_i) = \mu_i(1 + \gamma)$. To distinguish between the two, the model (32.17) is termed "Type 2". Gretl implements model 1 via the option `--model1`.
- A script which exemplifies the above models is included among gretl's sample scripts, under the name `camtriv.inp`.

FIXME: expand.

32.10 Duration models

In some contexts we wish to apply econometric methods to measurements of the duration of certain states. Classic examples include the following:

- From engineering, the "time to failure" of electronic or mechanical components: how long do, say, computer hard drives last until they malfunction?
- From the medical realm: how does a new treatment affect the time from diagnosis of a certain condition to exit from that condition (where "exit" might mean death or full recovery)?
- From economics: the duration of strikes, or of spells of unemployment.

In each case we may be interested in how the durations are distributed, and how they are affected by relevant covariates. There are several approaches to this problem; the one we discuss here—which is currently the only one supported by gretl—is estimation of a parametric model by means of Maximum Likelihood. In this approach we hypothesize that the durations follow some definite probability law and we seek to estimate the parameters of that law, factoring in the influence of covariates.

We may express the density (PDF) of the durations as $f(t, X, \theta)$, where t is the length of time in the state in question, X is a matrix of covariates, and θ is a vector of parameters. The likelihood for a sample of n observations indexed by i is then

$$L = \prod_{i=1}^{n} f(t_i, x_i, \theta)$$

Rather than working with the density directly, however, it is standard practice to factor $f(\cdot)$ into two components, namely a *hazard function*, λ, and a *survivor function*, S. The survivor function gives the probability that a state lasts at least as long as t; it is therefore $1 - F(t, X, \theta)$ where F is the CDF corresponding to the density $f(\cdot)$. The hazard function addresses this question: given that a state has persisted as long as t, what is the likelihood that it ends within a short increment of time beyond t—that is, it ends between t and $t + \Delta$? Taking the limit as Δ goes to zero, we end up with the ratio of the density to the survivor function:[4]

$$\lambda(t, X, \theta) = \frac{f(t, X, \theta)}{S(t, X, \theta)} \tag{32.18}$$

so the log-likelihood can be written as

$$\ell = \sum_{i=1}^{n} \log f(t_i, x_i, \theta) = \sum_{i=1}^{n} \log \lambda(t_i, x_i, \theta) + \log S(t_i, x_i, \theta) \tag{32.19}$$

[4]For a fuller discussion see, for example, Davidson and MacKinnon (2004).

One point of interest is the shape of the hazard function, in particular its dependence (or not) on time since the state began. If λ does not depend on t we say the process in question exhibits *duration independence*: the probability of exiting the state at any given moment neither increases nor decreases based simply on how long the state has persisted to date. The alternatives are positive duration dependence (the likelihood of exiting the state rises, the longer the state has persisted) or negative duration dependence (exit becomes less likely, the longer it has persisted). Finally, the behavior of the hazard with respect to time need not be monotonic; some parameterizations allow for this possibility and some do not.

Since durations are inherently positive the probability distribution used in modeling must respect this requirement, giving a density of zero for $t \leq 0$. Four common candidates are the exponential, Weibull, log-logistic and log-normal, the Weibull being the most common choice. The table below displays the density and the hazard function for each of these distributions as they are commonly parameterized, written as functions of t alone. (ϕ and Φ denote, respectively, the Gaussian PDF and CDF.)

	density, $f(t)$	hazard, $\lambda(t)$
Exponential	$\gamma \exp(-\gamma t)$	γ
Weibull	$\alpha\gamma^{\alpha}t^{\alpha-1}\exp[-(\gamma t)^{\alpha}]$	$\alpha\gamma^{\alpha}t^{\alpha-1}$
Log-logistic	$\gamma\alpha\frac{(\gamma t)^{\alpha-1}}{[1+(\gamma t)^{\alpha}]^2}$	$\gamma\alpha\frac{(\gamma t)^{\alpha-1}}{[1+(\gamma t)^{\alpha}]}$
Log-normal	$\frac{1}{\sigma t}\phi[(\log t-\mu)/\sigma]$	$\frac{1}{\sigma t}\frac{\phi[(\log t-\mu)/\sigma]}{\Phi[-(\log t-\mu)/\sigma]}$

The hazard is constant for the exponential distribution. For the Weibull, it is monotone increasing in t if $\alpha > 1$, or monotone decreasing for $\alpha < 1$. (If $\alpha = 1$ the Weibull collapses to the exponential.) The log-logistic and log-normal distributions allow the hazard to vary with t in a non-monotonic fashion.

Covariates are brought into the picture by allowing them to govern one of the parameters of the density, so that the durations are not identically distributed across cases. For example, when using the log-normal distribution it is natural to make μ, the expected value of $\log t$, depend on the covariates, X. This is typically done via a linear index function: $\mu = X\beta$.

Note that the expressions for the log-normal density and hazard contain the term $(\log t - \mu)/\sigma$. Replacing μ with $X\beta$ this becomes $(\log t - X\beta)/\sigma$. It turns out that this constitutes a useful simplifying change of variables for all of the distributions discussed here. As in Kalbfleisch and Prentice (2002), we define

$$w_i \equiv (\log t_i - x_i\beta)/\sigma$$

The interpretation of the scale factor, σ, in this expression depends on the distribution. For the log-normal, σ represents the standard deviation of $\log t$; for the Weibull and the log-logistic it corresponds to $1/\alpha$; and for the exponential it is fixed at unity. For distributions other than the log-normal, $X\beta$ corresponds to $-\log\gamma$, or in other words $\gamma = \exp(-X\beta)$.

With this change of variables, the density and survivor functions may be written compactly as follows (the exponential is the same as the Weibull).

	density, $f(w_i)$	survivor, $S(w_i)$
Weibull	$\exp(w_i - e^{w_i})$	$\exp(-e^{w_i})$
Log-logistic	$e^{w_i}(1+e^{w_i})^{-2}$	$(1+e^{w_i})^{-1}$
Log-normal	$\phi(w_i)$	$\Phi(-w_i)$

In light of the above we may think of the generic parameter vector θ, as in $f(t, X, \theta)$, as composed of the coefficients on the covariates, β, plus (in all cases apart from the exponential) the additional parameter σ.

A complication in estimation of θ is posed by "incomplete spells". That is, in some cases the state in question may not have ended at the time the observation is made (e.g. some workers remain unemployed, some components have not yet failed). If we use t_i to denote the time from entering the state to either (a) exiting the state or (b) the observation window closing, whichever comes first, then all we know of the "right-censored" cases (b) is that the duration was at least as long as t_i. This can be handled by rewriting the the log-likelihood (compare 32.19) as

$$\ell_i = \sum_{i=1}^{n} \delta_i \log S(w_i) + (1 - \delta_i)\left[-\log \sigma + \log f(w_i)\right] \tag{32.20}$$

where δ_i equals 1 for censored cases (incomplete spells), and 0 for complete observations. The rationale for this is that the log-density equals the sum of the log hazard and the log survivor function, but for the incomplete spells only the survivor function contributes to the likelihood. So in (32.20) we are adding up the log survivor function alone for the incomplete cases, plus the full log density for the completed cases.

Implementation in gretl and illustration

The `duration` command accepts a list of series on the usual pattern: dependent variable followed by covariates. If right-censoring is present in the data this should be represented by a dummy variable corresponding to δ_i above, separated from the covariates by a semicolon. For example,

```
duration durat 0 X ; cens
```

where `durat` measures durations, `0` represents the constant (which is required for such models), `X` is a named list of regressors, and `cens` is the censoring dummy.

By default the Weibull distribution is used; you can substitute any of the other three distributions discussed here by appending one of the option flags `--exponential`, `--loglogistic` or `--lognormal`.

Interpreting the coefficients in a duration model requires some care, and we will work through an illustrative case. The example comes from section 20.3 of Wooldridge (2002a), and it concerns criminal recidivism.[5] The data (filename `recid.gdt`) pertain to a sample of 1,445 convicts released from prison between July 1, 1977 and June 30, 1978. The dependent variable is the time in months until they are again arrested. The information was gathered retrospectively by examining records in April 1984; the maximum possible length of observation is 81 months. Right-censoring is important: when the date were compiled about 62 percent had not been arrested. The dataset contains several covariates, which are described in the data file; we will focus below on interpretation of the `married` variable, a dummy which equals 1 if the respondent was married when imprisoned.

Example 32.7 shows the gretl commands for a Weibull model along with most of the output. Consider first the scale factor, σ. The estimate is 1.241 with a standard error of 0.048. (We don't print a z score and p-value for this term since $H_0 : \sigma = 0$ is not of interest.) Recall that σ corresponds to $1/\alpha$; we can be confident that α is less than 1, so recidivism displays negative duration dependence. This makes sense: it is plausible that if a past offender manages to stay out of trouble for an extended period his risk of engaging in crime again diminishes. (The exponential model would therefore not be appropriate in this case.)

On a priori grounds, however, we may doubt the monotonic decline in hazard that is implied by the Weibull specification. Even if a person is liable to return to crime, it seems relatively unlikely that he would do so straight out of prison. In the data, we find that only 2.6 percent of those followed were rearrested within 3 months. The log-normal specification, which allows the hazard to rise and then fall, may be more appropriate. Using the `duration` command again with the same covariates but the `--lognormal` flag, we get a log-likelihood of -1597 as against -1633 for the Weibull, confirming that the log-normal gives a better fit.

[5]Germán Rodríguez of Princeton University has a page discussing this example and displaying estimates from Stata at `http://data.princeton.edu/pop509a/recid1.html`.

Example 32.7: Weibull model for recidivism data

Input:

```
open recid.gdt
list X = workprg priors tserved felon alcohol drugs \
 black married educ age
duration durat 0 X ; cens
duration durat 0 X ; cens --lognormal
```

Partial output:

```
Model 1: Duration (Weibull), using observations 1-1445
Dependent variable: durat

             coefficient   std. error      z       p-value
  --------------------------------------------------------
  const       4.22167      0.341311      12.37    3.85e-35 ***
  workprg    -0.112785     0.112535      -1.002   0.3162
  priors     -0.110176     0.0170675     -6.455   1.08e-10 ***
  tserved    -0.0168297    0.00213029    -7.900   2.78e-15 ***
  felon       0.371623     0.131995       2.815   0.0049   ***
  alcohol    -0.555132     0.132243      -4.198   2.69e-05 ***
  drugs      -0.349265     0.121880      -2.866   0.0042   ***
  black      -0.563016     0.110817      -5.081   3.76e-07 ***
  married     0.188104     0.135752       1.386   0.1659
  educ        0.0289111    0.0241153      1.199   0.2306
  age         0.00462188   0.000664820    6.952   3.60e-12 ***
  sigma       1.24090      0.0482896

Chi-square(10)        165.4772   p-value              2.39e-30
Log-likelihood       -1633.032   Akaike criterion     3290.065

Model 2: Duration (log-normal), using observations 1-1445
Dependent variable: durat

             coefficient   std. error       z       p-value
  ---------------------------------------------------------
  const       4.09939      0.347535      11.80     4.11e-32 ***
  workprg    -0.0625693    0.120037      -0.5213   0.6022
  priors     -0.137253     0.0214587     -6.396    1.59e-10 ***
  tserved    -0.0193306    0.00297792    -6.491    8.51e-11 ***
  felon       0.443995     0.145087       3.060    0.0022   ***
  alcohol    -0.634909     0.144217      -4.402    1.07e-05 ***
  drugs      -0.298159     0.132736      -2.246    0.0247   **
  black      -0.542719     0.117443      -4.621    3.82e-06 ***
  married     0.340682     0.139843       2.436    0.0148   **
  educ        0.0229194    0.0253974      0.9024   0.3668
  age         0.00391028   0.000606205    6.450    1.12e-10 ***
  sigma       1.81047      0.0623022

Chi-square(10)        166.7361   p-value              1.31e-30
Log-likelihood       -1597.059   Akaike criterion     3218.118
```

Let us now focus on the `married` coefficient, which is positive in both specifications but larger and more sharply estimated in the log-normal variant. The first thing is to get the interpretation of the sign right. Recall that $X\beta$ enters negatively into the intermediate variable w. The Weibull hazard is $\lambda(w_i) = e^{w_i}$, so being married reduces the hazard of re-offending, or in other words lengthens the expected duration out of prison. The same qualitative interpretation applies for the log-normal.

To get a better sense of the married effect, it is useful to show its impact on the hazard across time. We can do this by plotting the hazard for two values of the index function $X\beta$: in each case the values of all the covariates other than `married` are set to their means (or some chosen values) while `married` is set first to 0 then to 1. Example 32.8 provides a script that does this, and the resulting plots are shown in Figure 32.1. Note that when computing the hazards we need to multiply by the Jacobian of the transformation from t_i to $w_i = \log(t_i - x_i\beta)/\sigma$, namely $1/t$. Note also that the estimate of σ is available via the accessor `$sigma`, but it is also present as the last element in the coefficient vector obtained via `$coeff`.

A further difference between the Weibull and log-normal specifications is illustrated in the plots. The Weibull is an instance of a *proportional hazard* model. This means that for any sets of values of the covariates, x_i and x_j, the ratio of the associated hazards is invariant with respect to duration. In this example the Weibull hazard for unmarried individuals is always 1.1637 times that for married. In the log-normal variant, on the other hand, this ratio gradually declines from 1.6703 at one month to 1.1766 at 100 months.

Alternative representations of the Weibull model

One point to watch out for with the Weibull duration model is that the estimates may be represented in different ways. The representation given by gretl is sometimes called the *accelerated failure-time* (AFT) metric. An alternative that one sometimes sees is the *log relative-hazard* metric; in fact this is the metric used in Wooldridge's presentation of the recidivism example. To get from AFT estimates to log relative-hazard form it is necessary to multiply the coefficients by $-\sigma^{-1}$. For example, the `married` coefficient in the Weibull specification as shown here is 0.188104 and $\hat{\sigma}$ is 1.24090, so the alternative value is −0.152, which is what Wooldridge shows (2002a, Table 20.1).

Fitted values and residuals

By default, gretl computes fitted values (accessible via `$yhat`) as the conditional mean of duration. The formulae are shown below (where Γ denotes the gamma function, and the exponential variant is just Weibull with $\sigma = 1$).

Weibull	Log-logistic	Log-normal
$\exp(X\beta)\Gamma(1+\sigma)$	$\exp(X\beta)\dfrac{\pi\sigma}{\sin(\pi\sigma)}$	$\exp(X\beta + \sigma^2/2)$

The expression given for the log-logistic mean, however, is valid only for $\sigma < 1$; otherwise the expectation is undefined, a point that is not noted in all software.[6]

Alternatively, if the `--medians` option is given, gretl's `duration` command will produce conditional medians as the content of `$yhat`. For the Weibull the median is $\exp(X\beta)(\log 2)^{\sigma}$; for the log-logistic and log-normal it is just $\exp(X\beta)$.

The values we give for the accessor `$uhat` are generalized (Cox-Snell) residuals, computed as the integrated hazard function, which equals the negative of the log of the survivor function:

$$\hat{u}_i = \Lambda(t_i, x_i, \theta) = -\log S(t_i, x_i, \theta)$$

Under the null of correct specification of the model these generalized residuals should follow the unit exponential distribution, which has mean and variance both equal to 1 and density e^{-1}. See Cameron and Trivedi (2005) for further discussion.

[6] The `predict` adjunct to the `streg` command in Stata 10, for example, gaily produces large negative values for the log-logistic mean in duration models with $\sigma > 1$.

Example 32.8: Create plots showing conditional hazards

```
open recid.gdt -q

# leave 'married' separate for analysis
list X = workprg priors tserved felon alcohol drugs \
 black educ age

# Weibull variant
duration durat 0 X married ; cens
# coefficients on all Xs apart from married
matrix beta_w = $coeff[1:$ncoeff-2]
# married coefficient
scalar mc_w = $coeff[$ncoeff-1]
scalar s_w = $sigma

# Log-normal variant
duration durat 0 X married ; cens --lognormal
matrix beta_n = $coeff[1:$ncoeff-2]
scalar mc_n = $coeff[$ncoeff-1]
scalar s_n = $sigma

list allX = 0 X
# evaluate X\beta at means of all variables except marriage
scalar Xb_w = meanc({allX}) * beta_w
scalar Xb_n = meanc({allX}) * beta_n

# construct two plot matrices
matrix mat_w = zeros(100, 3)
matrix mat_n = zeros(100, 3)

loop t=1..100 -q
  # first column, duration
  mat_w[t, 1] = t
  mat_n[t, 1] = t
  wi_w = (log(t) - Xb_w)/s_w
  wi_n = (log(t) - Xb_n)/s_n
  # second col: hazard with married = 0
  mat_w[t, 2] = (1/t) * exp(wi_w)
  mat_n[t, 2] = (1/t) * pdf(z, wi_n) / cdf(z, -wi_n)
  wi_w = (log(t) - (Xb_w + mc_w))/s_w
  wi_n = (log(t) - (Xb_n + mc_n))/s_n
  # third col: hazard with married = 1
  mat_w[t, 3] = (1/t) * exp(wi_w)
  mat_n[t, 3] = (1/t) * pdf(z, wi_n) / cdf(z, -wi_n)
endloop

colnames(mat_w, "months unmarried married")
colnames(mat_n, "months unmarried married")

gnuplot 2 3 1 --with-lines --supp --matrix=mat_w --output=weibull.plt
gnuplot 2 3 1 --with-lines --supp --matrix=mat_n --output=lognorm.plt
```

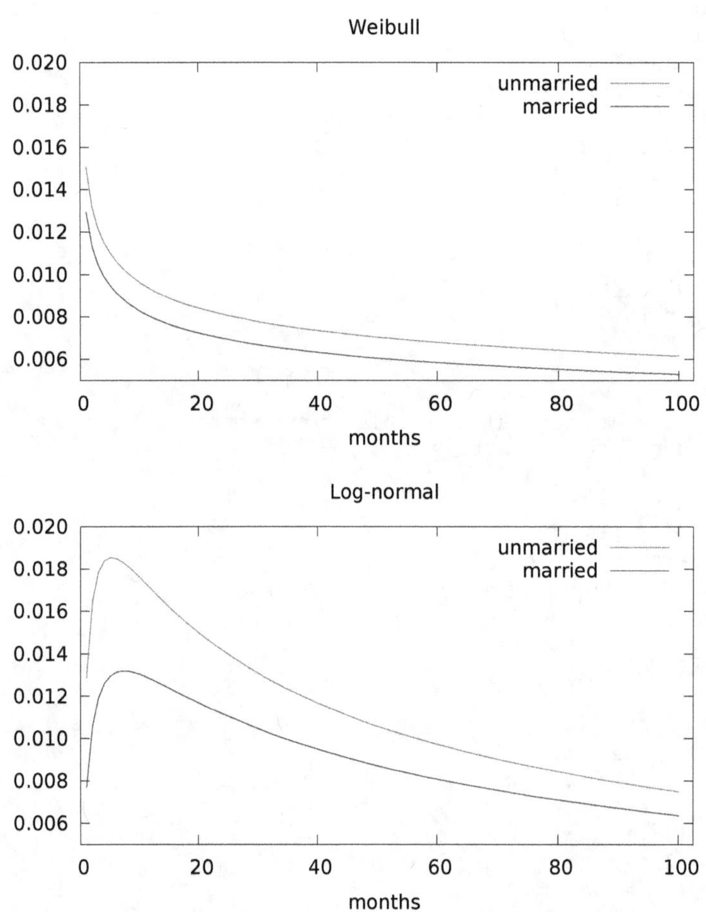

Figure 32.1: Recidivism hazard estimates for married and unmarried ex-convicts

Chapter 33

Quantile regression

33.1 Introduction

In Ordinary Least Squares (OLS) regression, the fitted values, $\hat{y}_i = X_i\hat{\beta}$, represent the *conditional mean* of the dependent variable—conditional, that is, on the regression function and the values of the independent variables. In median regression, by contrast and as the name implies, fitted values represent the *conditional median* of the dependent variable. It turns out that the principle of estimation for median regression is easily stated (though not so easily computed), namely, choose $\hat{\beta}$ so as to minimize the sum of absolute residuals. Hence the method is known as Least Absolute Deviations or LAD. While the OLS problem has a straightforward analytical solution, LAD is a linear programming problem.

Quantile regression is a generalization of median regression: the regression function predicts the conditional τ-quantile of the dependent variable—for example the first quartile ($\tau = .25$) or the ninth decile ($\tau = .90$).

If the classical conditions for the validity of OLS are satisfied—that is, if the error term is independently and identically distributed, conditional on X—then quantile regression is redundant: all the conditional quantiles of the dependent variable will march in lockstep with the conditional mean. Conversely, if quantile regression reveals that the conditional quantiles behave in a manner quite distinct from the conditional mean, this suggests that OLS estimation is problematic.

As of version 1.7.5, gretl offers quantile regression functionality (in addition to basic LAD regression, which has been available since early in gretl's history via the `lad` command).[1]

33.2 Basic syntax

The basic invocation of quantile regression is

`quantreg` *tau reglist*

where

- *reglist* is a standard gretl regression list (dependent variable followed by regressors, including the constant if an intercept is wanted); and
- *tau* is the desired conditional quantile, in the range 0.01 to 0.99, given either as a numerical value or the name of a pre-defined scalar variable (but see below for a further option).

Estimation is via the Frisch-Newton interior point solver (Portnoy and Koenker, 1997), which is substantially faster than the "traditional" Barrodale-Roberts (1974) simplex approach for large problems.

By default, standard errors are computed according to the asymptotic formula given by Koenker and Bassett (1978). Alternatively, if the `--robust` option is given, we use the sandwich estimator developed in Koenker and Zhao (1994).[2]

[1] We gratefully acknowledge our borrowing from the `quantreg` package for GNU R (version 4.17). The core of the `quantreg` package is composed of Fortran code written by Roger Koenker; this is accompanied by various driver and auxiliary functions written in the R language by Koenker and Martin Mächler. The latter functions have been re-worked in C for gretl. We have added some guards against potential numerical problems in small samples.

[2] These correspond to the `iid` and `nid` options in R's `quantreg` package, respectively.

33.3 Confidence intervals

An option `--intervals` is available. When this is given we print confidence intervals for the parameter estimates instead of standard errors. These intervals are computed using the rank inversion method and in general they are asymmetrical about the point estimates—that is, they are not simply "plus or minus so many standard errors". The specifics of the calculation are inflected by the `--robust` option: without this, the intervals are computed on the assumption of IID errors (Koenker, 1994); with it, they use the heteroskedasticity-robust estimator developed by Koenker and Machado (1999).

By default, 90 percent intervals are produced. You can change this by appending a confidence value (expressed as a decimal fraction) to the intervals option, as in

```
quantreg tau reglist --intervals=.95
```

When the confidence intervals option is selected, the parameter estimates are calculated using the Barrodale-Roberts method. This is simply because the Frisch-Newton code does not currently support the calculation of confidence intervals.

Two further details. First, the mechanisms for generating confidence intervals for quantile estimates require that the model has at least two regressors (including the constant). If the `--intervals` option is given for a model containing only one regressor, an error is flagged. Second, when a model is estimated in this mode, you can retrieve the confidence intervals using the accessor `$coeff_ci`. This produces a $k \times 2$ matrix, where k is the number of regressors. The lower bounds are in the first column, the upper bounds in the second. See also section 33.5 below.

33.4 Multiple quantiles

As a further option, you can give *tau* as a matrix—either the name of a predefined matrix or in numerical form, as in `{.05, .25, .5, .75, .95}`. The given model is estimated for all the τ values and the results are printed in a special form, as shown below (in this case the `--intervals` option was also given).

```
Model 1: Quantile estimates using the 235 observations 1-235
Dependent variable: foodexp
With 90 percent confidence intervals

     VARIABLE       TAU    COEFFICIENT       LOWER        UPPER

   const            0.05      124.880       98.3021      130.517
                    0.25      95.4835       73.7861      120.098
                    0.50      81.4822       53.2592      114.012
                    0.75      62.3966       32.7449      107.314
                    0.95      64.1040       46.2649      83.5790

   income           0.05     0.343361      0.343327     0.389750
                    0.25     0.474103      0.420330     0.494329
                    0.50     0.560181      0.487022     0.601989
                    0.75     0.644014      0.580155     0.690413
                    0.95     0.709069      0.673900     0.734441
```

The gretl GUI has an entry for Quantile Regression (under /Model/Robust estimation), and you can select multiple quantiles there too. In that context, just give space-separated numerical values (as per the predefined options, shown in a drop-down list).

When you estimate a model in this way most of the standard menu items in the model window are disabled, but one extra item is available—graphs showing the τ sequence for a given coefficient in comparison with the OLS coefficient. An example is shown in Figure 33.1. This sort of graph provides a simple means of judging whether quantile regression is redundant (OLS is fine) or informative.

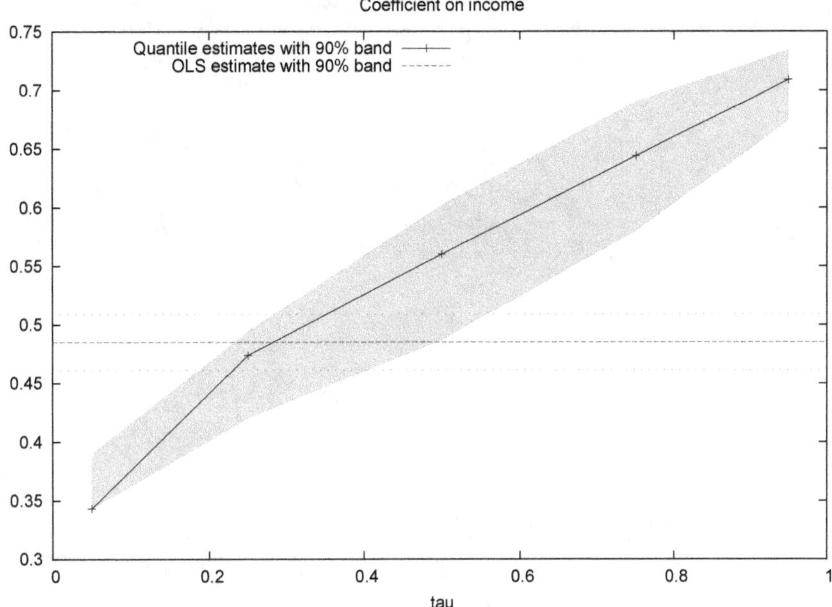

Figure 33.1: Regression of food expenditure on income; Engel's data

In the example shown—based on data on household income and food expenditure gathered by Ernst Engel (1821–1896)—it seems clear that simple OLS regression is potentially misleading. The "crossing" of the OLS estimate by the quantile estimates is very marked.

However, it is not always clear what implications should be drawn from this sort of conflict. With the Engel data there are two issues to consider. First, Engel's famous "law" claims an income-elasticity of food consumption that is less than one, and talk of elasticities suggests a logarithmic formulation of the model. Second, there are two apparently anomalous observations in the data set: household 105 has the third-highest income but unexpectedly low expenditure on food (as judged from a simple scatter plot), while household 138 (which also has unexpectedly low food consumption) has much the highest income, almost twice that of the next highest.

With $n = 235$ it seems reasonable to consider dropping these observations. If we do so, and adopt a log-log formulation, we get the plot shown in Figure 33.2. The quantile estimates still cross the OLS estimate, but the "evidence against OLS" is much less compelling: the 90 percent confidence bands of the respective estimates overlap at all the quantiles considered.

33.5 Large datasets

As noted above, when you give the `--intervals` option with the `quantreg` command, which calls for estimation of confidence intervals via rank inversion, gretl switches from the default Frisch–Newton algorithm to the Barrodale–Roberts simplex method.

This is OK for moderately large datasets (up to, say, a few thousand observations) but on very large problems the simplex algorithm may become seriously bogged down. For example, Koenker and Hallock (2001) present an analysis of the determinants of birth weights, using 198377 observations and with 15 regressors. Generating confidence intervals via Barrodale–Roberts for a single value of τ took about half an hour on a Lenovo Thinkpad T60p with 1.83GHz Intel Core 2 processor.

If you want confidence intervals in such cases, you are advised not to use the `--intervals` option, but to compute them using the method of "plus or minus so many standard errors". (One Frisch–Newton run took about 8 seconds on the same machine, showing the superiority of the interior point method.) The script below illustrates:

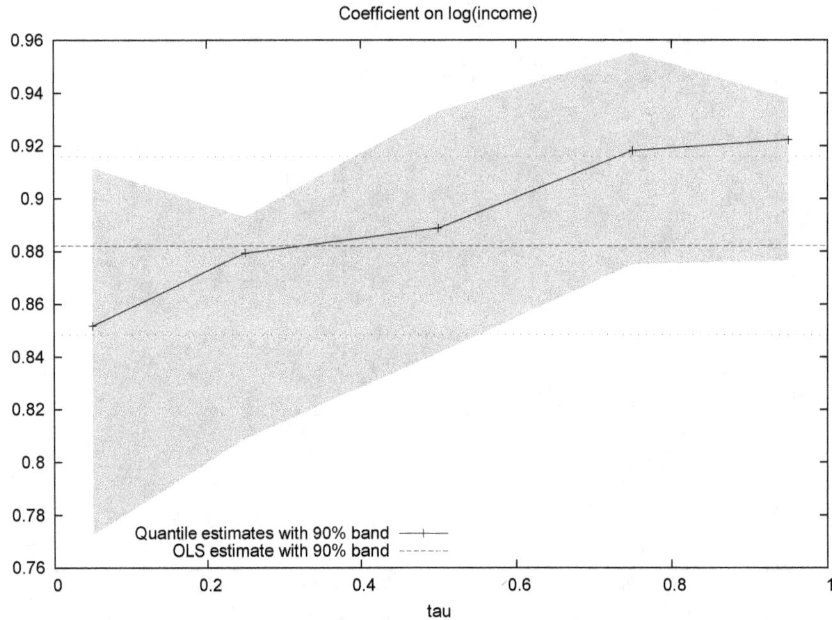

Figure 33.2: Log-log regression; 2 observations dropped from full Engel data set.

```
quantreg .10 y 0 xlist
scalar crit = qnorm(.95)
matrix ci = $coeff - crit * $stderr
ci = ci~($coeff + crit * $stderr)
print ci
```

The matrix ci will contain the lower and upper bounds of the (symmetrical) 90 percent confidence intervals.

To avoid a situation where gretl becomes unresponsive for a very long time we have set the maximum number of iterations for the Borrodale–Roberts algorithm to the (somewhat arbitrary) value of 1000. We will experiment further with this, but for the meantime if you really want to use this method on a large dataset, and don't mind waiting for the results, you can increase the limit using the set command with parameter rq_maxiter, as in

```
set rq_maxiter 5000
```

Chapter 34

Nonparametric methods

The main focus of gretl is on parametric estimation, but we offer a selection of nonparametric methods. The most basic of these

- various tests for difference in distribution (Sign test, Wilcoxon rank-sum test, Wilcoxon signed-rank test);
- the Runs test for randomness; and
- nonparametric measures of association: Spearman's rho and Kendall's tau.

Details on the above can be found by consulting the help for the commands `difftest`, `runs`, `corr` and `spearman`. In the GUI program these items are found under the Tools menu and the Robust estimation item under the Model menu.

In this chapter we concentrate on two relatively complex methods for nonparametric curve-fitting and prediction, namely William Cleveland's "loess" (also known as "lowess") and the Nadaraya–Watson estimator.

34.1 Locally weighted regression (loess)

Loess (Cleveland, 1979) is a nonparametric smoother employing locally weighted polynomial regression. It is intended to yield an approximation to $g(\cdot)$ when the dependent variable, y, can be expressed as

$$y_i = g(x_i) + \epsilon_i$$

for some smooth function $g(\cdot)$.

Given a sample of n observations on the variables y and x, the procedure is to run a weighted least squares regression (a polynomial of order $d = 0$, 1 or 2 in x) localized to each data point, i. In each such regression the sample consists of the r nearest neighbors (in the x dimension) to the point i, with weights that are inversely related to the distance $|x_i - x_k|$, $k = 1, \ldots, r$. The predicted value $\hat{y}_i$ is then obtained by evaluating the estimated polynomial at x_i. The most commonly used order is $d = 1$.

A bandwidth parameter $0 < q \leq 1$ controls the proportion of the total number of data points used in each regression; thus $r = qn$ (rounded up to an integer). Larger values of q lead to a smoother fitted series, smaller values to a series that tracks the actual data more closely; $0.25 \leq q \leq 0.5$ is often a suitable range.

In gretl's implementation of loess the weighting scheme is that given by Cleveland, namely,

$$w_k(x_i) = W(h_i^{-1}(x_k - x_i))$$

where h_i is the distance between x_i and its rth nearest neighbor, and $W(\cdot)$ is the tricube function,

$$W(x) = \begin{cases} (1 - |x|^3)^3 & \text{for } |x| < 1 \\ 0 & \text{for } |x| \geq 1 \end{cases}$$

The local regression can be made robust via an adjustment based on the residuals, $e_i = y_i - \hat{y}_i$. Robustness weights, δ_k, are defined by

$$\delta_k = B(e_k/6s)$$

where s is the median of the $|e_i|$ and $B(\cdot)$ is the bisquare function,

$$B(x) = \begin{cases} (1-x^2)^2 & \text{for } |x| < 1 \\ 0 & \text{for } |x| \geq 1 \end{cases}$$

The polynomial regression is then re-run using weight $\delta_k w_k(x_i)$ at (x_k, y_k).

The `loess()` function in gretl takes up to five arguments as follows: the y series, the x series, the order d, the bandwidth q, and a boolean switch to turn on the robust adjustment. The last three arguments are optional: if they are omitted the default values are $d = 1$, $q = 0.5$ and no robust adjustment. An example of a full call to `loess()` is shown below; in this case a quadratic in x is specified, three quarters of the data points will be used in each local regression, and robustness is turned on:

```
series yh = loess(y, x, 2, 0.75, 1)
```

An illustration of loess is provided in Example 34.1: we generate a series that has a deterministic sine wave component overlaid with noise uniformly distributed on $(-1, 1)$. Loess is then used to retrieve a good approximation to the sine function. The resulting graph is shown in Figure 34.1.

Example 34.1: Loess script

```
nulldata 120
series x = index
scalar n = $nobs
series y = sin(2*pi*x/n) + uniform(-1, 1)
series yh = loess(y, x, 2, 0.75, 0)
gnuplot y yh x --output=display --with-lines=yh
```

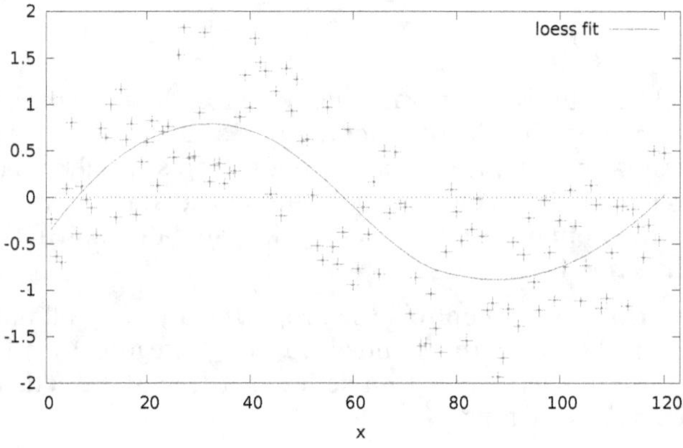

Figure 34.1: Loess: retrieving a sine wave

34.2 The Nadaraya-Watson estimator

The Nadaraya-Watson nonparametric estimator (Nadaraya, 1964; Watson, 1964) is an estimator for the conditional mean of a variable Y, available in a sample of size n, for a given value of a conditioning variable X, and is defined as

$$m(X) = \frac{\sum_{j=1}^{n} y_j \cdot K_h(X - x_j)}{\sum_{j=1}^{n} K_h(X - x_j)}$$

where $K_h(\cdot)$ is the so-called *kernel function*, which is usually some simple transform of a density function that depends on a scalar called the *bandwidth*. The one gretl uses is given by

$$K_h(x) = \exp\left(-\frac{x^2}{2h}\right)$$

for $|x| < \tau$ and zero otherwise. The scalar τ is used to prevent numerical problems when the kernel function is evaluated too far away from zero and is called the trim parameter.

Example 34.2: Nadaraya-Watson example

```
# Nonparametric regression example: husband's age on wife's age
open mroz87.gdt

# initial value for the bandwidth
scalar h = $nobs^(-0.2)
# three increasingly smooth estimates
series m0 = nadarwat(HA, WA, h)
series m1 = nadarwat(HA, WA, h * 5)
series m2 = nadarwat(HA, WA, h * 10)

# produce the graph
dataset sortby WA
gnuplot HA m0 m1 m2 WA --output=display --with-lines=m0,m1,m2
```

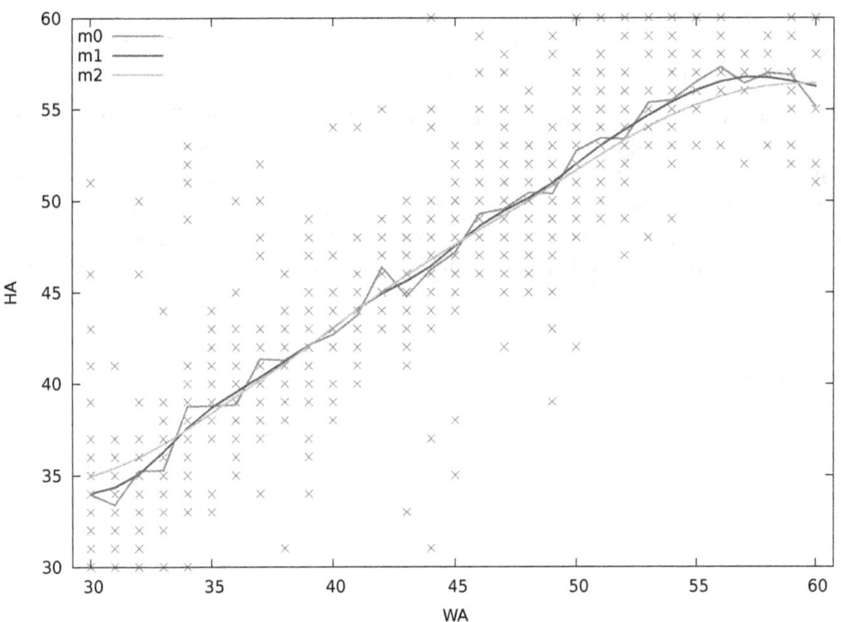

Figure 34.2: Nadaraya-Watson example for several choices of the bandwidth parameter

Example 34.2 produces the graph shown in Figure 34.2 (after some slight editing).

The choice of the bandwidth is up to the user: larger values of h lead to a smoother $m(\cdot)$ function; smaller values make the $m(\cdot)$ function follow the y_i values more closely, so that the function appears more "jagged". In fact, as $h \to \infty$, $m(x_i) \to \bar{Y}$; on the contrary, if $h \to 0$, observations for which $x_i \neq X$ are not taken into account at all when computing $m(X)$.

Also, the statistical properties of $m(\cdot)$ vary with h: its variance can be shown to be decreasing in

h, while its squared bias is increasing in h. It can be shown that choosing $h \sim n^{-1/5}$ minimizes the RMSE, so that value is customarily taken as a reference point.

Note that the kernel function has its tails "trimmed". The scalar τ, which controls the level at which trimming occurs is set by default at $4 \cdot h$; this setting, however, may be changed via the `set` command. For example,

```
set nadarwat_trim 10
```

sets $\tau = 10 \cdot h$. This may at times produce more sensible results in regions of X with sparse support; however, you should be aware that in those same cases machine precision (division by numerical zero) may render your results spurious. The default is relatively safe, but experimenting with larger values may be a sensible strategy in some cases.

A common variant of the Nadaraya-Watson estimator is the so-called "leave-one-out" estimator: this is a variant of the estimator that does not use the i-th observation for evaluating $m(x_i)$. This makes the estimator more robust numerically and its usage is often advised for inference purposes. In formulae, the leave-one-out estimator is

$$m(x_i) = \frac{\sum_{j \neq i} y_j \cdot K_h(x_i - x_j)}{\sum_{j \neq i} K_h(x_i - x_j)}$$

In order to have gretl compute the leave-one-out estimator, just reverse the sign of h: if we changed example 34.2 by substituting

```
scalar h = $nobs^(-0.2)
```

with

```
scalar h = -($nobs^(-0.2))
```

the rest of the example would have stayed unchanged, the only difference being the usage of the leave-one-out estimator.

Although X could be, in principle, any value, in the typical usage of this estimator you want to compute $m(X)$ for X equal to one or more values actually observed in your sample, that is $m(x_i)$. If you need a point estimate of $m(X)$ for some value of X which is not present among the valid observations of your dependent variable, you may want to add some "fake" observations to your dataset in which y is missing and x contains the values you want $m(x)$ evaluated at. For example, the following script evaluates $m(x)$ at regular intervals between -2.0 and 2.0:

```
nulldata 120
set seed 120496

# first part of the sample: actual data
smpl 1 100
x = normal()
y = x^2 + sin(x) + normal()

# second part of the sample: fake x data
smpl 101 120
x = (obs-110) / 5

# compute the Nadaraya-Watson estimate
# with bandwidth equal to 0.4 (note that
# 100^(-0.2) = 0.398)
smpl full
m = nadarwat(y, x, 0.4)

# show m(x) for the fake x values only
smpl 101 120
print x m -o
```

and running it produces

```
                  x           m

 101           -1.8    1.165934
 102           -1.6    0.730221
 103           -1.4    0.314705
 104           -1.2    0.026057
 105           -1.0   -0.131999
 106           -0.8   -0.215445
 107           -0.6   -0.269257
 108           -0.4   -0.304451
 109           -0.2   -0.306448
 110            0.0   -0.238766
 111            0.2   -0.038837
 112            0.4    0.354660
 113            0.6    0.908178
 114            0.8    1.485178
 115            1.0    2.000003
 116            1.2    2.460100
 117            1.4    2.905176
 118            1.6    3.380874
 119            1.8    3.927682
 120            2.0    4.538364
```

Part III

Technical details

Chapter 35

Gretl and TeX

35.1 Introduction

TeX — initially developed by Donald Knuth of Stanford University and since enhanced by hundreds of contributors around the world — is the gold standard of scientific typesetting. Gretl provides various hooks that enable you to preview and print econometric results using the TeX engine, and to save output in a form suitable for further processing with TeX.

This chapter explains the finer points of gretl's TeX-related functionality. The next section describes the relevant menu items; section 35.3 discusses ways of fine-tuning TeX output; and section 35.4 gives some pointers on installing (and learning) TeX if you do not already have it on your computer. (Just to be clear: TeX is not included with the gretl distribution; it is a separate package, including several programs and a large number of supporting files.)

Before proceeding, however, it may be useful to set out briefly the stages of production of a final document using TeX. For the most part you don't have to worry about these details, since, in regard to previewing at any rate, gretl handles them for you. But having some grasp of what is going on behind the scences will enable you to understand your options better.

The first step is the creation of a plain text "source" file, containing the text or mathematics to be typset, interspersed with mark-up that defines how it should be formatted. The second step is to run the source through a processing engine that does the actual formatting. Typically this a program called pdflatex that generates PDF output.[1] (In times gone by it was a program called latex that generated so-called DVI (device-independent) output.)

So gretl calls pdflatex to process the source file. On MS Windows and Mac OS X, gretl expects the operating system to find the default viewer for PDF output. On GNU/Linux you can specify your preferred PDF viewer via the menu item "Tools, Preferences, General," under the "Programs" tab.

35.2 TeX-related menu items

The model window

The fullest TeX support in gretl is found in the GUI model window. This has a menu item titled "LaTeX" with sub-items "View", "Copy", "Save" and "Equation options" (see Figure 35.1).

The first three sub-items have branches titled "Tabular" and "Equation". By "Tabular" we mean that the model is represented in the form of a table; this is the fullest and most explicit presentation of the results. See Table 35.1 for an example; this was pasted into the manual after using the "Copy, Tabular" item in gretl (a few lines were edited out for brevity).

The "Equation" option is fairly self-explanatory—the results are written across the page in equation format, as below:

$$\widehat{\text{ENROLL}} = \underset{(0.066022)}{0.241105} + \underset{(0.04597)}{0.223530}\,\text{CATHOL} - \underset{(0.0027196)}{0.00338200}\,\text{PUPIL} - \underset{(0.040706)}{0.152643}\,\text{WHITE}$$

$$T = 51 \quad \bar{R}^2 = 0.4462 \quad F(3,47) = 14.431 \quad \hat{\sigma} = 0.038856$$

$$\text{(standard errors in parentheses)}$$

[1] Experts will be aware of something called "plain TeX", which is processed using the program tex. The great majority of TeX users, however, use the LaTeX macros, initially developed by Leslie Lamport. gretl does not support plain TeX.

Figure 35.1: LATEX menu in model window

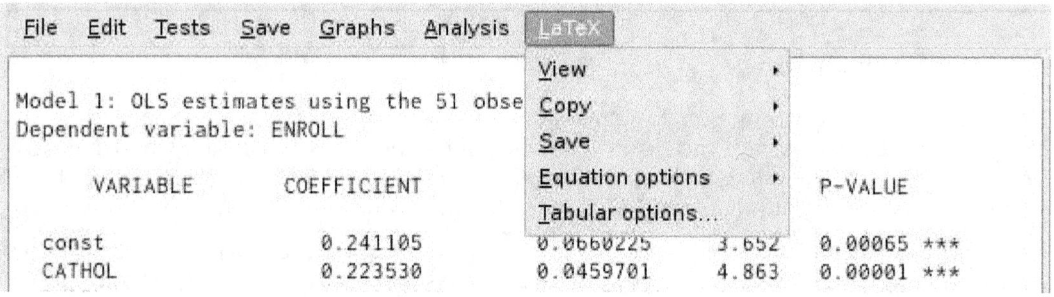

File Edit Tests Save Graphs Analysis LaTeX

View ▸
Copy ▸
Save ▸
Equation options ▸
Tabular options...

Model 1: OLS estimates using the 51 obse
Dependent variable: ENROLL

VARIABLE	COEFFICIENT			P-VALUE
const	0.241105	0.0660225	3.652	0.00065 ***
CATHOL	0.223530	0.0459701	4.863	0.00001 ***

Table 35.1: Example of LATEX tabular output

Model 1: OLS estimates using the 51 observations 1–51
Dependent variable: ENROLL

Variable	Coefficient	Std. Error	t-statistic	p-value
const	0.241105	0.0660225	3.6519	0.0007
CATHOL	0.223530	0.0459701	4.8625	0.0000
PUPIL	−0.00338200	0.00271962	−1.2436	0.2198
WHITE	−0.152643	0.0407064	−3.7499	0.0005

Mean of dependent variable	0.0955686
S.D. of dependent variable	0.0522150
Sum of squared residuals	0.0709594
Standard error of residuals ($\hat{\sigma}$)	0.0388558
Unadjusted R^2	0.479466
Adjusted $\bar{R}^2$	0.446241
$F(3, 47)$	14.4306

The distinction between the "Copy" and "Save" options (for both tabular and equation) is twofold. First, "Copy" puts the TEX source on the clipboard while with "Save" you are prompted for the name of a file into which the source should be saved. Second, with "Copy" the material is copied as a "fragment" while with "Save" it is written as a complete file. The point is that a well-formed TEX source file must have a header that defines the `documentclass` (article, report, book or whatever) and tags that say `\begin{document}` and `\end{document}`. This material is included when you do "Save" but not when you do "Copy", since in the latter case the expectation is that you will paste the data into an existing TEX source file that already has the relevant apparatus in place.

The items under "Equation options" should be self-explanatory: when printing the model in equation form, do you want standard errors or t-ratios displayed in parentheses under the parameter estimates? The default is to show standard errors; if you want t-ratios, select that item.

Other windows

Several other sorts of output windows also have TEX preview, copy and save enabled. In the case of windows having a graphical toolbar, look for the TEX button. Figure 35.2 shows this icon (second from the right on the toolbar) along with the dialog that appears when you press the button.

Figure 35.2: TEX icon and dialog

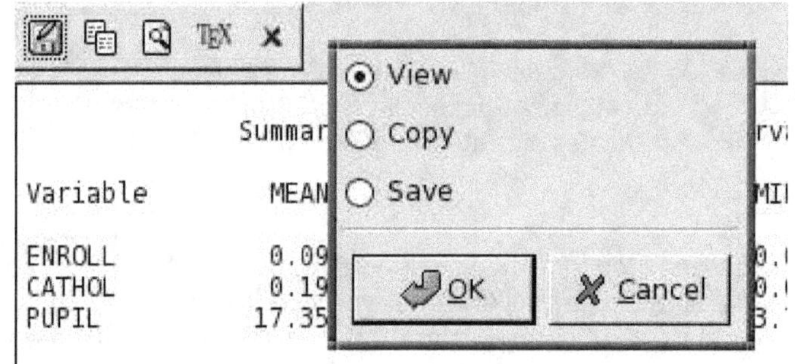

One aspect of gretl's TEX support that is likely to be particularly useful for publication purposes is the ability to produce a typeset version of the "model table" (see section 3.4). An example of this is shown in Table 35.2.

35.3 Fine-tuning typeset output

There are three aspects to this: adjusting the appearance of the output produced by gretl in LATEX preview mode; adjusting the formatting of gretl's tabular output for models when using the `tabprint` command; and incorporating gretl's output into your own TEX files.

Previewing in the GUI

As regards *preview mode*, you can control the appearance of gretl's output using a file named `gretlpre.tex`, which should be placed in your gretl user directory (see the *Gretl Command Reference*). If such a file is found, its contents will be used as the "preamble" to the TEX source. The default value of the preamble is as follows:

```
\documentclass[11pt]{article}
\usepackage[utf8]{inputenc}
\usepackage{amsmath}
\usepackage{dcolumn,longtable}
\begin{document}
\thispagestyle{empty}
```

Table 35.2: Example of model table output

OLS estimates
Dependent variable: ENROLL

	Model 1	Model 2	Model 3
const	0.2907**	0.2411**	0.08557
	(0.07853)	(0.06602)	(0.05794)
CATHOL	0.2216**	0.2235**	0.2065**
	(0.04584)	(0.04597)	(0.05160)
PUPIL	−0.003035	−0.003382	−0.001697
	(0.002727)	(0.002720)	(0.003025)
WHITE	−0.1482**	−0.1526**	
	(0.04074)	(0.04071)	
ADMEXP	−0.1551		
	(0.1342)		
n	51	51	51
$\bar{R}^2$	0.4502	0.4462	0.2956
ℓ	96.09	95.36	88.69

Standard errors in parentheses
* indicates significance at the 10 percent level
** indicates significance at the 5 percent level

Note that the `amsmath` and `dcolumn` packages are required. (For some sorts of output the `longtable` package is also needed.) Beyond that you can, for instance, change the type size or the font by altering the `documentclass` declaration or including an alternative font package.

In addition, if you wish to typeset gretl output in more than one language, you can set up per-language preamble files. A "localized" preamble file is identified by a name of the form `gretlpre_xx.tex`, where xx is replaced by the first two letters of the current setting of the LANG environment variable. For example, if you are running the program in Polish, using LANG=pl_PL, then gretl will do the following when writing the preamble for a TEX source file.

1. Look for a file named `gretlpre_pl.tex` in the gretl user directory. If this is not found, then
2. look for a file named `gretlpre.tex` in the gretl user directory. If this is not found, then
3. use the default preamble.

Conversely, suppose you usually run gretl in a language other than English, and have a suitable `gretlpre.tex` file in place for your native language. If on some occasions you want to produce TEX output in English, then you could create an additional file `gretlpre_en.tex`: this file will be used for the preamble when gretl is run with a language setting of, say, en_US.

Command-line options

After estimating a model via a script—or interactively via the gretl console or using the command-line program gretlcli—you can use the commands `tabprint` or `eqnprint` to print the model to file in tabular format or equation format respectively. These options are explained in the *Gretl Command Reference*.

If you wish alter the appearance of gretl's tabular output for models in the context of the `tabprint` command, you can specify a custom row format using the `--format` flag. The format string must be enclosed in double quotes and must be tied to the flag with an equals sign. The pattern for the format string is as follows. There are four fields, representing the coefficient, standard error, t-ratio and p-value respectively. These fields should be separated by vertical bars; they may contain a `printf`-type specification for the formatting of the numeric value in question, or may be left blank to suppress the printing of that column (subject to the constraint that you can't leave all the columns blank). Here are a few examples:

```
--format="%.4f|%.4f|%.4f|%.4f"
--format="%.4f|%.4f|%.3f|"
--format="%.5f|%.4f||%.4f"
--format="%.8g|%.8g||%.4f"
```

The first of these specifications prints the values in all columns using 4 decimal places. The second suppresses the p-value and prints the t-ratio to 3 places. The third omits the t-ratio. The last one again omits the t, and prints both coefficient and standard error to 8 significant figures.

Once you set a custom format in this way, it is remembered and used for the duration of the gretl session. To revert to the default formatting you can use the special variant `--format=default`.

Further editing

Once you have pasted gretl's TEX output into your own document, or saved it to file and opened it in an editor, you can of course modify the material in any wish you wish. In some cases, machine-generated TEX is hard to understand, but gretl's output is intended to be human-readable and -editable. In addition, it does not use any non-standard style packages. Besides the standard LATEX document classes, the only files needed are, as noted above, the `amsmath`, `dcolumn` and `longtable` packages. These should be included in any reasonably full TEX implementation.

35.4 Installing and learning TEX

This is not the place for a detailed exposition of these matters, but here are a few pointers.

So far as we know, every GNU/Linux distribution has a package or set of packages for TEX, and in fact these are likely to be installed by default. Check the documentation for your distribution. For MS Windows, several packaged versions of TEX are available: one of the most popular is MiKTEX at http://www.miktex.org/. For Mac OS X a nice implementation is iTEXMac, at http://itexmac.sourceforge.net/. An essential starting point for online TEX resources is the Comprehensive TEX Archive Network (CTAN) at http://www.ctan.org/.

As for learning TEX, many useful resources are available both online and in print. Among online guides, Tony Roberts' "LATEX: from quick and dirty to style and finesse" is very helpful, at

http://www.sci.usq.edu.au/staff/robertsa/LaTeX/latexintro.html

An excellent source for advanced material is *The LATEX Companion* (Goossens *et al.*, 2004).

Chapter 36

Gretl and R

36.1 Introduction

R is, by far, the largest free statistical project.[1] Like gretl, it is a GNU project and the two have a lot in common; however, gretl's approach focuses on ease of use much more than R, which instead aims to encompass the widest possible range of statistical procedures.

As is natural in the free software ecosystem, we don't view ourselves as competitors to R,[2] but rather as projects sharing a common goal who should support each other whenever possible. For this reason, gretl provides a way to interact with R and thus enable users to pool the capabilities of the two packages.

In this chapter, we will explain how to exploit R's power from within gretl. We assume that the reader has a working installation of R available and a basic grasp of R's syntax.[3]

Despite several valiant attempts, no graphical shell has gained wide acceptance in the R community: by and large, the standard method of working with R is by writing scripts, or by typing commands at the R prompt, much in the same way as one would write gretl scripts or work with the gretl console. In this chapter, the focus will be on the methods available to execute R commands without leaving gretl.

36.2 Starting an interactive R session

The easiest way to use R from gretl is in interactive mode. Once you have your data loaded in gretl, you can select the menu item "Tools, Start GNU R" and an interactive R session will be started, with your dataset automatically pre-loaded.

A simple example: OLS on cross-section data

For this example we use Ramanathan's dataset `data4-1`, one of the sample files supplied with gretl. We first run, in gretl, an OLS regression of `price` on `sqft`, `bedrms` and `baths`. The basic results are shown in Table 36.1.

Table 36.1: OLS house price regression via gretl

Variable	Coefficient	Std. Error	t-statistic	p-value
const	129.062	88.3033	1.4616	0.1746
sqft	0.154800	0.0319404	4.8465	0.0007
bedrms	−21.587	27.0293	−0.7987	0.4430
baths	−12.192	43.2500	−0.2819	0.7838

We will now replicate the above results using R. Select the menu item "Tools, Start GNU R". A window similar to the one shown in figure 36.1 should appear.

[1] R's homepage is at `http://www.r-project.org/`.

[2] OK, who are we kidding? But it's *friendly* competition!

[3] The main reference for R documentation is `http://cran.r-project.org/manuals.html`. In addition, R tutorials abound on the Net; as always, Google is your friend.

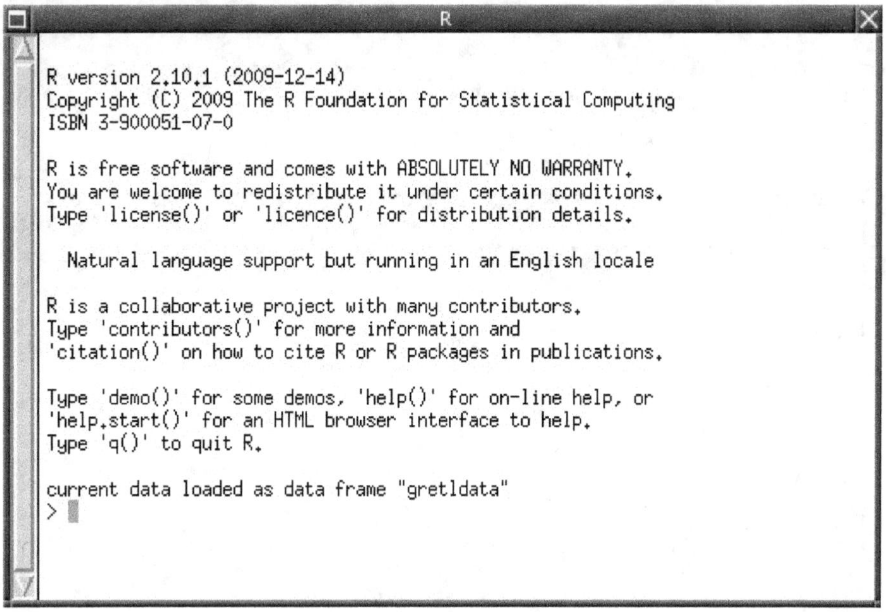

```
R version 2.10.1 (2009-12-14)
Copyright (C) 2009 The R Foundation for Statistical Computing
ISBN 3-900051-07-0

R is free software and comes with ABSOLUTELY NO WARRANTY.
You are welcome to redistribute it under certain conditions.
Type 'license()' or 'licence()' for distribution details.

  Natural language support but running in an English locale

R is a collaborative project with many contributors.
Type 'contributors()' for more information and
'citation()' on how to cite R or R packages in publications.

Type 'demo()' for some demos, 'help()' for on-line help, or
'help.start()' for an HTML browser interface to help.
Type 'q()' to quit R.

current data loaded as data frame "gretldata"
>
```

Figure 36.1: R window

The actual look of the R window may be somewhat different from what you see in Figure 36.1 (especially for Windows users), but this is immaterial. The important point is that you have a window where you can type commands to R. If the above procedure doesn't work and no R window opens, it means that gretl was unable to launch R. You should ensure that R is installed and working on your system and that gretl knows where it is. The relevant settings can be found by selecting the "Tools, Preferences, General" menu entry, under the "Programs" tab.

Assuming R was launched successfully, you will see notification that the data from gretl are available. In the background, gretl has arranged for two R commands to be executed, one to load the gretl dataset in the form of a *data frame* (one of several forms in which R can store data) and one to *attach* the data so that the variable names defined in the gretl workspace are available as valid identifiers within R.

In order to replicate gretl's OLS estimation, go into the R window and type at the prompt

```
model <- lm(price ~ sqft + bedrms + baths)
summary(model)
```

You should see something similar to Figure 36.2. Surprise—the estimates coincide! To get out, just close the R window or type q() at the R prompt.

Time series data

We now turn to an example which uses time series data: we will compare gretl's and R's estimates of Box and Jenkins' immortal "airline" model. The data are contained in the bjg sample dataset. The following gretl code

```
open bjg
arima 0 1 1 ; 0 1 1 ; lg --nc
```

produces the estimates shown in Table 36.2.

If we now open an R session as described in the previous subsection, the data-passing mechanism is slightly different. Since our data were defined in gretl as time series, we use an R *time-series* object (*ts* for short) for the transfer. In this way we can retain in R useful information such as the periodicity of the data and the sample limits. The downside is that the names

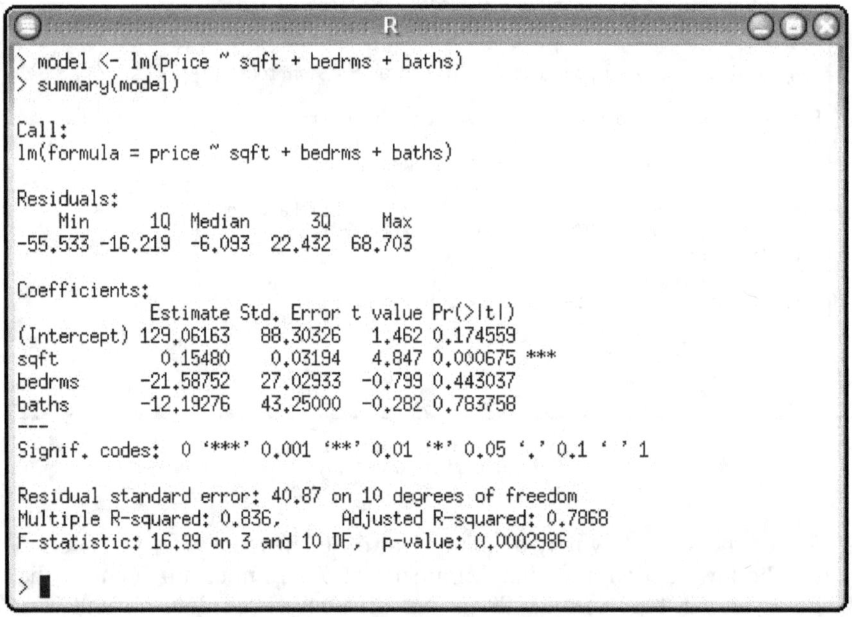

```
R
> model <- lm(price ~ sqft + bedrms + baths)
> summary(model)

Call:
lm(formula = price ~ sqft + bedrms + baths)

Residuals:
    Min      1Q  Median      3Q     Max
-55.533 -16.219  -6.093  22.432  68.703

Coefficients:
             Estimate Std. Error t value Pr(>|t|)
(Intercept) 129.06163   88.30326   1.462 0.174559
sqft          0.15480    0.03194   4.847 0.000675 ***
bedrms      -21.58752   27.02933  -0.799 0.443037
baths       -12.19276   43.25000  -0.282 0.783758
---
Signif. codes:  0 '***' 0.001 '**' 0.01 '*' 0.05 '.' 0.1 ' ' 1

Residual standard error: 40.87 on 10 degrees of freedom
Multiple R-squared: 0.836,     Adjusted R-squared: 0.7868
F-statistic: 16.99 on 3 and 10 DF,  p-value: 0.0002986

>
```

Figure 36.2: OLS regression on house prices via R

Table 36.2: Airline model from Box and Jenkins (1976) – selected portion of gretl's estimates

Variable	Coefficient	Std. Error	t-statistic	p-value
θ_1	−0.401824	0.0896421	−4.4825	0.0000
Θ_1	−0.556936	0.0731044	−7.6184	0.0000

Variance of innovations	0.00134810
Log-likelihood	244.696
Akaike information criterion	−483.39

of individual series, as defined in gretl, are not valid identifiers. In order to extract the variable `lg`, one needs to use the syntax `lg <- gretldata[, "lg"]`.

ARIMA estimation can be carried out by issuing the following two R commands:

```
lg <- gretldata[, "lg"]
arima(lg, c(0,1,1), seasonal=c(0,1,1))
```

which yield

```
Coefficients:
          ma1     sma1
      -0.4018  -0.5569
s.e.   0.0896   0.0731

sigma^2 estimated as 0.001348:  log likelihood = 244.7,  aic = -483.4
```

Happily, the estimates again coincide.

36.3 Running an R script

Opening an R window and keying in commands is a convenient method when the job is small. In some cases, however, it would be preferable to have R execute a script prepared in advance. One way to do this is via the `source()` command in R. Alternatively, gretl offers the facility to edit an R script and run it, having the current dataset pre-loaded automatically. This feature can be accessed via the "File, Script Files" menu entry. By selecting "User file", one can load a pre-existing R script; if you want to create a new script instead, select the "New script, R script" menu entry.

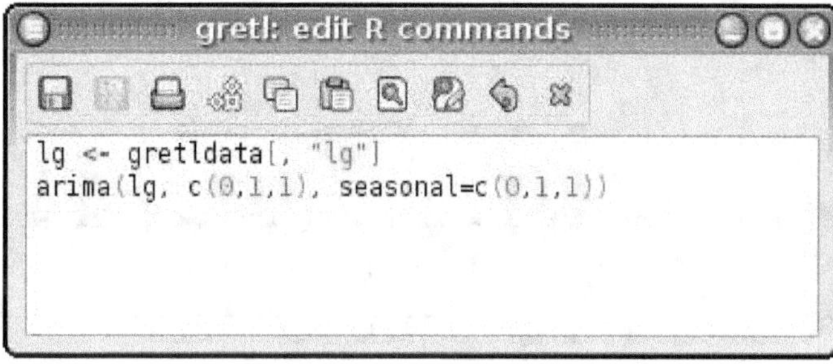

Figure 36.3: Editing window for R scripts

In either case, you are presented with a window very similar to the editor window used for ordinary gretl scripts, as in Figure 36.3.

There are two main differences. First, you get syntax highlighting for R's syntax instead of gretl's. Second, clicking on the Execute button (the gears icon), launches an instance of R in which your commands are executed. Before R is actually run, you are asked if you want to run R interactively or not (see Figure 36.4).

An interactive run opens an R instance similar to the one seen in the previous section: your data will be pre-loaded (if the "pre-load data" box is checked) and your commands will be executed. Once this is done, you will find yourself at the R prompt, where you can enter more commands.

A non-interactive run, on the other hand, will execute your script, collect the output from R and present it to you in an output window; R will be run in the background. If, for example, the script in Figure 36.3 is run non-interactively, a window similar to Figure 36.5 will appear.

36.4 Taking stuff back and forth

As regards the passing of data between the two programs, so far we have only considered passing series from gretl to R. In order to achieve a satisfactory degree of interoperability, more

gretl: R
R mode
Non-interactive (just get output)
Interactive R session
pre-load data
Cancel
OK

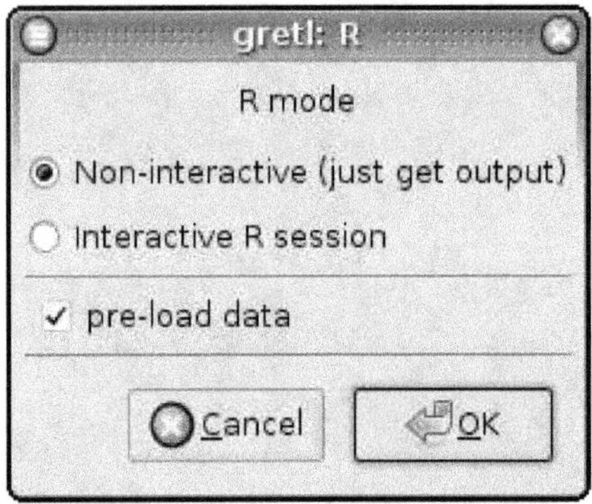

Figure 36.4: Editing window for R scripts

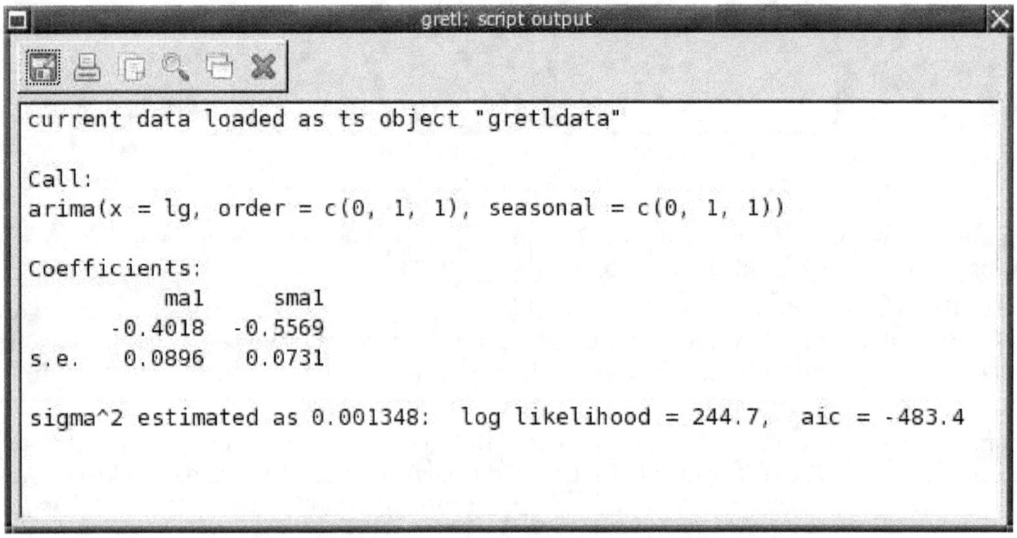

gretl: script output
current data loaded as ts object "gretldata"
Call:
arima(x = lg, order = c(0, 1, 1), seasonal = c(0, 1, 1))
Coefficients:
ma1 sma1
-0.4018 -0.5569
s.e. 0.0896 0.0731
sigma^2 estimated as 0.001348: log likelihood = 244.7, aic = -483.4

Figure 36.5: Output from a non-interactive R run

is needed. In the following sub-sections we see how matrices can be exchanged, and how data can be passed from R back to gretl.

Passing matrices from gretl to R

For passing matrices from gretl to R, you can use the mwrite matrix function described in section 15.6. For example, the following gretl code fragment generates the matrix

$$A = \begin{bmatrix} 3 & 7 & 11 \\ 4 & 8 & 12 \\ 5 & 9 & 13 \\ 6 & 10 & 14 \end{bmatrix}$$

and stores it into the file mymatfile.mat.

```
matrix A = mshape(seq(3,14),4,3)
err = mwrite(A, "mymatfile.mat")
```

The R code to import such a matrix is

```
A <- as.matrix(read.table("mymatfile.mat", skip=1))
```

Although in principle you can give your matrix file any valid filename, a couple of conventions may prove useful. First, you may want to use an informative file suffix such as ".mat", but this is a matter of taste. More importantly, the exact location of the file created by mwrite could be an issue. By default, if no path is specified in the file name, gretl writes matrix files in the current working directory. However, it may be wise for the purpose at hand to use a directory that is known to be writable by the user, namely the user's "dotdir"(see section 14.2). This can be achieved on the gretl side by passing a non-zero integer as the (optional) third argument to mwrite; on the R side you can use the function gretl.loadmat, which is pre-defined when R is called from gretl.[4] So the above example may be rewritten as

Gretl side:

```
mwrite(A, "mymatfile.mat", 1)
```

R side:

```
A <- gretl.loadmat("mymatfile.mat")
```

Passing data from R to gretl

For passing data in the opposite direction, gretl defines a special function that can be used in the R environment. An R object will be written as a temporary file in gretl's dotdir directory, from where it can be easily retrieved from within gretl.

The name of this function is gretl.export(); it takes one required argument, the object to be exported. At present, the objects that can be exported with this method are matrices, data frames and time-series objects. The function creates a text file, by default with the same name as the exported object (plus an appropriate suffix), in gretl's temporary directory. Data frames and time-series objects are stored as CSV files, and can be retrieved by using gretl's append command. Matrices are stored in a special text format that is understood by gretl (see section 15.6); the file suffix is in this case .mat, and to read the matrix in gretl you must use the mread() function.

This function also has an optional second argument, namely a string which specifies a basename for the export file, in case you want to use a name other than that attached to the object within R. As in the default case an appropriate suffix, .csv or .mat, will be added to the basename.

[4] In case you need to do this sort of thing manually, you can use the built-in string variable dotdir in gretl, while in R you have the same variable under the name gretl.dotdir.

As an example, we take the airline data and use them to estimate a structural time series model à la Harvey (1989). The model we will use is the *Basic Structural Model* (BSM), in which a time series is decomposed into three terms:

$$y_t = \mu_t + \gamma_t + \varepsilon_t$$

where μ_t is a trend component, γ_t is a seasonal component and ε_t is a noise term. In turn, the following is assumed to hold:

$$\begin{aligned} \Delta\mu_t &= \beta_{t-1} + \eta_t \\ \Delta\beta_t &= \zeta_t \\ \Delta_s\gamma_t &= \Delta\omega_t \end{aligned}$$

where Δ_s is the seasonal differencing operator, $(1 - L^s)$, and η_t, ζ_t and ω_t are mutually uncorrelated white noise processes. The object of the analysis is to estimate the variances of the noise components (which may be zero) and to recover estimates of the latent processes μ_t (the "level"), β_t (the "slope") and γ_t.

Gretl does not provide (yet) a command for estimating this class of models, so we will use R's `StructTS` command and import the results back into gretl. Once the `bjg` dataset is loaded in gretl, we pass the data to R and execute the following script:

```
# extract the log series
y <- gretldata[, "lg"]
# estimate the model
strmod <- StructTS(y)
# save the fitted components (smoothed)
compon <- as.ts(tsSmooth(strmod))
# save the estimated variances
vars <- as.matrix(strmod$coef)

# export into gretl's temp dir
gretl.export(compon)
gretl.export(vars)
```

Running this script via gretl produces minimal output:

```
current data loaded as ts object "gretldata"
wrote /home/cottrell/.gretl/compon.csv
wrote /home/cottrell/.gretl/vars.mat
```

However, we are now able to pull the results back into gretl by executing the following commands, either from the console or by creating a small script:

```
string fname
sprintf fname "%s/compon.csv", $dotdir
append @fname
vars = mread("vars.mat", 1)
```

The first command reads the estimated time-series components from a CSV file, which is the format that the passing mechanism employs for series. The matrix `vars` is read from the file `vars.mat`.

After the above commands have been executed, three new series will have appeared in the gretl workspace, namely the estimates of the three components; by plotting them together with the original data, you should get a graph similar to Figure 36.6. The estimates of the variances can be seen by printing the `vars` matrix, as in

```
? print vars
vars (4 x 1)
```

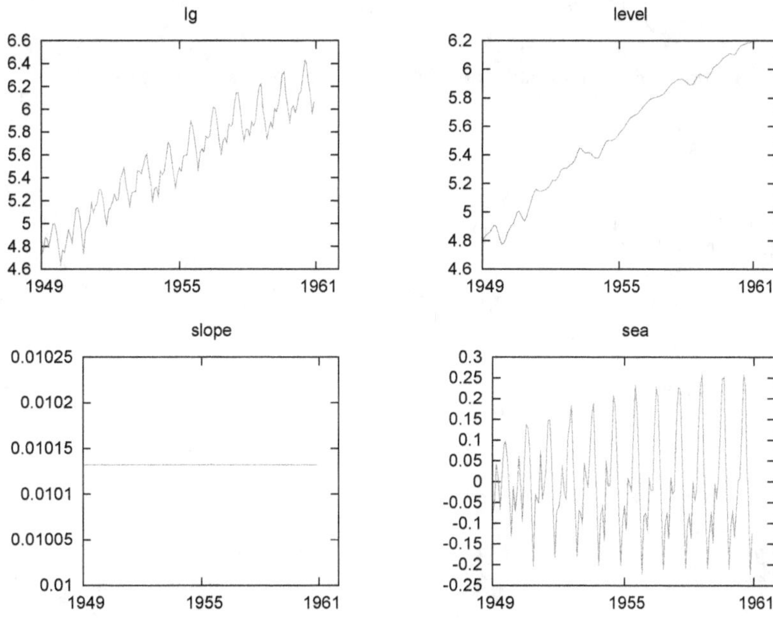

Figure 36.6: Estimated components from BSM

```
0.00077185
    0.0000
 0.0013969
    0.0000
```

That is,

$$\hat{\sigma}^2_\eta = 0.00077185, \quad \hat{\sigma}^2_\zeta = 0, \quad \hat{\sigma}^2_\omega = 0.0013969, \quad \hat{\sigma}^2_\varepsilon = 0$$

Notice that, since $\hat{\sigma}^2_\zeta = 0$, the estimate for β_t is constant and the level component is simply a random walk with a drift.

36.5 Interacting with R from the command line

Up to this point we have spoken only of interaction with R via the GUI program. In order to do the same from the command line interface, gretl provides the `foreign` command. This enables you to embed non-native commands within a gretl script.

A "foreign" block takes the form

```
foreign language=R [--send-data[=list]] [--quiet]
    ... R commands ...
end foreign
```

and achieves the same effect as submitting the enclosed R commands via the GUI in the non-interactive mode (see section 36.3 above). The `--send-data` option arranges for auto-loading of the data present in the gretl session, or a subset thereof specified via a named list. The `--quiet` option prevents the output from R from being echoed in the gretl output.

Using this method, replicating the example in the previous subsection is rather easy: basically, all it takes is encapsulating the content of the R script in a `foreign...end foreign` block; see example 36.1.

The above syntax, despite being already quite useful by itself, shows its full power when it is used in conjunction with user-written functions. Example 36.2 shows how to define a gretl function that calls R internally.

Example 36.1: Estimation of the Basic Structural Model – simple

```
open bjg.gdt

foreign language=R --send-data
    y <- gretldata[, "lg"]
    strmod <- StructTS(y)
    compon <- as.ts(tsSmooth(strmod))
    vars <- as.matrix(strmod$coef)

    gretl.export(compon)
    gretl.export(vars)
end foreign

append @dotdir/compon.csv
rename level lg_level
rename slope lg_slope
rename sea lg_seas

vars = mread("vars.mat", 1)
```

Example 36.2: Estimation of the Basic Structural Model – via a function

```
function list RStructTS(series myseries)

    smpl ok(myseries) --restrict
    sx = argname(myseries)

    foreign language=R --send-data --quiet
        @sx <- gretldata[, "myseries"]
        strmod <- StructTS(@sx)
        compon <- as.ts(tsSmooth(strmod))
        gretl.export(compon)
    end foreign

    append @dotdir/compon.csv
    rename level @sx_level
    rename slope @sx_slope
    rename sea @sx_seas

    list ret = @sx_level @sx_slope @sx_seas
    return ret
end function

# ------------ main -------------------------

open bjg.gdt
list X = RStructTS(lg)
```

36.6 Performance issues with R

R is a large and complex program, which takes an appreciable time to initialize itself.[5] In interactive use this not a significant problem, but if you have a gretl script that calls R repeatedly the cumulated start-up costs can become bothersome. To get around this, gretl calls the R shared library by preference; in this case the start-up cost is borne only once, on the first invocation of R code from within gretl.

Support for the R shared library is built into the gretl packages for MS Windows and OS X—but the advantage is realized only if the library is in fact available at run time. If you are building gretl yourself on Linux and wish to make use of the R library, you should ensure (a) that R has been built with the shared library enabled (specify `--enable-R-shlib` when configuring your build of R), and (b) that the `pkg-config` program is able to detect your R installation. We do not link to the R library at build time, rather we open it dynamically on demand. The gretl GUI has an item under the Tools/Preferences menu which enables you to select the path to the library, if it is not detected automatically.

If you have the R shared library installed but want to force gretl to call the R executable instead, you can do

```
set R_lib off
```

36.7 Further use of the R library

Besides improving performance, as noted above, use of the R shared library makes possible a further refinement. That is, you can define functions in R, within a `foreign` block, then call those functions later in your script much as if they were gretl functions. This is illustrated below.

```
set R_functions on
foreign language=R
  plus_one <- function(q) {
     z = q+1
     invisible(z)
  }
end foreign

scalar b=R.plus_one(2)
```

The R function `plus_one` is obviously trivial in itself, but the example shows a couple of points. First, for this mechanism to work you need to enable `R_functions` via the `set` command. Second, to avoid collision with the gretl function namespace, calls to functions defined in this way must be prefixed with "`R.`", as in `R.plus_one`.

Built-in R functions may also be called in this way, once `R_functions` is set `on`. For example one can invoke R's `choose` function, which computes binomial coefficients:

```
set R_functions on
scalar b=R.choose(10,4)
```

Note, however, that the possibilities for use of built-in R functions are limited; only functions whose arguments and return values are sufficiently generic (basically scalars or matrices) will work.

[5]About one third of a second on an Intel Core Duo machine of 2009 vintage.

Chapter 37

Gretl and Ox

37.1 Introduction

Ox, written by Jurgen A. Doornik (see Doornik, 2007), is described by its author as "an object-oriented statistical system. At its core is a powerful matrix language, which is complemented by a comprehensive statistical library. Among the special features of Ox are its speed [and] well-designed syntax.... Ox comes in two versions: Ox Professional and Ox Console. Ox is available for Windows, Linux, Mac (OS X), and several Unix platforms." (www.doornik.com)

Ox is proprietary, closed-source software. The command-line version of the program is, however, available free of change for academic users. Quoting again from Doornik's website: "The Console (command line) versions may be used freely for academic research and teaching purposes only.... The Ox syntax is public, and, of course, you may do with your own Ox code whatever you wish." If you wish to use Ox in conjunction with gretl please refer to doornik.com for further details on licensing.

As the reader will no doubt have noticed, most other software that we discuss in this Guide is open-source and freely available for all users. We make an exception for Ox on the grounds that it is indeed fast and well designed, and that its statistical library—along with various add-on packages that are also available—has exceptional coverage of cutting-edge techniques in econometrics. The gretl authors have used Ox for benchmarking some of gretl's more advanced features such as dynamic panel models and state space models.[1]

37.2 Ox support in gretl

The support offered for Ox in gretl is similar to that offered for R, as discussed in chapter 36.

☞ To enable support for Ox, go to the Tools/Preferences/General menu item and look under the Programs tab. Find the entry for the path to the `oxl` executable, that is, the program that runs Ox files (on MS Windows it is called `oxl.exe`). Adjust the path if it's not already right for your system and you should be ready to go.

With support enabled, you can open and edit Ox programs in the gretl GUI. Clicking the "execute" icon in the editor window will send your code to Ox for execution. Figures 37.1 and Figure 37.2 show an Ox program and part of its output.

In addition you can embed Ox code within a gretl script using a `foreign` block, as described in connection with R. A trivial example, which simply prints the gretl data matrix within Ox, is shown below:

```
open data4-1
matrix m = { dataset }
mwrite(m, "gretl.mat", 1)

foreign language=Ox
#include <oxstd.h>
main()
{
   decl gmat = gretl_loadmat("gretl.mat");
   print(gmat);
```

[1]For a review of Ox, see Cribari-Neto and Zarkos (2003) and for a (somewhat dated) comparison of Ox with other matrix-oriented packages such as GAUSS, see Steinhaus (1999).

```
#include <oxstd.h>
#include <packages/ssfpack/ssfpack.h>

main()
{
    decl mphi, momega, msigma;
    GetSsfArma(<0.6,0.2>, <-0.2>, 0.9, &mphi, &momega, &msigma);
    print("Phi =", mphi, "Omega =", momega, "Sigma =", msigma);

    decl myt = loadmat("arma21.mat")';

    decl mkf = KalmanFil(myt, mphi, momega, msigma);
    print(mkf');

    decl sm = SsfCondDens(ST_SMO, myt, mphi, momega, msigma);
    print(sm');

    decl dLL, dVar;
    SsfLik(&dLL, &dVar, myt, mphi, momega, msigma);
    print("dLL = ", dLL, "\n");
    print("dvar = ", dVar, "\n");
}
```

Figure 37.1: Ox editing window

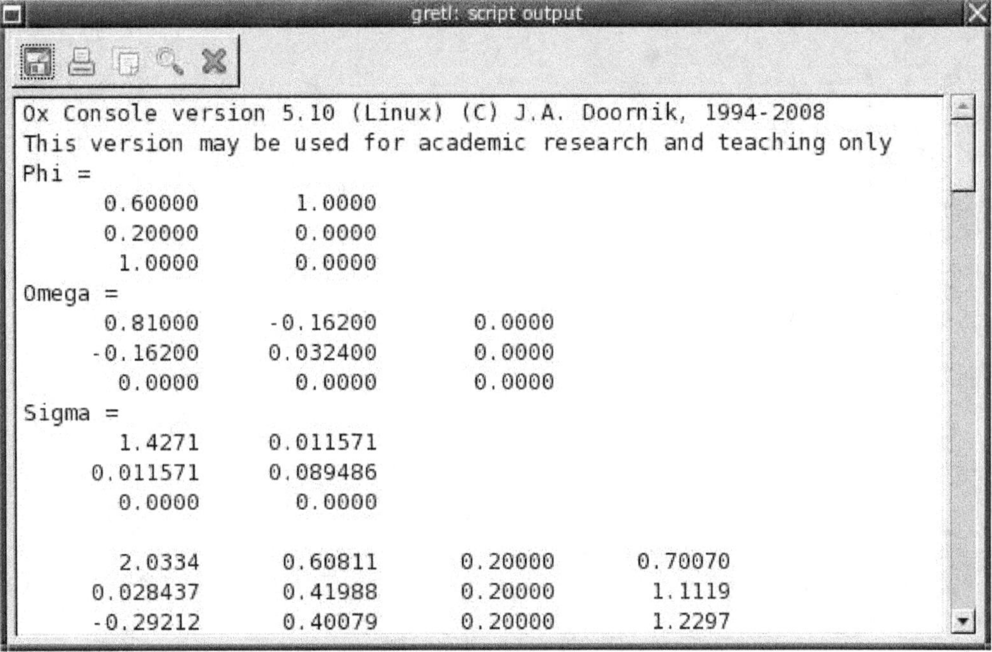

```
Ox Console version 5.10 (Linux) (C) J.A. Doornik, 1994-2008
This version may be used for academic research and teaching only
Phi =
      0.60000       1.0000
      0.20000       0.0000
       1.0000       0.0000
Omega =
      0.81000     -0.16200       0.0000
     -0.16200     0.032400       0.0000
       0.0000       0.0000       0.0000
Sigma =
       1.4271     0.011571
     0.011571     0.089486
       0.0000       0.0000

       2.0334      0.60811      0.20000      0.70070
     0.028437      0.41988      0.20000       1.1119
     -0.29212      0.40079      0.20000       1.2297
```

Figure 37.2: Output from Ox

```
}
end foreign
```

The above example illustrates how a matrix can be passed from gretl to Ox. We use the `mwrite` function to write a matrix into the user's "dotdir" (see section 14.2), then in Ox we use the function `gretl_loadmat` to retrieve the matrix.

How does `gretl_loadmat` come to be defined? When gretl writes out the Ox program corresponding to your `foreign` block it does two things in addition. First, it writes a small utility file named `gretl_io.ox` into your dotdir. This contains a definition for `gretl_loadmat` and also for the function `gretl_export` (see below). Second, gretl interpolates into your Ox code a line which includes this utility file (it is inserted right after the inclusion of `oxstd.h`, which is needed in all Ox programs). Note that `gretl_loadmat` expects to find the named file in the user's dotdir.

37.3 Illustration: replication of DPD model

Example 37.1 shows a more ambitious case. This script replicates one of the dynamic panel data models in Arellano and Bond (1991), first using gretl and then using Ox; we then check the relative differences between the parameter estimates produced by the two programs (which turn out to be reassuringly small).

Unlike the previous example, in this case we pass the dataset from gretl to Ox as a CSV file in order to preserve the variable names. Note the use of the internal variable `csv_na` to get the right representation of missing values for use with Ox—and also note that the `--send-data` option for the `foreign` command is not available in connection with Ox.

We get the parameter estimates back from Ox using `gretl_export` on the Ox side and `mread` on the gretl side. The `gretl_export` function takes two arguments, a matrix and a file name. The file is written into the user's dotdir, from where it can be picked up using `mread`. The final portion of the output from Example 37.1 is shown below:

```
? matrix oxparm = mread("oxparm.mat", 1)
Generated matrix oxparm
? eval abs((parm - oxparm) ./ oxparm)
  1.4578e-13
  3.5642e-13
  5.0672e-15
  1.6091e-13
  8.9808e-15
  2.0450e-14
  1.0218e-13
  2.1048e-13
  9.5898e-15
  1.8658e-14
  2.1852e-14
  2.9451e-13
  1.9398e-13
```

Example 37.1: Estimation of dynamic panel data model via gretl and Ox

```
open abdata.gdt

# Take first differences of the independent variables
genr Dw = diff(w)
genr Dk = diff(k)
genr Dys = diff(ys)

# 1-step GMM estimation
arbond 2 ; n Dw Dw(-1) Dk Dys Dys(-1) 0 --time-dummies
matrix parm = $coeff

# Write CSV file for Ox
set csv_na .NaN
store @dotdir/abdata.csv

# Replicate using the Ox DPD package
foreign language=Ox
#include <oxstd.h>
#import <packages/dpd/dpd>

main ()
{
    decl dpd = new DPD();
    dpd.Load("@dotdir/abdata.csv");
    dpd.SetYear("YEAR");

    dpd.Select(Y_VAR, {"n", 0, 2});
    dpd.Select(X_VAR, {"w", 0, 1, "k", 0, 0, "ys", 0, 1});
    dpd.Select(I_VAR, {"w", 0, 1, "k", 0, 0, "ys", 0, 1});

    dpd.Gmm("n", 2, 99);  // GMM-type instrument
    dpd.SetDummies(D_CONSTANT + D_TIME);
    dpd.SetTest(2, 2); // Sargan, AR 1-2 tests
    dpd.Estimate();    // 1-step estimation
    decl parm = dpd.GetPar();
    gretl_export(parm, "oxparm.mat");

    delete dpd;
}
end foreign

# Compare the results
matrix oxparm = mread("oxparm.mat", 1)
eval abs((parm - oxparm) ./ oxparm)
```

Chapter 38

Gretl and Octave

38.1 Introduction

GNU Octave, written by John W. Eaton and others, is described as "a high-level language, primarily intended for numerical computations." The program is oriented towards "solving linear and nonlinear problems numerically" and "performing other numerical experiments using a language that is mostly compatible with Matlab." (www.gnu.org/software/octave) Octave is available in source-code form (naturally, for GNU software) and also in the form of binary packages for MS Windows and Mac OS X. Numerous contributed packages that extend Octave's functionality in various ways can be found at octave.sf.net.

38.2 Octave support in gretl

The support offered for Octave in gretl is similar to that offered for R (chapter 36). For example, you can open and edit Octave scripts in the gretl GUI. Clicking the "execute" icon in the editor window will send your code to Octave for execution. Figures 38.1 and Figure 38.2 show an Octave script and its output; in this example we use the function `logistic_regression` to replicate some results from Greene (2000).

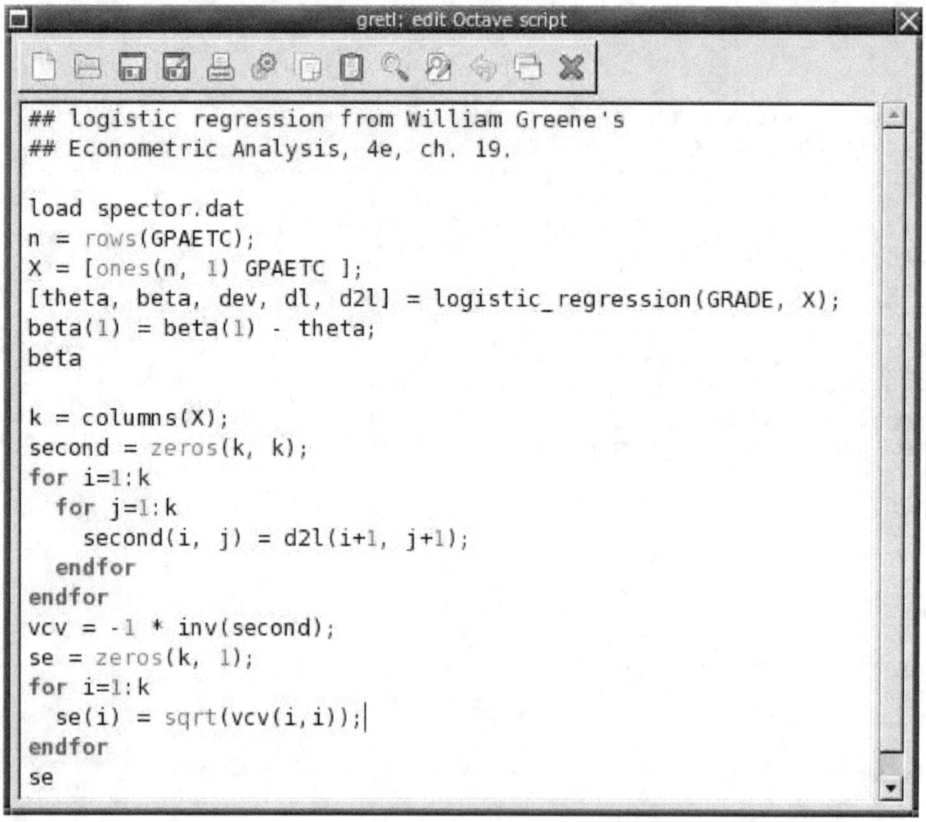

gretl: edit Octave script

```
## logistic regression from William Greene's
## Econometric Analysis, 4e, ch. 19.

load spector.dat
n = rows(GPAETC);
X = [ones(n, 1) GPAETC ];
[theta, beta, dev, dl, d2l] = logistic_regression(GRADE, X);
beta(1) = beta(1) - theta;
beta

k = columns(X);
second = zeros(k, k);
for i=1:k
  for j=1:k
    second(i, j) = d2l(i+1, j+1);
  endfor
endfor
vcv = -1 * inv(second);
se = zeros(k, 1);
for i=1:k
  se(i) = sqrt(vcv(i,i));
endfor
se
```

Figure 38.1: Octave editing window

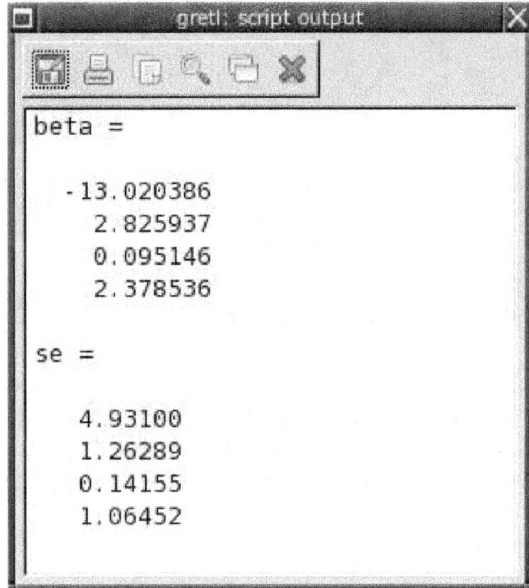

Figure 38.2: Output from Octave

In addition you can embed Octave code within a gretl script using a `foreign` block, as described in connection with R. A trivial example, which simply loads and prints the gretl data matrix within Octave, is shown below. (Note that in Octave, appending "`;`" to a line suppresses verbose output; leaving off the semicolon results in printing of the object that is produced, if any.)

```
open data4-1
matrix m = { dataset }
mwrite(m, "gretl.mat", 1)

foreign language=Octave
   gmat = gretl_loadmat("gretl.mat")
end foreign
```

We use the `mwrite` function to write a matrix into the user's "dotdir" (see section 14.2), then in Octave we use the function `gretl_loadmat` to retrieve the matrix. The "magic" behind `gretl_loadmat` works in essentially the same way as for Ox (chapter 37).

38.3 Illustration: spectral methods

We now present a more ambitious example which exploits Octave's handling of the frequency domain (and also its ability to use code written for MATLAB), namely estimation of the spectral coherence of two time series. For this illustration we require two extra Octave packages from `octave.sf.net`, namely those supporting spectral functions (`specfun`) and signal processing (`signal`). After downloading the packages you can install them from within Octave as follows (using version numbers as of March 2010):

```
pkg install specfun-1.0.8.tar.gz
pkg install signal-1.0.10.tar.gz
```

In addition we need some specialized MATLAB files made available by Mario Forni of the University of Modena, at `http://morgana.unimore.it/forni_mario/matlab.htm`. The files needed are `coheren2.m`, `coheren.m`, `coher.m`, `cospec.m`, `crosscov.m`, `crosspec.m`, `crosspe.m` and `spec.m`. These are in a form appropriate for MS Windows. On Linux you could run the following shell script to get the files and remove the Windows end-of-file character (which prevents the files from running under Octave):

```
SITE=http://morgana.unimore.it/forni_mario/MYPROG
# download files and delete trailing Ctrl-Z
```

```
for f in \
  coheren2.m \
  coheren.m \
  coher.m \
  cospec.m \
  crosscov.m \
  crosspec.m \
  crosspe.m \
  spec.m ; do
    wget $SITE/$f && \
    cat $f | tr -d \\032 > tmp.m && mv tmp.m $f
done
```

The Forni files should be placed in some appropriate directory, and you should tell Octave where to find them by adding that directory to Octave's `loadpath`. On Linux this can be done via an entry in one's `~/.octaverc` file. For example

```
addpath("~/stats/octave/forni");
```

Alternatively, an addpath directive can be written into the Octave script that calls on these files.

With everything set up on the Octave side we now write a gretl script (see Example 38.1) which opens a time-series dataset, constructs and writes a matrix containing two series, and defines a `foreign` block containing the Octave statements needed to produce the spectral coherence matrix. The matrix is exported via the `gretl_export` function, which is automatically defined for you; this function takes two arguments, a matrix and a file name. The file is written into the user's "dotdir", from where it can be picked up using `mread`. Finally, we produce a graph from the matrix in gretl. In the script this is sent to the screen; Figure 38.3 shows the same graph in PDF format.

Example 38.1: Estimation of spectral coherence via Octave

```
open data9-7
matrix xy = { PRIME, UNEMP }
mwrite(xy, "xy.mat", 1)

foreign language=Octave
 # uncomment and modify the following if necessary
 # addpath("~/stats/octave/forni");
 xy = gretl_loadmat("xy.mat");
 x = xy(:,1);
 y = xy(:,2);
 # note: the last parameter is the Bartlett window size
 h = coher(x, y, 8);
 gretl_export(h, "h.mat");
end foreign

h = mread("h.mat", 1)
colnames(h, "coherence")
gnuplot 1 --time-series --with-lines --matrix=h --output=display
```

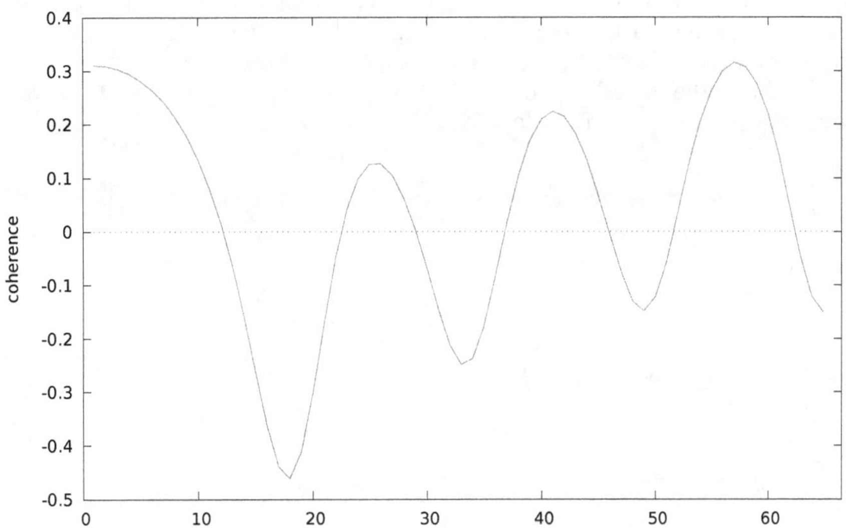

Figure 38.3: Spectral coherence estimated via Octave

Chapter 39

Gretl and Stata

Stata (www.stata.com) is closed-source, proprietary (and expensive) software and as such is not a natural companion to gretl. Nonetheless, given Stata's popularity it is desirable to have a convenient way of comparing results across the two programs, and to that end we provide some support for Stata code under the `foreign` command.

☞ To enable support for Stata, go to the Tools/Preferences/General menu item and look under the Programs tab. Find the entry for the path to the Stata executable. Adjust the path if it's not already right for your system and you should be ready to go.

The following example illustrates what's available. You can send the current gretl dataset to Stata using the `--send-data` flag. And having defined a matrix within Stata you can export it for use with gretl via the `gretl_export` command: this takes two arguments, the name of the matrix to export and the filename to use; the file is written to the user's "dotdir", from where it can be retrieved using the `mread()` function.[1] To suppress printed output from Stata you can add the `--quiet` flag to the `foreign` block.

Example 39.1: Comparison of clustered standard errors with Stata

```
function matrix stata_reorder (matrix se)
  # stata puts the intercept last, but gretl puts it first
  scalar n = rows(se)
  return se[n] | se[1:n-1]
end function

open data4-1
ols 1 0 2 3 --cluster=bedrms
matrix se = $stderr

foreign language=stata --send-data
  regress price sqft bedrms, vce(cluster bedrms)
  matrix vcv = e(V)
  gretl_export vcv "vcv.mat"
end foreign

matrix stata_vcv = mread("vcv.mat", 1)
stata_se = stata_reorder(sqrt(diag(stata_vcv)))
matrix check = se - stata_se
print check
```

In addition you can edit "pure" Stata scripts in the gretl GUI and send them for execution as with native gretl scripts.

Note that Stata coerces all variable names to lower-case on data input, so even if series names in gretl are upper-case, or of mixed case, it's necessary to use all lower-case in Stata. Also note that when opening a data file within Stata via the `use` command it will be necessary to provide the full path to the file.

[1]We do not currently offer the complementary functionality of `gretl_loadmat`, which enables reading of matrices written by gretl's `mwrite()` function in Ox and Octave. This is not at all easy to implement in Stata code.

Chapter 40

Gretl and Python

40.1 Introduction

According to www.python.org, Python is "an easy to learn, powerful programming language. It has efficient high-level data structures and a simple but effective approach to object-oriented programming. Python's elegant syntax and dynamic typing, together with its interpreted nature, make it an ideal language for scripting and rapid application development in many areas on most platforms."

Indeed, Python is widely used in a great variety of contexts. Numerous add-on modules are available; the ones likely to be of greatest interest to econometricians include NumPy ("the fundamental package for scientific computing with Python"—see www.numpy.org); SciPy (which builds on NumPy—see www.scipy.org); and Statsmodels (http://statsmodels.sourceforge.net/).

40.2 Python support in gretl

The support offered for Python in gretl is similar to that offered for Octave (chapter 38). You can open and edit Python scripts in the gretl GUI. Clicking the "execute" icon in the editor window will send your code to Python for execution. In addition you can embed Python code within a gretl script using a `foreign` block, as described in connection with R.

When you launch Python from within gretl one variable and two convenience functions are pre-defined, as follows.

```
gretl_dotdir
gretl_loadmat(filename, autodot=1)
gretl_export(M, filename, autodot=1)
```

The variable `gretl_dotdir` holds the path to the user's "dot directory." The first function loads a matrix of the given `filename` as written by gretl's `mwrite` function, and the second writes matrix `M`, under the given `filename`, in the format wanted by gretl.

By default the traffic in matrices goes via the dot directory on the Python side; that is, the name of this directory is prepended to `filename` for both reading and writing. (This is complementary to use of the *export* and *import* parameters with gretl's `mwrite` and `mread` functions, respectively.) However, if you wish to take control over the reading and writing locations you can supply a zero value for `autodot` when calling `gretl_loadmat` and `gretl_export`: in that case the `filename` argument is used as is.

Note that `gretl_loadmat` and `gretl_export` depend on NumPy; they make use of the functions `loadtxt` and `savetxt` respectively. Nonetheless, the presence of NumPy is not an absolute requirement if you don't need to use these two functions.

40.3 Illustration: linear regression with multicollinearity

Example 40.1 compares the numerical accuracy of gretl's `ols` command with that of NumPy's `linalg.lstsq`, using the notorious Longley test data which exhibit extreme multicollinearity. Unlike some econometrics packages, NumPy does a good job on these data. The script computes and prints the log-relative error in estimation of the regression coefficients, using the NIST-certified values as a benchmark;[1] the error values correspond to the number of correct digits

[1] See http://www.itl.nist.gov/div898/strd/lls/data/Longley.shtml.

(with a maximum of 15). The results will differ somewhat by computer architecture; the output shown was obtained on a 32-bit Linux Intel i5 system.

Example 40.1: Comparing regression results with Python

```
set echo off
set messages off

function matrix logrel_error (matrix est, matrix true)
  return -log10(abs(est - true) ./ abs(true))
end function

open longley.gdt -q
list LX = prdefl .. year
# gretl's regular OLS
ols employ 0 LX -q
matrix b = $coeff

mwrite({employ}, "y.mat", 1)
mwrite({const} ~ {LX}, "X.mat", 1)

foreign language=python
   import numpy as np
   y = gretl_loadmat('y.mat', 1)
   X = gretl_loadmat('X.mat', 1)
   # NumPy's OLS
   b = np.linalg.lstsq(X, y)[0]
   gretl_export(np.transpose(np.matrix(b)), 'py_b.mat', 1)
end foreign

# NIST's certified coefficient values
matrix nist_b = {-3482258.63459582, 15.0618722713733,
    -0.358191792925910E-01, -2.02022980381683,
    -1.03322686717359, -0.511041056535807E-01,
     1829.15146461355}'

matrix py_b = mread("py_b.mat", 1)
matrix errs = logrel_error(b, nist_b) ~ logrel_error(py_b, nist_b)
colnames(errs, "gretl python")
printf "Log-relative errors, Longley coefficients:\n\n%12.5g\n", errs
printf "Column means\n%12.5g\n", meanc(errs)
```

Output:

```
  Log-relative errors, Longley coefficients:

         gretl      python
        12.844       12.85
        11.528      11.414
        12.393      12.401
        13.135      13.121
        13.738      13.318
        12.587      12.363
        12.848      12.852

  Column means
        12.725      12.617
```

Chapter 41

Gretl and Julia

41.1 Introduction

According to julialang.org, Julia is "a high-level, high-performance dynamic programming language for technical computing, with syntax that is familiar to users of other technical computing environments. It provides a sophisticated compiler, distributed parallel execution, numerical accuracy, and an extensive mathematical function library." Julia is well known for being very fast.

41.2 Julia support in gretl

The support offered for Julia in gretl is similar to that offered for Octave (chapter 38). You can open and edit Julia scripts in the gretl GUI. Clicking the "execute" icon in the editor window will send your code to Julia for execution. In addition you can embed Julia code within a gretl script using a `foreign` block, as described in connection with R.

When you launch Julia from within gretl one variable and two convenience functions are predefined, as follows.

```
gretl_dotdir
gretl_loadmat(filename, autodot=true)
gretl_export(M, filename, autodot=true)
```

The variable `gretl_dotdir` holds the path to the user's "dot directory." The first function loads a matrix of the given `filename` as written by gretl's `mwrite` function, and the second writes matrix `M`, under the given `filename`, in the format wanted by gretl.

By default the traffic in matrices goes via the dot directory on the Julia side; that is, the name of this directory is prepended to `filename` for both reading and writing. (This is complementary to use of the *export* and *import* parameters with gretl's `mwrite` and `mread` functions, respectively.) However, if you wish to take control over the reading and writing locations you can supply a zero value for `autodot` when calling `gretl_loadmat` and `gretl_export`: in that case the `filename` argument is used as is.

41.3 Illustration

TO BE WRITTEN!

Chapter 42

Troubleshooting gretl

42.1 Bug reports

Bug reports are welcome—well, if not exactly welcome then useful and appreciated. Hopefully, you are unlikely to find bugs in the actual calculations done by gretl (although this statement does not constitute any sort of warranty). You may, however, come across bugs or oddities in the behavior of the graphical interface. Please remember that the usefulness of bug reports is greatly enhanced if you can be as specific as possible: what *exactly* went wrong, under what conditions, and on what operating system? If you saw an error message, what precisely did it say?

One way of making a bug report more useful is to run the program in such a way that you can see (and copy) any additional information that gets printed to the `stderr` output stream. On Linux and Mac OS X that's just a matter of launching gretl from the command prompt in a terminal window. On MS Windows it's a bit more complicated since `stderr` is by default "invisble." However, you can quite easily set up a special gretl shortcut that does the job. On the Windows desktop, right-click and select "New shortcut." In the dialog box that appears, browse to find `gretl.exe` and append the `--debug` flag, as shown in Figure 42.1. Note that there are two dashes before "debug".

Figure 42.1: Creating a debugging shortcut

When you start gretl in this mode, a "console window" appears as well as the gretl window, and stderr output goes to the console. To copy this output, click at the top left of the console window for a menu (Figure 42.2): first do "Select all", then "Copy." You can paste the results into Notepad or similar.

Figure 42.2: The program with console window

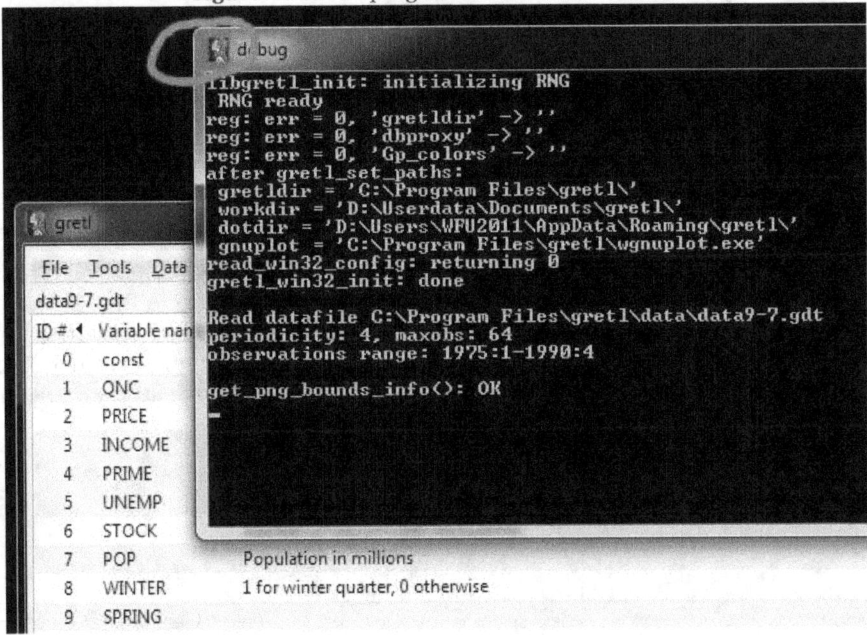

42.2 Auxiliary programs

As mentioned above, gretl calls some other programs to accomplish certain tasks (gnuplot for graphing, LaTeX for high-quality typesetting of regression output, GNU R). If something goes wrong with such external links, it is not always easy for gretl to produce an informative error message. If such a link fails when accessed from the gretl graphical interface, you may be able to get more information by starting gretl from the command prompt rather than via a desktop menu entry or icon. On the X window system, start gretl from the shell prompt in an xterm; on MS Windows, start the program gretl.exe from a console window or "DOS box" using the -g or --debug option flag. Additional error messages may be displayed on the terminal window.

Also please note that for most external calls, gretl assumes that the programs in question are available in your "path"—that is, that they can be invoked simply via the name of the program, without supplying the program's full location.[1] Thus if a given program fails, try the experiment of typing the program name at the command prompt, as shown below.

	Graphing	*Typesetting*	*GNU R*
X window system	gnuplot	pdflatex	R
MS Windows	wgnuplot.exe	pdflatex	RGui.exe

If the program fails to start from the prompt, it's not a gretl issue but rather that the program's home directory is not in your path, or the program is not installed (properly). For details on modifying your path please see the documentation or online help for your operating system or shell.

[1]The exception to this rule is the invocation of gnuplot under MS Windows, where a full path to the program is given.

Chapter 43

The command line interface

The gretl package includes the command-line program gretlcli. On Linux it can be run from a terminal window (xterm, rxvt, or similar), or at the text console. Under MS Windows it can be run in a console window (sometimes inaccurately called a "DOS box"). gretlcli has its own help file, which may be accessed by typing "help" at the prompt. It can be run in batch mode, sending output directly to a file (see also the *Gretl Command Reference*).

If gretlcli is linked to the readline library (this is automatically the case in the MS Windows version; also see Appendix C), the command line is recallable and editable, and offers command completion. You can use the Up and Down arrow keys to cycle through previously typed commands. On a given command line, you can use the arrow keys to move around, in conjunction with Emacs editing keystrokes.[1] The most common of these are:

Keystroke	*Effect*
`Ctrl-a`	go to start of line
`Ctrl-e`	go to end of line
`Ctrl-d`	delete character to right

where "`Ctrl-a`" means press the "a" key while the "`Ctrl`" key is also depressed. Thus if you want to change something at the beginning of a command, you *don't* have to backspace over the whole line, erasing as you go. Just hop to the start and add or delete characters. If you type the first letters of a command name then press the Tab key, readline will attempt to complete the command name for you. If there's a unique completion it will be put in place automatically. If there's more than one completion, pressing Tab a second time brings up a list.

Probably the most useful mode for heavy-duty work with gretlcli is batch (non-interactive) mode, in which the program reads and processes a script, and sends the output to file. For example

```
gretlcli -b scriptfile > outputfile
```

Note that *scriptfile* is treated as a program argument; only the output file requires redirection (>). Don't forget the `-b` (batch) switch, otherwise the program will wait for user input after executing the script (and if output is redirected, the program will appear to "hang").

[1] Actually, the key bindings shown below are only the defaults; they can be customized. See the readline manual.

Part IV

Appendices

Appendix A

Data file details

A.1 Basic native format

In gretl's basic native data format-for which we use the suffix gdt—a data set is stored in XML (extensible mark-up language). Data files correspond to the simple DTD (document type definition) given in gretldata.dtd, which is supplied with the gretl distribution and is installed in the system data directory (e.g. /usr/share/gretl/data on Linux.) Such files may be plain text (uncompressed) or gzipped. They contain the actual data values plus additional information such as the names and descriptions of variables, the frequency of the data, and so on.

In a gdt file the actual data values are written to 17 significant figures (for generated data such as logs or pseudo-random numbers) or to a maximum of 15 figures for primary data. The C printf format "%.*g" is used (for * = 17 or 15) so that trailing zeros are not printed.

Most users will probably not have need to read or write such files other than via gretl itself, but if you want to manipulate them using other software tools you should examine the DTD and also take a look at a few of the supplied practice data files: data4-1.gdt gives a simple example; data4-10.gdt is an example where observation labels are included.

A.2 Binary data file format

As of gretl 1.9.15, an alternative, binary format is available for data storage. Files of this sort have suffix gdtb, and they take the form of a PKZIP archive containing two files, data.xml and data.bin, with the following characteristics.

- data.xml is an XML file conforming to the gretldata DTD mentioned above, holding all the metadata.
- data.bin starts with a header composed of one of the strings

  gretl-bin:little-endian or
  gretl-bin:big-endian

  padded to 24 bytes with nul characters. This is followed by a binary dump of the data series, by variable, as double-precision floating-point values.

Binary values are saved in the endianness of the machine on which they're written; the header information enables gretl to convert to the endianness of the host on which the data are read if need be.

The rationale for introducing the binary gdtb format is that for very large datasets it is a lot faster to write and read data in this form rather than as text. For small to moderately sized datasets (say, up to 10 megabytes or so) there is no appreciable advantage in the binary format and we recommend use of plain gdt.

Some illustrative timings are shown in Table A.1; these were obtained on a Lenovo ThinkPad X1 Carbon running gretl 1.9.15. The datasets contained a mixture of random normal series and random binary (dummy) series. The largest comprised 50000 observations on 1000 series and the smallest 5000 observations on 250 series. As can be seen, there is a big time saving from the binary format when writing (and to a lesser extent, when reading) a dataset in the hundreds of megabytes range. At a size of around 10 megabytes, however, gdt files can be both written and read in under a second, surely fast enough for most purposes.

	381 MB		38 MB		19 MB		10 MB	
format	write	read	write	read	write	read	write	read
gdt	21.76	4.91	2.20	0.48	1.19	0.32	0.59	0.16
gdtb	3.82	1.10	0.39	0.11	0.25	0.06	0.13	0.03

Table A.1: Data write and read timings in seconds for datasets of various sizes in megabytes. The MB numbers represent the size of the datasets in memory; files of both formats are substantially smaller when compression is applied.

A.3 Native database format

A gretl database consists of two parts: an ASCII index file (with filename suffix `.idx`) containing information on the series, and a binary file (suffix `.bin`) containing the actual data. Two examples of the format for an entry in the `idx` file are shown below:

```
G0M910  Composite index of 11 leading indicators (1987=100)
M 1948.01 - 1995.11   n = 575
currbal Balance of Payments: Balance on Current Account; SA
Q 1960.1 - 1999.4 n = 160
```

The first field is the series name. The second is a description of the series (maximum 128 characters). On the second line the first field is a frequency code: `M` for monthly, `Q` for quarterly, `A` for annual, `B` for business-daily (daily with five days per week) and `D` for daily (seven days per week). No other frequencies are accepted at present. Then comes the starting date (N.B. with two digits following the point for monthly data, one for quarterly data, none for annual), a space, a hyphen, another space, the ending date, the string "`n = `" and the integer number of observations. In the case of daily data the starting and ending dates should be given in the form `YYYY/MM/DD`. This format must be respected exactly.

Optionally, the first line of the index file may contain a short comment (up to 64 characters) on the source and nature of the data, following a hash mark. For example:

```
# Federal Reserve Board (interest rates)
```

The corresponding binary database file holds the data values, represented as "floats", that is, single-precision floating-point numbers, typically taking four bytes apiece. The numbers are packed "by variable", so that the first n numbers are the observations of variable 1, the next m the observations on variable 2, and so on.

Appendix B

Data import via ODBC

Since version 1.7.5, gretl provides a method for retrieving data from databases which support the Open Database Connectivity (ODBC) standard. Most users won't be interested in this, but there may be some for whom this feature matters a lot — typically, those who work in an environment where huge data collections are accessible via a Data Base Management System (DBMS).

In the following section we explain what is needed for ODBC support in gretl. We provide some background information on how ODBC works in section B.2, and explain the details of getting gretl to retrieve data from a database in section B.3.

B.1 ODBC support

The piece of software that bridges between gretl and the ODBC system is a dynamically loaded "plugin". This is included in the gretl packages for MS Windows and Mac OS X (on OS X support was added in gretl 1.9.0). On other unix-type platforms (notably Linux) you will have to build gretl from source to get ODBC support. This is because the gretl plugin depends on having unixODBC installed, which we cannot assume to be the case on typical Linux systems. To enable the ODBC plugin when building gretl, you must pass the option `--with-odbc` to gretl's `configure` script. In addition, if unixODBC is installed in a non-standard location you will have to specify its installation prefix using `--with-ODBC-prefix`, as in (for example)

```
./configure --with-odbc --with-ODBC-prefix=/opt/ODBC
```

B.2 ODBC base concepts

ODBC is short for *Open DataBase Connectivity*, a group of software methods that enable a *client* to interact with a database *server*. The most common operation is when the client fetches some data from the server. ODBC acts as an intermediate layer between client and server, so the client "talks" to ODBC rather than accessing the server directly (see Figure B.1).

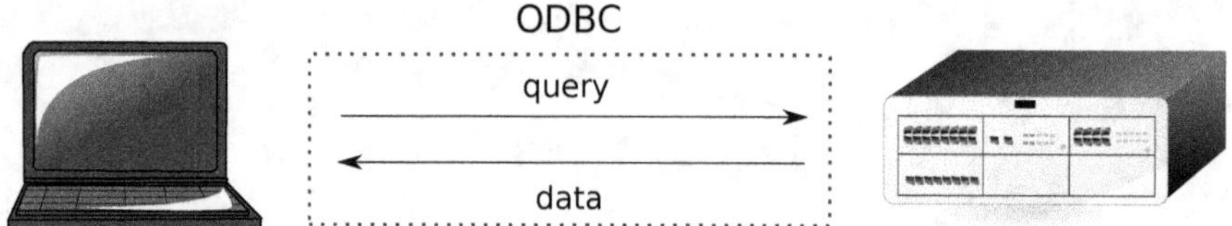

Figure B.1: Retrieving data via ODBC

For the above mechanism to work, it is necessary that the relevant ODBC software is installed and working on the client machine (contact your DB administrator for details). At this point, the database (or databases) that the server provides will be accessible to the client as a *data source* with a specific identifier (a Data Source Name or DSN); in most cases, a username and a password are required to connect to the data source.

Once the connection is established, the user sends a *query* to ODBC, which contacts the database manager, collects the results and sends them back to the user. The query is almost invariably formulated in a special language used for the purpose, namely SQL.[1] We will not provide here

[1] See http://en.wikipedia.org/wiki/SQL.

an SQL tutorial: there are many such tutorials on the Net; besides, each database manager tends to support its own SQL dialect so the precise form of an SQL query may vary slightly if the DBMS on the other end is Oracle, MySQL, PostgreSQL or something else.

Suffice it to say that the main statement for retrieving data is the SELECT statement. Within a DBMS, data are organized in *tables*, which are roughly equivalent to spreadsheets. The SELECT statement returns a subset of a table, which is itself a table. For example, imagine that the database holds a table called "NatAccounts", containing the data shown in Table B.1.

year	qtr	gdp	consump	tradebal
1970	1	584763	344746.9	−5891.01
1970	2	597746	350176.9	−7068.71
1970	3	604270	355249.7	−8379.27
1970	4	609706	361794.7	−7917.61
1971	1	609597	362490	−6274.3
1971	2	617002	368313.6	−6658.76
1971	3	625536	372605	−4795.89
1971	4	630047	377033.9	−6498.13

Table B.1: The "NatAccounts" table

The SQL statement

```
SELECT qtr, tradebal, gdp FROM NatAccounts WHERE year=1970;
```

produces the subset of the original data shown in Table B.2.

qtr	tradebal	gdp
1	−5891.01	584763
2	−7068.71	597746
3	−8379.27	604270
4	−7917.61	609706

Table B.2: Result of a SELECT statement

Gretl provides a mechanism for forwarding your query to the DBMS via ODBC and including the results in your currently open dataset.

B.3 Syntax

At present gretl does not offer a graphical interface for ODBC import; this must be done via the command line interface. The two commands used for fetching data via an ODBC connection are open and data.

The open command is used for connecting to a DBMS: its syntax is

```
open dsn=database [user=username] [password=password] --odbc
```

The user and password items are optional; the effect of this command is to initiate an ODBC connection. It is assumed that the machine gretl runs on has a working ODBC client installed.

In order to actually retrieve the data, the data command is used. Its syntax is:

```
data series [obs-format=format-string] query=query-string --odbc
```

where:

series is a list of names of gretl series to contain the incoming data, separated by spaces. Note that these series need not exist pior to the ODBC import.

format-string is an optional parameter, used to handle cases when a "rectangular" organisation of the database cannot be assumed (more on this later);

query-string is a string containing the SQL statement used to extract the data.[2]

There should be no spaces around the equals signs in the `obs-format` and `query` fields in the `data` command.

The *query-string* can, in principle, contain any valid SQL statement which results in a table. This string may be specified directly within the command, as in

```
data x query="SELECT foo FROM bar" --odbc
```

which will store into the gretl variable `x` the content of the column `foo` from the table `bar`. However, since in a real-life situation the string containing the SQL statement may be rather long, it may be best to store it in a string variable. For example:

```
string SqlQry = "SELECT foo1, foo2 FROM bar"
data x y query=SqlQry --odbc
```

The observation format specifier

If the optional parameter `obs-format` is absent, as in the above example, the SQL query should return k columns of data, where k is the number of series names listed in the `data` command. It may be necessary to include a `smpl` command before the `data` command to set up the right "window" for the incoming data. In addition, if one cannot assume that the data will be delivered in the correct order (typically, chronological order), the SQL query should contain an appropriate `ORDER BY` clause.

The optional format string is used for those cases when there is no certainty that the data from the query will arrive in the same order as the gretl dataset. This may happen when missing values are interspersed within a column, or with data that do not have a natural ordering, e.g. cross-sectional data. In this case, the SQL statement should return a table with $m + k$ columns, where the first m columns are used to identify the observation or row in the gretl dataset into which the actual data values in the final k columns should be placed. The `obs-format` string is used to translate the first m fields into a string which matches the string gretl uses to identify observations in the currently open dataset. Up to three columns can be used for this purpose ($m \leq 3$).

Note that the strings gretl uses to identify observations can be seen by printing any variable "by observation", as in

```
print index --byobs
```

(The series named `index` is automatically added to a dataset created via the `nulldata` command.)

The format specifiers available for use with `obs-format` are as follows:

%d	print an integer value
%s	print an string value
%g	print a floating-point value

In addition the format can include literal characters to be passed through, such as slashes or colons, to make the resulting string compatible with gretl's observation identifiers.

For example, consider the following fictitious case: we have a 5-days-per-week dataset, to which we want to add the stock index for the Verdurian market;[3] it so happens that in Verduria

[2]Prior to gretl 1.8.8, the tag "query=" was not required (or accepted) before the query string, and only one series could be imported at a time. This variant is still accepted for the sake of backward compatibility.

[3]See http://www.almeopedia.com/index.php/Verduria.

Saturdays are working days but Wednesdays are not. We want a column which does *not* contain data on Saturdays, because we wouldn't know where to put them, but at the same time we want to place missing values on all the Wednesdays.

In this case, the following syntax could be used

```
string QRY="SELECT year,month,day,VerdSE FROM AlmeaIndexes"
data y obs-format="%d/%d/%d" query=QRY --odbc
```

The column VerdSE holds the data to be fetched, which will go into the gretl series y. The first three columns are used to construct a string which identifies the day. Daily dates take the form YYYY/MM/DD in gretl. If a row from the DBMS produces the observation string 2008/04/01 this will match OK (it's a Tuesday), but 2008/04/05 will not match since it is a Saturday; the corresponding row will therefore be discarded. On the other hand, since no string 2008/04/23 will be found in the data coming from the DBMS (it's a Wednesday), that entry is left blank in our series y.

B.4 Examples

Table Consump

Field	Type
time	decimal(7,2)
income	decimal(16,6)
consump	decimal(16,6)

Table DATA

Field	Type
year	decimal(4,0)
qtr	decimal(1,0)
varname	varchar(16)
xval	decimal(20,10)

Table B.3: Example AWM database - structure

Table Consump

1970.00	424278.975500	344746.944000
1970.25	433218.709400	350176.890400
1970.50	440954.219100	355249.672300
1970.75	446278.664700	361794.719900
1971.00	447752.681800	362489.970500
1971.25	453553.860100	368313.558500
1971.50	460115.133100	372605.015300
...		

Table DATA

1970	1	CAN	−517.9085000000
1970	2	CAN	662.5996000000
1970	3	CAN	1130.4155000000
1970	4	CAN	467.2508000000
1970	1	COMPR	18.4000000000
1970	2	COMPR	18.6341000000
1970	3	COMPR	18.3000000000
1970	4	COMPR	18.2663000000
1970	1	D1	1.0000000000
1970	2	D1	0.0000000000
...			

Table B.4: Example AWM database — data

In the following examples, we will assume that access is available to a database known to ODBC with the data source name "AWM", with username "Otto" and password "Bingo". The database "AWM" contains quarterly data in two tables (see B.3 and B.4):

The table Consump is the classic "rectangular" dataset; that is, its internal organization is the same as in a spreadsheet or econometrics package: each row is a data point and each column is a variable. The structure of the DATA table is different: each record is one figure, stored in the column xval, and the other fields keep track of which variable it belongs to, for which date.

Example B.1 shows a query for two series: first we set up an empty quarterly dataset. Then we connect to the database using the open statement. Once the connection is established we

Example B.1: Simple query from a rectangular table

```
nulldata 160
setobs 4 1970:1 --time
open dsn=AWM user=Otto password=Bingo --odbc

string Qry = "SELECT consump, income FROM Consump"
data cons inc query=Qry --odbc
```

retrieve two columns from the Consump table. No observation string is required because the data already have a suitable structure; we need only import the relevant columns.

Example B.2: Simple query from a non-rectangular table

```
string S = "select year, qtr, xval from DATA \
       where varname='WLN' ORDER BY year, qtr"
data wln obs-format="%d:%d" query=S --odbc
```

In example B.2, by contrast, we make use of the observation string since we are drawing from the DATA table, which is not rectangular. The SQL statement stored in the string S produces a table with three columns. The ORDER BY clause ensures that the rows will be in chronological order, although this is not strictly necessary in this case.

Example B.3: Handling of missing values for a non-rectangular table

```
string foo = "select year, qtr, xval from DATA \
      where varname='STN' AND qtr>1"
data bar obs-format="%d:%d" query=foo --odbc
print bar --byobs
```

Example B.3 shows what happens if the rows in the outcome from the SELECT statement do not match the observations in the currently open gretl dataset. The query includes a condition which filters out all the data from the first quarter. The query result (invisible to the user) would be something like

```
+------+------+---------------+
| year | qtr  | xval          |
+------+------+---------------+
| 1970 |    2 |  7.8705000000 |
| 1970 |    3 |  7.5600000000 |
| 1970 |    4 |  7.1892000000 |
| 1971 |    2 |  5.8679000000 |
| 1971 |    3 |  6.2442000000 |
| 1971 |    4 |  5.9811000000 |
| 1972 |    2 |  4.6883000000 |
| 1972 |    3 |  4.6302000000 |
...
```

Internally, gretl fills the variable bar with the corresponding value if it finds a match; otherwise, NA is used. Printing out the variable bar thus produces

```
      Obs            bar

   1970:1
   1970:2         7.8705
   1970:3         7.5600
   1970:4         7.1892
   1971:1
   1971:2         5.8679
   1971:3         6.2442
   1971:4         5.9811
   1972:1
   1972:2         4.6883
   1972:3         4.6302
...
```

Appendix C

Building gretl

Here we give instructions detailed enough to allow a user with only a basic knowledge of a Unix-type system to build gretl. These steps were tested on a fresh installation of Debian Etch. For other Linux distributions (especially Debian-based ones, like Ubuntu and its derivatives) little should change. Other Unix-like operating systems such as Mac OS X and BSD would probably require more substantial adjustments.

In this guided example, we will build gretl complete with documentation. This introduces a few more requirements, but gives you the ability to modify the documentation files as well, like the help files or the manuals.

C.1 Installing the prerequisites

We assume that the basic GNU utilities are already installed on the system, together with these other programs:

- some TEX/LATEXsystem (`texlive` will do beautifully)
- Gnuplot
- ImageMagick

We also assume that the user has administrative privileges and knows how to install packages. The examples below are carried out using the `apt-get` shell command, but they can be performed with menu-based utilities like `aptitude`, `dselect` or the GUI-based program `synaptic`. Users of Linux distributions which employ rpm packages (e.g. Red Hat/Fedora, Mandriva, SuSE) may want to refer to the dependencies page on the gretl website.

The first step is installing the C compiler and related basic utilities, if these are not already in place. On a Debian (or derivative) system, these are contained in a bunch of packages that can be installed via the command

```
apt-get install gcc autoconf automake1.9 libtool flex bison gcc-doc \
libc6-dev libc-dev gfortran gettext pkgconfig
```

Then it is necessary to install the "development" (`dev`) packages for the libraries that gretl uses:

Library	*command*
GLIB	`apt-get install libglib2.0-dev`
GTK 3.0	`apt-get install libgtk3.0-dev`
PNG	`apt-get install libpng12-dev`
XSLT	`apt-get install libxslt1-dev`
LAPACK	`apt-get install liblapack-dev`
FFTW	`apt-get install libfftw3-dev`
READLINE	`apt-get install libreadline-dev`
ZLIB	`apt-get install zlib1g-dev`
XML	`apt-get install libxml2-dev`
GMP	`apt-get install libgmp-dev`
CURL	`apt-get install libcurl4-gnutls-dev`
MPFR	`apt-get install libmpfr-dev`

MPFR is optional, but recommended. It is possible to substitute GTK 2.0 for GTK 3.0. The dev packages for these libraries are necessary to *compile* gretl—you'll also need the plain, non-dev library packages to *run* gretl, but most of these should already be part of a standard installation. In order to enable other optional features, like audio support, you may need to install more libraries.

☞ The above steps can be much simplified on Linux systems that provide deb-based package managers, such as Debian and its derivatives (Ubuntu, Knoppix and other distributions). The command

```
apt-get build-dep gretl
```

will download and install all the necessary packages for building the version of gretl that is currently present in your APT sources. Techincally, this does not guarantee that all the software necessary to build the git version is included, because the version of gretl on your repository may be quite old and build requirements may have changed in the meantime. However, the chances of a mismatch are rather remote for a reasonably up-to-date system, so in most cases the above command should take care of everything correctly.

C.2 Getting the source: release or git

At this point, it is possible to build from the source. You have two options here: obtain the latest released source package, or retrieve the current git version of gretl (git = the version control software currently in use for gretl). The usual caveat applies to the git version, namely, that it may not build correctly and may contain "experimental" code; on the other hand, git often contains bug-fixes relative to the released version. If you want to help with testing and to contribute bug reports, we recommend using git gretl.

To work with the released source:

1. Download the gretl source package from gretl.sourceforge.net.
2. Unzip and untar the package. On a system with the GNU utilities available, the command would be `tar xvfJ gretl-N.tar.xz` (replace `N` with the specific version number of the file you downloaded at step 1).
3. Change directory to the gretl source directory created at step 2 (e.g. `gretl-1.10.2`).
4. Proceed to the next section, "Configure and make".

To work with git you'll first need to install the git client program if it's not already on your system. Relevant resources you may wish to consult include the main git website at git-scm.com and instructions specific to gretl: gretl git basics.

When grabbing the git sources *for the first time*, you should first decide where you want to store the code. For example, you might create a directory called `git` under your home directory. Open a terminal window, `cd` into this directory, and type the following commands:

```
git clone git://git.code.sf.net/p/gretl/git gretl-git
```

At this point git should create a subdirectory named `gretl-git` and fill it with the current sources.

When you want to *update the source*, this is very simple: just move into the `gretl-git` directory and type

```
git pull
```

Assuming you're now in the `gretl-git` directory, you can proceed in the same manner as with the released source package.

C.3 Configure the source

The next command you need is `./configure`; this is a complex script that detects which tools you have on your system and sets things up. The `configure` command accepts many options; you may want to run

```
./configure --help
```

first to see what options are available. One option you way wish to tweak is `--prefix`. By default the installation goes under `/usr/local` but you can change this. For example

```
./configure --prefix=/usr
```

will put everything under the `/usr` tree.

If you have a multi-core machine you may want to activate support for OpenMP, which permits the parallelization of matrix multiplication and some other tasks. This requires adding the `configure` flag

```
--enable-openmp
```

By default the gretl GUI is built using version 3.0 of the GTK library, if available, otherwise version 2.0. If you have both versions installed and prefer to use GTK 2.0, use the flag

```
--enable-gtk2
```

In order to have the documentation built, we need to pass the relevant option to `configure`, as in

```
--enable-build-doc
```

But please note that this option will work only if you are using the git source.

☞ In order to build the documentation, there is the possibility that you will have to install some extra software on top of the packages mentioned in the previous section. For example, you may need some extra LaTeX packages to compile the manuals. Two of the required packages, that not every standard LaTeX installation include, are typically pifont.sty and appendix.sty. You could install the corresponding packages from your distribution or you could simply download them from CTAN and install them by hand.

This, for example, if you want to install under `/usr`, with OpenMP support, and also build the documentation, you would do

```
./configure --prefix=/usr \
 --enable-openmp \
 --enable-build-doc
```

You will see a number of checks being run, and if everything goes according to plan, you should see a summary similar to that displayed in Example C.1.

☞ If you're using git, it's a good idea to re-run the `configure` script after doing an update. This is not always necessary, but sometimes it is, and it never does any harm. For this purpose, you may want to write a little shell script that calls `configure` with any options you want to use.

C.4 Build and install

We are now ready to undertake the compilation proper: this is done by running the `make` command, which takes care of compiling all the necessary source files in the correct order. All you need to do is type

```
make
```

Example C.1: Sample output from `./configure`

```
Configuration:

  Installation path:                      /usr
  Use readline library:                   yes
  Use gnuplot for graphs:                 yes
  Use LaTeX for typesetting output:       yes
  Use libgsf for zip/unzip:               no
  MPFR support:                           yes
  sse2 support for RNG:                   yes
  OpenMP support:                         yes
  MPI support:                            no
  AVX support for arithmetic:             no
  Build with GTK version:                 2.0
  Build gretl documentation:              yes
  Use Lucida fonts:                       no
  Build message catalogs:                 yes
  X-12-ARIMA support:                     yes
  TRAMO/SEATS support:                    yes
  libR support:                           yes
  ODBC support:                           no
  Experimental audio support:             no
  Use xdg-utils in installation:          if DESTDIR not set
  LAPACK libraries:
    -llapack -lblas -lgfortran

Now type 'make' to build gretl.
You can also do 'make pdfdocs' to build the PDF documentation.
```

This step will likely take several minutes to complete; a lot of output will be produced on screen. Once this is done, you can install your freshly baked copy of gretl on your system via

```
make install
```

On most systems, the `make install` command requires you to have administrative privileges. Hence, either you log in as `root` before launching `make install` or you may want to use the `sudo` utility, as in:

```
sudo make install
```

Now try if everything works: go back to your home directory and run gretl

```
cd ~
gretl &
```

If all is well, you ought to see gretl start, at which point just exit the program in the usual way. On the other hand, there is the possibility that gretl doesn't start and instead you see a message like

```
/usr/local/bin/gretl_x11: error while loading shared libraries:
libgretl-1.0.so.0: cannot open shared object file: No such file or directory
```

In this case, just run

```
sudo ldconfig
```

The problem should be fixed once and for all.

Appendix D

Numerical accuracy

Gretl uses double-precision arithmetic throughout—except for the multiple-precision plugin invoked by the menu item "Model, Other linear models, High precision OLS" which represents floating-point values using a number of bits given by the environment variable `GRETL_MP_BITS` (default value 256).

The normal equations of Least Squares are by default solved via Cholesky decomposition, which is highly accurate provided the matrix of cross-products of the regressors, $X'X$, is not very ill conditioned. If this problem is detected, gretl automatically switches to use QR decomposition.

The program has been tested rather thoroughly on the statistical reference datasets provided by NIST (the U.S. National Institute of Standards and Technology) and a full account of the results may be found on the gretl website (follow the link "Numerical accuracy").

To date, two published reviews have discussed gretl's accuracy: Giovanni Baiocchi and Walter Distaso (2003), and Talha Yalta and Yasemin Yalta (2007). We are grateful to these authors for their careful examination of the program. Their comments have prompted several modifications including the use of Stephen Moshier's `cephes` code for computing p-values and other quantities relating to probability distributions (see netlib.org), changes to the formatting of regression output to ensure that the program displays a consistent number of significant digits, and attention to compiler issues in producing the MS Windows version of gretl (which at one time was slighly less accurate than the Linux version).

Gretl now includes a "plugin" that runs the NIST linear regression test suite. You can find this under the "Tools" menu in the main window. When you run this test, the introductory text explains the expected result. If you run this test and see anything other than the expected result, please send a bug report to `cottrell@wfu.edu`.

All regression statistics are printed to 6 significant figures in the current version of gretl (except when the multiple-precision plugin is used, in which case results are given to 12 figures). If you want to examine a particular value more closely, first save it (for example, using the `genr` command) then print it using `printf`, to as many digits as you like (see the *Gretl Command Reference*).

Appendix E

Related free software

Gretl's capabilities are substantial, and are expanding. Nonetheless you may find there are some things you can't do in gretl, or you may wish to compare results with other programs. If you are looking for complementary functionality in the realm of free, open-source software we recommend the following programs. The self-description of each program is taken from its website.

- **GNU R** r-project.org: "R is a system for statistical computation and graphics. It consists of a language plus a run-time environment with graphics, a debugger, access to certain system functions, and the ability to run programs stored in script files... It compiles and runs on a wide variety of UNIX platforms, Windows and MacOS." Comment: There are numerous add-on packages for R covering most areas of statistical work.
- **GNU Octave** www.octave.org: "GNU Octave is a high-level language, primarily intended for numerical computations. It provides a convenient command line interface for solving linear and nonlinear problems numerically, and for performing other numerical experiments using a language that is mostly compatible with Matlab. It may also be used as a batch-oriented language."
- **Julia** julialang.org: "Julia is a high-level, high-performance dynamic programming language for technical computing, with syntax that is familiar to users of other technical computing environments. It provides a sophisticated compiler, distributed parallel execution, numerical accuracy, and an extensive mathematical function library."
- **JMulTi** www.jmulti.de: "JMulTi was originally designed as a tool for certain econometric procedures in time series analysis that are especially difficult to use and that are not available in other packages, like Impulse Response Analysis with bootstrapped confidence intervals for VAR/VEC modelling. Now many other features have been integrated as well to make it possible to convey a comprehensive analysis." Comment: JMulTi is a java GUI program: you need a java run-time environment to make use of it.

As mentioned above, gretl offers the facility of exporting data in the formats of both Octave and R. In the case of Octave, the gretl data set is saved as a single matrix, X. You can pull the `X` matrix apart if you wish, once the data are loaded in Octave; see the Octave manual for details. As for R, the exported data file preserves any time series structure that is apparent to gretl. The series are saved as individual structures. The data should be brought into R using the `source()` command.

In addition, gretl has a convenience function for moving data quickly into R. Under gretl's "Tools" menu, you will find the entry "Start GNU R". This writes out an R version of the current gretl data set (in the user's gretl directory), and sources it into a new R session. The particular way R is invoked depends on the internal gretl variable `Rcommand`, whose value may be set under the "Tools, Preferences" menu. The default command is `RGui.exe` under MS Windows. Under X it is `xterm -e R`. Please note that at most three space-separated elements in this command string will be processed; any extra elements are ignored.

Appendix F

Listing of URLs

Below is a listing of the full URLs of websites mentioned in the text.

Estima (RATS) http://www.estima.com/

FFTW3 http://www.fftw.org/

Gnome desktop homepage http://www.gnome.org/

GNU Multiple Precision (GMP) library http://gmplib.org/

CURL library http://curl.haxx.se/libcurl/

GNU Octave homepage http://www.octave.org/

GNU R homepage http://www.r-project.org/

GNU R manual http://cran.r-project.org/doc/manuals/R-intro.pdf

Gnuplot homepage http://www.gnuplot.info/

Gnuplot manual http://ricardo.ecn.wfu.edu/gnuplot.html

Gretl data page http://gretl.sourceforge.net/gretl_data.html

Gretl homepage http://gretl.sourceforge.net/

GTK+ homepage http://www.gtk.org/

GTK+ port for win32 http://www.gimp.org/~tml/gimp/win32/

InfoZip homepage http://www.info-zip.org/pub/infozip/zlib/

JMulTi homepage http://www.jmulti.de/

JRSoftware http://www.jrsoftware.org/

Julia homepage http://julialang.org/

Mingw (gcc for win32) homepage http://www.mingw.org/

Minpack http://www.netlib.org/minpack/

Penn World Table http://pwt.econ.upenn.edu/

Readline homepage http://cnswww.cns.cwru.edu/~chet/readline/rltop.html

Readline manual http://cnswww.cns.cwru.edu/~chet/readline/readline.html

Xmlsoft homepage http://xmlsoft.org/

Bibliography

Akaike, H. (1974) 'A new look at the statistical model identification', *IEEE Transactions on Automatic Control* AC-19: 716–723.

Anderson, B. and J. Moore (1979) *Optimal Filtering*, Upper Saddle River, NJ: Prentice-Hall.

Anderson, T. W. and C. Hsiao (1981) 'Estimation of dynamic models with error components', *Journal of the American Statistical Association* 76: 598–606.

Andrews, D. W. K. and J. C. Monahan (1992) 'An improved heteroskedasticity and autocorrelation consistent covariance matrix estimator', *Econometrica* 60: 953–966.

Arellano, M. (2003) *Panel Data Econometrics*, Oxford: Oxford University Press.

Arellano, M. and S. Bond (1991) 'Some tests of specification for panel data: Monte carlo evidence and an application to employment equations', *The Review of Economic Studies* 58: 277–297.

Baiocchi, G. and W. Distaso (2003) 'GRETL: Econometric software for the GNU generation', *Journal of Applied Econometrics* 18: 105–110.

Baltagi, B. H. (1995) *Econometric Analysis of Panel Data*, New York: Wiley.

Barrodale, I. and F. D. K. Roberts (1974) 'Solution of an overdetermined system of equations in the ℓ_l norm', *Communications of the ACM* 17: 319–320.

Baxter, M. and R. G. King (1999) 'Measuring business cycles: Approximate band-pass filters for economic time series', *The Review of Economics and Statistics* 81(4): 575–593.

Beck, N. and J. N. Katz (1995) 'What to do (and not to do) with time-series cross-section data', *The American Political Science Review* 89: 634–647.

Berndt, E., B. Hall, R. Hall and J. Hausman (1974) 'Estimation and inference in nonlinear structural models', *Annals of Economic and Social Measurement* 3(4): 653–665.

Blundell, R. and S. Bond (1998) 'Initial conditions and moment restrictions in dynamic panel data models', *Journal of Econometrics* 87: 115–143.

Bond, S., A. Hoeffler and J. Temple (2001) 'GMM estimation of empirical growth models'. Economics Papers from Economics Group, Nuffield College, University of Oxford, No 2001-W21.

Boswijk, H. P. (1995) 'Identifiability of cointegrated systems'. Tinbergen Institute Discussion Paper 95-78. URL `http://www.ase.uva.nl/pp/bin/258fulltext.pdf`.

Boswijk, H. P. and J. A. Doornik (2004) 'Identifying, estimating and testing restricted cointegrated systems: An overview', *Statistica Neerlandica* 58(4): 440–465.

Box, G. E. P. and G. Jenkins (1976) *Time Series Analysis: Forecasting and Control*, San Franciso: Holden-Day.

Brand, C. and N. Cassola (2004) 'A money demand system for euro area M3', *Applied Economics* 36(8): 817–838.

Butterworth, S. (1930) 'On the theory of filter amplifiers', *Experimental Wireless & The Wireless Engineer* 7: 536–541.

Byrd, R. H., P. Lu, J. Nocedal and C. Zhu (1995) 'A limited memory algorithm for bound constrained optimization', *SIAM Journal on Scientific Computing* 16: 1190–1208.

Cameron, A. C. and P. K. Trivedi (2005) *Microeconometrics, Methods and Applications*, Cambridge: Cambridge University Press.

_______ (2013) *Regression Analysis of Count Data*, Cambridge University Press.

Caselli, F., G. Esquivel and F. Lefort (1996) 'Reopening the convergence debate: A new look at cross-country growth empirics', *Journal of Economic Growth* 1(3): 363–389.

Chesher, A. and M. Irish (1987) 'Residual analysis in the grouped and censored normal linear model', *Journal of Econometrics* 34: 33–61.

Choi, I. (2001) 'Unit root tests for panel data', *Journal of International Money and Finance* 20(2): 249–272.

Cleveland, W. S. (1979) 'Robust locally weighted regression and smoothing scatterplots', *Journal of the American Statistical Association* 74(368): 829–836.

Cribari-Neto, F. and S. G. Zarkos (2003) 'Econometric and statistical computing using Ox', *Computational Economics* 21: 277–295.

Davidson, R. and E. Flachaire (2001) 'The wild bootstrap, tamed at last'. GREQAM Document de Travail 99A32. URL http://russell.vcharite.univ-mrs.fr/GMMboot/wild5-euro.pdf.

Davidson, R. and J. G. MacKinnon (1993) *Estimation and Inference in Econometrics*, New York: Oxford University Press.

_______ (2004) *Econometric Theory and Methods*, New York: Oxford University Press.

Doornik, J. A. (1995) 'Testing general restrictions on the cointegrating space'. Discussion Paper, Nuffield College. URL http://www.doornik.com/research/coigen.pdf.

_______ (1998) 'Approximations to the asymptotic distribution of cointegration tests', *Journal of Economic Surveys* 12: 573–593. Reprinted with corrections in McAleer and Oxley (1999).

_______ (2007) *Object-Oriented Matrix Programming Using Ox*, London: Timberlake Consultants Press, third edn. URL http://www.doornik.com.

Doornik, J. A., M. Arellano and S. Bond (2006) *Panel Data estimation using DPD for Ox*.

Elliott, G., T. J. Rothenberg and J. H. Stock (1996) 'Efficient tests for an autoregressive unit root', *Econometrica* 64: 813–836.

Engle, R. F. and C. W. J. Granger (1987) 'Co-integration and error correction: Representation, estimation, and testing', *Econometrica* 55: 251–276.

Fiorentini, G., G. Calzolari and L. Panattoni (1996) 'Analytic derivatives and the computation of GARCH estimates', *Journal of Applied Econometrics* 11: 399–417.

Frigo, M. and S. G. Johnson (2005) 'The design and implementation of FFTW3', *Proceedings of the IEEE 93* 2: 216–231.

Goossens, M., F. Mittelbach and A. Samarin (2004) *The LaTeX Companion*, Boston: Addison-Wesley, second edn.

Gould, W. (2013) 'Interpreting the intercept in the fixed-effects model'. URL http://www.stata.com/support/faqs/statistics/intercept-in-fixed-effects-model/.

Gourieroux, C. and A. Monfort (1996) *Simulation-Based Econometric Methods*, Oxford: Oxford University Press.

Gourieroux, C., A. Monfort, E. Renault and A. Trognon (1987) 'Generalized residuals', *Journal of Econometrics* 34: 5–32.

Greene, W. H. (2000) *Econometric Analysis*, Upper Saddle River, NJ: Prentice-Hall, fourth edn.

_______ (2003) *Econometric Analysis*, Upper Saddle River, NJ: Prentice-Hall, fifth edn.

Hall, A. D. (2005) *Generalized Method of Moments*, Oxford: Oxford University Press.

Hamilton, J. D. (1994) *Time Series Analysis*, Princeton, NJ: Princeton University Press.

Hannan, E. J. and B. G. Quinn (1979) 'The determination of the order of an autoregression', *Journal of the Royal Statistical Society, B* 41: 190–195.

Hansen, L. P. (1982) 'Large sample properties of generalized method of moments estimation', *Econometrica* 50: 1029–1054.

Hansen, L. P. and K. J. Singleton (1982) 'Generalized instrumental variables estimation of non-linear rational expectations models', *Econometrica* 50: 1269–1286.

Harvey, A. C. (1989) *Forecasting, structural time series models and the Kalman filter*, Cambridge: Cambridge University Press.

Harvey, A. C. and T. Proietti (2005) *Readings in Unobserved Component Models*, Oxford: Oxford University Press.

Hausman, J. A. (1978) 'Specification tests in econometrics', *Econometrica* 46: 1251–1271.

Heckman, J. (1979) 'Sample selection bias as a specification error', *Econometrica* 47: 153–161.

Hodrick, R. and E. C. Prescott (1997) 'Postwar U.S. business cycles: An empirical investigation', *Journal of Money, Credit and Banking* 29: 1–16.

Im, K. S., M. H. Pesaran and Y. Shin (2003) 'Testing for unit roots in heterogeneous panels', *Journal of Econometrics* 115: 53–74.

Johansen, S. (1995) *Likelihood-Based Inference in Cointegrated Vector Autoregressive Models*, Oxford: Oxford University Press.

de Jong, P. (1991) 'The diffuse Kalman filter', *The Annals of Statistics* 19: 1073–1083.

Kalbfleisch, J. D. and R. L. Prentice (2002) *The Statistical Analysis of Failure Time Data*, New York: Wiley, second edn.

Kalman, R. E. (1960) 'A new approach to linear filtering and prediction problems', *Transactions of the ASME–Journal of Basic Engineering* 82(Series D): 35–45.

Keane, M. P. and K. I. Wolpin (1997) 'The career decisions of young men', *Journal of Political Economy* 105: 473–522.

Koenker, R. (1994) 'Confidence intervals for regression quantiles'. In P. Mandl and M. Huskova (eds.), *Asymptotic Statistics*, pp. 349–359. New York: Springer-Verlag.

Koenker, R. and G. Bassett (1978) 'Regression quantiles', *Econometrica* 46: 33–50.

Koenker, R. and K. Hallock (2001) 'Quantile regression', *Journal of Economic Perspectives* 15(4): 143–156.

Koenker, R. and J. Machado (1999) 'Goodness of fit and related inference processes for quantile regression', *Journal of the American Statistical Association* 94: 1296–1310.

Koenker, R. and Q. Zhao (1994) 'L-estimation for linear heteroscedastic models', *Journal of Nonparametric Statistics* 3: 223–235.

Koopman, S. J. (1997) 'Exact initial Kalman filtering and smoothing for nonstationary time series models', *Journal of the American Statistical Association* 92: 1630–1638.

Koopman, S. J., N. Shephard and J. A. Doornik (1999) 'Statistical algorithms for models in state space using SsfPack 2.2', *Econometrics Journal* 2: 113–166.

Kwiatkowski, D., P. C. B. Phillips, P. Schmidt and Y. Shin (1992) 'Testing the null of stationarity against the alternative of a unit root: How sure are we that economic time series have a unit root?', *Journal of Econometrics* 54: 159–178.

Levin, A., C.-F. Lin and J. Chu (2002) 'Unit root tests in panel data: asymptotic and finite-sample properties', *Journal of Econometrics* 108: 1–24.

Lucchetti, R., L. Papi and A. Zazzaro (2001) 'Banks' inefficiency and economic growth: A micro macro approach', *Scottish Journal of Political Economy* 48: 400–424.

Lütkepohl, H. (2005) *Applied Time Series Econometrics*, Springer.

MacKinnon, J. G. (1996) 'Numerical distribution functions for unit root and cointegration tests', *Journal of Applied Econometrics* 11: 601–618.

Magnus, J. R. and H. Neudecker (1988) *Matrix Differential Calculus with Applications in Statistics and Econometrics*, John Wiley & Sons.

McAleer, M. and L. Oxley (1999) *Practical Issues in Cointegration Analysis*, Oxford: Blackwell.

McCullough, B. D. and C. G. Renfro (1998) 'Benchmarks and software standards: A case study of GARCH procedures', *Journal of Economic and Social Measurement* 25: 59–71.

Mroz, T. (1987) 'The sensitivity of an empirical model of married women's hours of work to economic and statistical assumptions', *Econometrica* 5: 765–799.

Nadaraya, E. A. (1964) 'On estimating regression', *Theory of Probability and its Applications* 9: 141–142.

Nash, J. C. (1990) *Compact Numerical Methods for Computers: Linear Algebra and Function Minimisation*, Bristol: Adam Hilger, second edn.

Nerlove, M. (1971) 'Further evidence on the estimation of dynamic economic relations from a time series of cross sections', *Econometrica* 39: 359–382.

_______ (1999) 'Properties of alternative estimators of dynamic panel models: An empirical analysis of cross-country data for the study of economic growth'. In C. Hsiao, K. Lahiri, L.-F. Lee and M. H. Pesaran (eds.), *Analysis of Panels and Limited Dependent Variable Models*. Cambridge: Cambridge University Press.

Newey, W. K. and K. D. West (1987) 'A simple, positive semi-definite, heteroskedasticity and autocorrelation consistent covariance matrix', *Econometrica* 55: 703–708.

_______ (1994) 'Automatic lag selection in covariance matrix estimation', *Review of Economic Studies* 61: 631–653.

Okui, R. (2009) 'The optimal choice of moments in dynamic panel data models', *Journal of Econometrics* 151(1): 1–16.

Pollock, D. S. G. (1999) *A Handbook of Time-Series Analysis, Signal Processing and Dynamics*, New York: Academic Press.

_______ (2000) 'Trend estimation and de-trending via rational square-wave filters', *Journal of Econometrics* 99(2): 317–334.

Portnoy, S. and R. Koenker (1997) 'The Gaussian hare and the Laplacian tortoise: computability of squared-error versus absolute-error estimators', *Statistical Science* 12(4): 279–300.

Ramanathan, R. (2002) *Introductory Econometrics with Applications*, Fort Worth: Harcourt, fifth edn.

Roodman, D. (2006) 'How to do xtabond2: An introduction to "difference" and "system" GMM in Stata'. Center for Global Development, Working Paper Number 103.

Schwarz, G. (1978) 'Estimating the dimension of a model', *Annals of Statistics* 6: 461–464.

Sephton, P. S. (1995) 'Response surface estimates of the KPSS stationarity test', *Economics Letters* 47: 255–261.

Sims, C. A. (1980) 'Macroeconomics and reality', *Econometrica* 48: 1–48.

Steinhaus, S. (1999) 'Comparison of mathematical programs for data analysis (edition 3)'. University of Frankfurt. URL http://www.informatik.uni-frankfurt.de/~stst/ncrunch/.

Stock, J. H. and M. W. Watson (2003) *Introduction to Econometrics*, Boston: Addison-Wesley.

_______ (2008) 'Heteroskedasticity-robust standard errors for fixed effects panel data regression', *Econometrica* 76(1): 155–174.

Stokes, H. H. (2004) 'On the advantage of using two or more econometric software systems to solve the same problem', *Journal of Economic and Social Measurement* 29: 307–320.

Swamy, P. A. V. B. and S. S. Arora (1972) 'The exact finite sample properties of the estimators of coefficients in the error components regression models', *Econometrica* 40: 261–275.

Theil, H. (1961) *Economic Forecasting and Policy*, Amsterdam: North-Holland.

_______ (1966) *Applied Economic Forecasting*, Amsterdam: North-Holland.

Verbeek, M. (2004) *A Guide to Modern Econometrics*, New York: Wiley, second edn.

Watson, G. S. (1964) 'Smooth regression analysis', *Shankya Series A* 26: 359–372.

White, H. (1980) 'A heteroskedasticity-consistent covariance matrix astimator and a direct test for heteroskedasticity', *Econometrica* 48: 817–838.

Windmeijer, F. (2005) 'A finite sample correction for the variance of linear efficient two-step GMM estimators', *Journal of Econometrics* 126: 25–51.

Wooldridge, J. M. (2002a) *Econometric Analysis of Cross Section and Panel Data*, Cambridge, MA: MIT Press.

_______ (2002b) *Introductory Econometrics, A Modern Approach*, Mason, OH: South-Western, second edn.

Yalta, A. T. and A. Y. Yalta (2007) 'GRETL 1.6.0 and its numerical accuracy', *Journal of Applied Econometrics* 22: 849–854.

www.ingramcontent.com/pod-product-compliance
Lightning Source LLC
Chambersburg PA
CBHW062347220526
45472CB00008B/1732

* 9 7 8 9 8 8 8 4 0 6 2 7 2 *